The Yokohama System for Reporting Endometrial Cytology

Yasuo Hirai · Franco Fulciniti

Editors

The Yokohama System for Reporting Endometrial Cytology

Definitions, Criteria, and Explanatory Notes

Springer

Editors
Yasuo Hirai
Department of Obstetrics and Gynecology
Faculty of Medicine
Dokkyo Medical University
Tochigi, Japan

Franco Fulciniti
Clinical Cytology Service
Istituto Cantonale dì Patologia
Ente Ospedaliero Cantonale
Locarno, Switzerland

PCL Japan Pathology and Cytology Center
PCL Inc.
Saitama
Japan

ISBN 978-981-16-5010-9   ISBN 978-981-16-5011-6   (eBook)
https://doi.org/10.1007/978-981-16-5011-6

This Springer imprint is published by the registered company Springer Nature Singapore Pte Ltd.
The registered company address is: 152 Beach Road, #21-01/04 Gateway East, Singapore 189721,
Singapore

Foreword

It was a great pleasure and honor to be invited to write a foreword for this admirable book written and edited by experts, several of whom I met at Yokohama in 2016, who took that opportunity to work "seriously and fruitfully" to agree a reporting and clinical management system for an important but widely misunderstood and underused type of cytology. Although probably not amenable to population screening, in that context it would have an advantage over cervical screening by involving a limited age range of the female population (age 40 years and over) rather than almost the entire adult female population (age from 20 or 25 years). Endometrial sampling should be little more intrusive than a Pap smear. So it could even be carried out in women without risk factors or symptoms.

There are four special features of this book that I would like to emphasize: first, the long history of endometrial cytology dating from the origins of diagnostic cytology and culminating in a system that takes advantage of both its history and the most up-to-date techniques and understanding of endometrial carcinoma; second, the methodology that gets the best out of cytodiagnosis by using liquid-based cytology (LBC) in addition to cell blocks prepared from the same painlessly obtained sample; third, the tidy and beautifully presented algorithm leading to a logical classification named the Yokohama System (TYS) similar, but adapted for the endometrium, to the well-known Bethesda system for cervical cytology; and fourth, but certainly not least, cooperation from the outset with an international and interdisciplinary group of experts: gynecologists and oncologists, pathologists (histo-, cyto-, and both), and cytotechnologists.

Endometrial cytology has an almost unique history of development from as early as the 1940s when W.H. Carey and George Papanicolaou worked independently on endometrial cell sampling and diagnosis. During the early days and subsequent years, a technique was developed that did not damage the uterus or risk spreading malignant cells through the fallopian tubes to the peritoneum, nor did it compromise cytomorphology by damaging the cells or stroma. In the 1940s, Papanicolaou used cell culture to examine the differences between benign and neoplastic cells; now rapid LBC samples can be examined immediately allowing the same fixation method for immediate and permanent slides.

LBC is not always regarded as superior to directly spread cell samples, but there is no doubt that immediate fixation in a carefully modulated medium provides excellent preservation of cellular detail; in the methodology recommended for TYS, a cell block is prepared directly from cells and tissue fragments deposited on the filter through which cells are deposited on the microscope slide. Thus, the advantages of histological sections for examination of cellular and stromal architecture and immunocytochemistry are combined with optimal cell preservation; nevertheless, the illustrations published for TYS show remarkable preservation of architecture as well as cytomorphology with LBC.

The algorithm developed for TYS is designed for clinicians as much as pathologists, and the authors explain the process whereby a final classification was reached combining the terminology (and its diverse abbreviations) into a logical TYS 0 and 1–6 system. As the system uses terms widely used in cytology, it should be readily understood by modeling the unique variations in endometrial physiology and pathology; descriptive terms as well as TYS categories may be used, for example, to describe conditions such as endometrial hyperplasia and glandular and stromal breakdown. Gray areas such as "atypical endometrial hyperplasia" show considerable variation in outcome when audited with respect to a final outcome of endometrial carcinoma—but less so with TYS, which separates low- and high-risk types of "atypia."

The Yokohama System makes full use of interdisciplinary and international cooperation. Based on the long experience of endometrial cytology developed by experts in Japan and Greece and their national cytology societies, TYS has already received international approval (Jimenez-Ayala M. *Acta Cytol* 2013, 57: 113–114; Margari et al. *Diagn Cytopathol* 2016, 44: 888–901). TYS should have an important place in diagnosis and screening of high-risk women. Now that primary human papillomavirus testing is widely used as the initial test for cervical cancer and its precursors (this author's recommendation to combine it with cytology for the first couple of tests not having been heeded: Herbert A. *Cytopathology* 2017; 28:9–15.), its occasional but individually life-saving ability to detect asymptomatic endometrial cancer is no longer available. Thus, the future of an established noninvasive test for endometrial carcinoma should be assured and used more widely in the rest of the world.

Amanda Herbert, M.D., MRCS,
LRCP, FRCPath
Former Director Guy's and St Thomas'
NHS Foundation Trust
Honorary Senior Lecturer at Guy's, King's
and St Thomas' School of Medicine
London, UK

Foreword

Cytology, the study of cells, was born in 1665, when Hooke first observed dead cork under the microscope, and called its empty spaces "cells." Thus, discovered in dead tissue from the wrong, i.e., plant, kingdom, and named after its containers, not its contents, it would be another two centuries before animal cells would reach the center of biological sciences, where it remains to this day. But why go back to Schleiden's and Schwann's 1839 *Cell Theory* and Virchow's 1859 *Omnis Cellula e Cellula*, to introduce a book on diagnostic cytopathology? Because a similar series of serendipities would accompany the birth of modern "clinical cytology," attributed to Papanicolaou, in the mid-1950s. For one, a pathologist uncovered the genesis of neoplasia, by recognizing that cells only develop from preexisting cells. But Virchow did not foresee that by looking at cells, from live patients, he could be making the clinical diagnosis of cancer. And more, the Pap test was not discovered in a clinic but in a laboratory developed by an anatomist, not a physician; first used to study benign vaginal cells, in guinea pigs, not to detect cancer in women; and was first embraced by gynecologists, not by pathologists, some of whom still insist on obtaining a tissue biopsy for cancer diagnosis. In fact, the publication of Papanicolaou's *Atlas of Exfoliative Cytology* in 1954 still emphasized the determination of the "Maturation Index" of cervical cells, under the effect of hormones. The atlas nearly ignored the endometrium, which, under the effect of estrogens, may become cancerous, while endometrial cells remain largely undetected or studied in conventional Pap smears. As it is well analyzed in this new book, the Yokohama System (TYS) became possible only after the introduction of new endometrial sampling devices, which coincided with the peak of the utilization of cervico-vaginal cytology, for the early diagnosis of cervical cancer. Liquid-based cytology (LBC) improved the cytopathologist's ability to peer into the significance of premalignant lesions of uterine cancer and lay out the basis for elucidating the link between HPV and cervical cancer. This tome contributes toward the prospect of endometrial cytology playing a similar role in ushering endometrial oncology into the translational era. This transition is not a linear chain, with obvious links between one another, but more like a net, with threads that need a clinical as well as pathological context, to be fully appreciated. This book provides such a context.

Born from ideas discussed during a symposium of the International Academy of Cytology (IAC) Meeting, 2016, at Yokohama, Japan, the book is the fruit of multidisciplinary Working Groups, in Greece and Japan. Steeped in careful cytohistological correlation, the book emphasizes the detection or early endometrial neoplasia and the recognition of precursor lesions. Through the use of meticulous illustrations of architectural as well as individual cell cytomorphology, the book provides a rationale for meaningful phenotypical profiling and the judicious molecular testing applied to clinical samples. Covering from basic aspects of molecular histogenesis to practical aspects of specimen adequacy and cyto-preparatory techniques, the scope of the book is wide, as well as thorough in its depth. But it is at detailing specific diagnostic categories and cytomorphological criteria of TYS that the book shines brightest, in its purpose to ensure uniform and reliable reporting, via a standardized terminology with a global reach. Besides elaborating on the nomenclature used in TYS, separate chapters present its epidemiological and pathogenetic basis, along with useful interpretational algorithms, thus adding coherence to the diagnostic approach, with singular clarity. For the sake of explicitness, the authors add charts correlating TYS with the Bethesda System and WHO's histological classification, which increases its utility in day-to-day practices.

   The appearance of this text atlas is timely, given the fact that endometrial cancer now outnumbers cervical cancer somewhere between 4 and 5 to 1 in industrialized nations and the need for its early, accurate diagnosis by noninvasive means. A new title on endometrial cytology is justified, for the dearth of experience in dealing with this aspect of gynecological cytology, and for providing a valuable bridge between conventional smears, LBC preparations, cell blocks, and tissue biopsies. Cytohistological subtyping is currently routinely used to guide prognosis and treatment decisions for endometrial cancer patients, while ongoing studies evaluate the potential utility of subcellular, molecular subtyping. Although molecularly targeted therapies have been effective in some cancer types, relatively few have reached clinical actionability for use in endometrial cancer. This book provides a linchpin between current knowledge and this possible endeavor. A sound volume, laying out the pathological basis of endometrial disease, and illustrating its cytological and histopathological features, in light of clinical correlation, the book is a welcome addition to the library of oncologists, gynecologists, pathologists, and cytopathologists alike. Just the standardization of diagnostic terminology would be of great help to those practitioners interested in endometrial disease. The fact that discussions leading to the completion of this book took place in Japan as well as in the birthplace of Papanicolaou adds, serendipitously, to its historical significance. The richness of the other topics covered in this tome are a delight to be appreciated. The fact that it was expertly, conceived, planned, and written by long-time friends and colleagues is an added bonus.

Carlos W. M. Bedrossian, M.D.
Professor of Pathology
Former Director, Cytopathology Service
Northwestern University
Chicago, IL, USA

Preface

Cytopathology is one of the most fascinating areas of medicine. Just observing cells under the microscope, the cytopathologist is able not only to diagnose a wide range of diseases but also sometimes to give information about the evolution and management of these diseases. The first time that cytopathology was recognized as a science was with the description of the Papanicolaou system to report gynecological cytology. It is still amazing that this simple technique proved to be one of the most effective methods of cancer prevention. As a natural evolution, cytology was used as a diagnostic method in several organ systems that several body organisms such as the respiratory system, serous effusions, and urine, among others. With the establishment of fine needle aspiration, practically all organs became accessible to be studied by cytology. Today, this is a reality in the thyroid, salivary glands, lungs, breast, pancreas, etc. Shortly after the development of cervico-vaginal cytology, endometrial cytology began to be established as a method that could assist the gynecologist in the management of various uterine pathologies. However, as history shows, not only in the endometrium but also in other organs (e.g., prostate) and based on not entirely truly arguments, many centers have left cytology behind and replaced it totally with surgical biopsy. Luckily in some countries like Japan and due to the presence of extraordinary specialists in this area, some of them collaborators in this book, endometrial cytology has remained active and of clinical importance in certain areas of the world. Recently, the incorporation of a molecular classification of endometrial cancer, better recognition of its signaling pathways and precursor lesions, and the possibility of studying therapeutic targets in this type of material bring a new perspective to this field.

After a meticulous microscope examination, using complementary techniques, if necessary, the cytopathologist has another extremely important function: to transmit the result in a clear format to the clinician to allow a proper patient management. This book describes a standardized method for classifying and reporting invasive endometrial malignancies via direct endometrial sampling. Featuring a wealth of color illustrations, it provides specific diagnostic categories and cytomorphologic criteria to promote uniform and reliable diagnoses. It also describes the history of directly sampled endometrial cytology, reviews the sampling techniques and algorithmic approach, discusses specimen adequacy, and outlines challenges for the future. *The Yokohama System for Reporting Endometrial Cytology—Definitions, Criteria, and Explanatory Notes* offers a valuable resource for researchers at clinical cytopathological laboratories around the world whose work involves gyne-cological cytology, oncology, pathology, and cytopathology. It will also appeal to researchers in the fields of cytotechnology, basic science, pathology and related industries, medical residents, and clinicians.

At this moment, a worldwide effort is being developed to promote cytopathology reporting systems. All efforts to promote a systematization of the nomenclature are important, because despite different names, sponsoring organizations, and authors, they share the same advantages, such as they unify reporting of disease categories, reduce inter-observer variability, improve intra-observer reproducibility, better align patient management options with interpretations, and at the end of the day enhance patient care.

Now it is time to enjoy the text and the illustrations of this book that presents in a didactic way the updated knowledge in endometrial cytopathology.

Porto, Portugal   Fernando Schmitt

Preface: The Quest to Develop a Standardized Terminology for Endometrial Cytology

The main purpose of directly sampled endometrial cytology is to detect invasive endometrial malignancies. With this principle in mind, the Yokohama System Working Group, composed of cytopathologists, surgical pathologists, and gynecologic oncologists, met at the 2016 International Congress of Cytology, Yokohama, with the aim of publishing a standardized reporting system.

Currently, the Bethesda System for Reporting Cervical Cytology has also been widely adapted as the "Bethesda-style Cytologic Reporting System" to various anatomical compartments such as the vagina, vulva, anus, thyroid, breast [1], urinary system [2], effusions, pancreas [3], and other digestive organs. The "Bethesda-style Cytologic Reporting System" has now definitively decommissioned Papanicolaou's classification. However, no globally accepted reporting system or format has been established for endometrial cytology yet.

Internationally gathered experts on endometrial cytology have agreed to develop a globally acceptable consensus reporting system for endometrial cytology, named the Yokohama System (TYS) for Reporting Endometrial Cytology through the serious and fruitful discussion devoted to directly sampled endometrial cytology at the symposium held in ICC2016 in Yokohama, Japan. Dr. Hirai and Dr. Fulciniti were the main promoters of this work; they elicited the scientific collaboration that produced this work and article.

The TYS working group aimed to discuss ways to improve the reporting and performance of endometrial cytology. The value of ancillary tests in the screening and diagnosis of endometrial neoplasms was also included; the TYS adopted a "Bethesda-style" format to encourage better communication between cytopathologists and gynecologists by suggesting further clinicopathological steps (i.e., clinical handling, definitive histological examination, tissue biopsy, D&C, or follow-up). Moreover, the TYS has illustrated the respective diagnostic cytomorphologic criteria and evaluated reproducibility (inter- and intra-observer variabilities), the utility of each classification scheme, and the associated implied risks of malignancy for each diagnostic category; it also has a diagnostic algorithm.

In Japan, the examination of directly sampled endometrial cytology has been encouraged as a first-level screening method to be used before tissue examinations such as biopsy for women with an increased risk of endometrial cancer. Anyhow directly sampled endometrial cytology using conventional methods has not been globally recognized as a useful screening method, as some studies have shown that the diagnostic accuracy in detecting malignancy is too low. However, recent clinical results of directly sampled endometrial cytology using LBC in Japan and Europe have been remarkable.

Hence, we believe that directly sampled endometrial cytology using LBC needs global recognition, as its high performance and noninvasive nature as a screening examination are noteworthy. The proposed TYS and this book also cover the procedures, precautions, and detailed methods of examination required for clinicians to conduct directly sampled endometrial cytology using LBC in women with an increased risk of endometrial cancer. Endometrial cytology is highly accurate, safe, and less painful in detecting this type of cancer; it is the best method for cancer surveillance among women with Lynch syndrome. Furthermore, it is a useful first-step screening test for many cases in which vaginal bleeding or similar symptoms are recognized in daily clinical practice, but biopsy is questioned owing to its invasive nature.

Success of the TYS will depend on its acceptance by all the relevant pathology and gynecologic oncology communities, which, by their joint efforts, will adopt, critically evaluate, and optimize this method, aiming at further improving the impact of endometrial cytology on patient care.

Saitama, Japan   Yasuo Hirai
Locarno, Switzerland   Franco Fulciniti

References

1. Field AS. Breast FNA biopsy cytology: current problems and the international academy of cytology Yokohama standardized reporting system. Cancer Cytopathol. 2017;125:229–230.
2. Barkan GA, Wojcik EM, Nayar R, et al. The Paris system for reporting urinary cytology: the quest to develop a standardized terminology. Acta Cytol. 2016;60:185–197.
3. Layfield LJ, Pitman MB, DeMay RM, et al. Pancreaticobiliary tract cytology: journey toward "Bethesda" style guidelines from the Papanicolaou Society of Cytopathology. Cytojournal. 2014;11:18.

Introduction

Endometrial cytology has recently experienced a new era with the recent introduction of efficient uterine brushes in Japan, Italy, and Greece. This permitted the adoption of Liquid Based Cytology for these samples, with consequent marked improvement of excess blood contamination and fixation artifacts that were so common in traditional smears from the endometrial cavity. Moreover, thanks to the LBC method, the number of unsatisfactory samples has been significantly reduced and standardization of sample preparation and interpretation has made it possible to adopt this technique as a screening procedure in many regions of Japan. During the International Congress of Cytology held in Yokohama in 2016, the conceptual bases of a new reporting method for reporting directly sampled endometrial brushings based on cytomorphological criteria on LBC samples and the definition of Risk of Malignancy (ROM) in the various diagnostic categories, according to the corresponding cytologic features were put forward in a Symposium by a panel of dedicated cytopathologists, gynecologists, and oncologists "The Yokohama System for reporting endometrial cytology" (TYS) was hence born [1].

In analogy to the Bethesda reporting system for cervical cytology (TBS), the diagnostic categories of TYS imply several risk categories associated with the diagnostic classes. Basically, two indeterminate categories are present in the TYS: one with a low risk of malignancy (ATEC-US) roughly equivalent to the ASC-US class in the TBS and the other one with a higher level of clinical risk (ATEC-AE), roughly equivalent to the ASC-H category of TBS. The various diagnostic categories of the TYS along with their corresponding cytologic criteria may be seen in Fig. 5.1 of Chap. 5 of this book.

The diagnostic categories of the TYS reflect the state of the art and the relatively limited degree of diffusion of the adoption of direct endometrial brushings for the screening of endometrial cancer. We wish that this system could be progressively adopted in more and more countries of the world, as it is a relatively cheap, reproducible, and fairly accurate system that could replace more invasive and expensive screening systems. Moreover, traditional endometrial biopsy with Pipelle or diagnostic curettage may always be performed in these cases.

We expect TYS to be tested by a growing number of cytopathologists and clinicians worldwide in order to test its performance and its efficiency in the definition

of the follow-up strategies after the test. It is well possible that some of these categories might be changed in the future, depending on whether new risk categories could
be defined also by the application of supplemental or ancillary methods (like immunocytochemistry, or molecular pathology performed on cytologic material). In this
way, we sincerely hope that the study of endometrial cytology could live a new era
due to the significant technical advantages represented by LBC and ancillary techniques applied to the cytological sample.

This seems to be reflected by an increased interest in this method in Japan and in
several European countries, like Greece, Switzerland, Italy, and Spain, as demonstrated by a number of excellent publications.

This book will accompany its readers in the discovery of the history and newer
developments in endometrial cytology, which include its technical and cytomorphological aspects that take now a significant advantage by the study of several pathological variables, as they can be determined by the application of immunocytochemical
and molecular pathological study of cell blocks obtained by LBC samples.

All of the authors of this book warmly hope that our work be of interest to all the
professional figures (cytotechnicians, cytopathologists, gynecologists, gynecological oncologists) involved in the important task of securing the best possible woman's health (Fig. 1).

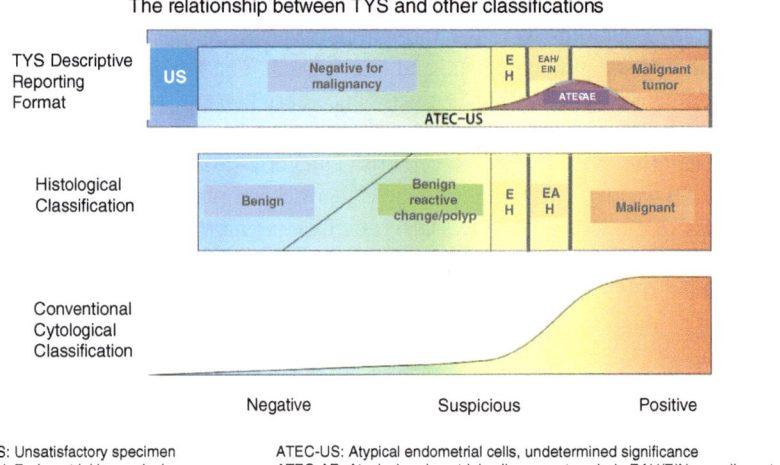

Fig. 1 The relationship between TYS and other classifications

Reference

1. Fulciniti F, Yanoh K, Karakitsos P, et al. The Yokohama system for reporting directly sampled endometrial cytology: The quest to develop a standardized terminology. Diagn Cytopathol. 2018;400–412.

Addendum

This figure shows the relationship between The Yokohama System (TYS) and other cytological and pathological classifications. In the classical cytological classification system, more than half of cases diagnosed as 'suspicious' correspond to pathologically benign endometrium. Furthermore, in the classical cytological classification system, the histological distinction between 'negative' and 'suspicious' appears to be ambiguous. In contrast, there is consistency between TYS classification and histological classification.

TYS includes an 'unsatisfactory for evaluation (TYS 0)' category under the heading of specimen adequacy criteria. Typically, a cytological diagnosis would not be reported in the case of TYS 0. However, flexible clinical management is permitted for the patient's benefit. For example, if the specimen has to be estimated as TYS 0 due to scant cellularity, designation as 'negative for malignancy' may be permitted when clinical information reveals that the case is a postmenopausal woman with no abnormal genital bleeding and no abnormal ultrasonographic findings. When atypical cells are observed even in cases estimated as TYS 0, the suspected cytological diagnosis must be reported.

When the cytological result is 'negative for malignancy (TYS 1)', subsequent endometrial histological evaluation is not necessarily required. However, in cases of irregular endometrial ultrasonographic findings or accompanying abnormal uterine bleeding, endometrial biopsy or curettage must be considered even in TYS 1 cases. Because ATEC-US (TYS 2) is expected to encompass a spectrum ranging from benign endometrium to neoplastic changes, it is difficult (at present) to define a triage method. Endometrial biopsy or repeat endometrial cytological assessment after 2 or 3 months is required for ATEC-US (TYS 2), whereas histological estimation is necessary in cases of ATEC-AE (TYS 4). When the cytological result is other than 'negative for malignancy (TYS 1)' or 'ATEC-US (TYS 2)', endometrial biopsy or curettage is required to confirm the endometrial diagnosis.

Appropriate clinical management, especially for ATEC-US (TYS 2) and ATEC-AE (TYS 4), will be defined on the basis of strong evidence obtained in further clinical studies and using new technologies.

Kenji Yanoh
Department of Obstetrics and Gynecology
Suzuka Chuo General Hospital
Suzuka, Mie, Japan

Contents

Contributors

Jun Akiba, MD, PhD Department of Diagnostic Pathology, Kurume University Hospital, Kurume, Japan

Arrigo Capitanio, MD, MSc Department of Clinical Pathology, Linköping University Hospital, Linköping, Sweden

Alessia Di Lorito, MD, PhD Pathology Department, SS Annunziata Hospital, Chieti, Italy

Milo Frattini, PhD Laboratory of Molecular Pathology, Institute of Pathology, Locarno, Switzerland

Istituto Cantonale di Patologia, Service of Molecular Pathology, Locarno, Switzerland

Franco Fulciniti, MD, PhD, MIAC Clinical Cytology Service, Istituto Cantonale dì Patologia, Ente Ospedaliero Cantonale, Locarno, Switzerland

Yasuo Hirai, MD, PhD, FIAC Department of Obstetrics and Gynecology, Faculty of Medicine, Dokkyo Medical University, Tochigi, Japan

PCL Japan Pathology and Cytology Center, PCL Inc., Saitama, Japan

Akihiko Kawahara, PhD, CFIAC Department of Diagnostic Cytopathology, Kurume University Hospital, Fukuoka, Japan

Maki Kihara, MD, PhD Department of Obstetrics and Gynecology, International University of Health and Welfare Narita Hospital, Chiba, Japan

Tadao K. Kobayashi, PhD, CFIAC Division of Health Sciences, Cancer Education and Research Center, Osaka University Graduate School of Medicine, Osaka, Japan

Tetsuji Kurokawa, MD, PhD Department of Gynecology and Obstetrics, Faculty of Medical Sciences, University of Fukui, Fukui, Japan

Alarice C. Lowe, MD Department of Pathology, Stanford University Hospital, Stanford, CA, USA

Yoshinobu Maeda, MD, PhD Department of Diagnostic Pathology, Toyama Red Cross Hospital, Toyama, Japan

Natalia Malara, MD, PhD Department of Experimental and Clinical Medicine, University Magna Graecia, Bionem Lab, Catanzaro, Italy

Niki Margari, MD, PhD Private Cytopathology Laboratory, Ex Scientific Collaborator of Department of Cytopathology at National and Kapodistrian University of Athens, Athens, Greece

Diana Martins, PhD I3S, Instituto de Investigação e Inovação em Saúde, University of Porto, Porto, Portugal

Polytechnic Institute of Coimbra, ESTESC-Coimbra Health School, Department of Biomedical Laboratory Sciences, Coimbra, Portugal

University of Coimbra, Coimbra Institute for Clinical and Biomedical Research (iCBR) Area of Environment Genetics and Oncobiology (CIMAGO), Biophysics Institute of Faculty of Medicine, Coimbra, Portugal

Luca Mazzucchelli, MD Istituto Cantonale di Patologia, Locarno, Switzerland

Akira Mitsuhashi, MD, PhD Department of Reproductive Medicine, Graduate School of Medicine, Chiba University, Chiba, Japan

Takeshi Nishikawa, PhD, CMIAC Department of Diagnostic Pathology, Nara Medical University Hospital, Kashihara, Nara, Japan

Yoshiaki Norimatsu, PhD, CFIAC Department of Medical Technology, Faculty of Health Sciences, Ehime Prefectural University of Health Sciences, Tobe-cho, Iyo-gun, Ehime, Japan

Toshimichi Onuma, MD Department of Gynecology and Obstetrics, Faculty of Medical Sciences, University of Fukui, Fukui, Japan

Ioannis G. Panayiotides, MD, PhD Professor of Pathology at University of Athens, Chair, 2nd Department of Pathology UoA, "Attikon" University Hospital, Athens, Greece

Fernando Schmitt, MD, PhD, FIAC Department of Pathology, Medical Faculty of Porto University, Porto, Portugal

Unit of Molecular Pathology, Institute of Molecular Pathology and Immunology, Porto University, Porto, Portugal

RISE, Health Research Network, Porto, Portugal

Akiko Shinagawa, MD Department of Gynecology and Obstetrics, Faculty of Medical Sciences, University of Fukui, Fukui, Japan

Jun Watanabe, MD, PhD, FIAC Department of Bioscience and Laboratory Medicine, Hirosaki University Graduate School of Health Science, Hirosaki, Japan

Kenji Yanoh, MD, PhD Department of Obstetrics and Gynecology, Suzuka Chuo General Hospital, Suzuka, Mie, Japan

Yoshio Yoshida, MD, PhD Department of Gynecology and Obstetrics, Faculty of Medical Sciences, University of Fukui, Fukui, Japan

Endometrial Cytology in Historical Perspective

Tadao K. Kobayashi

1.1   General Concepts

The prevalence of squamous cell carcinoma of the uterine cervix has been decreasing because of the early detection of its precursor lesions by cervical cytology and early treatment. The human papillomavirus (HPV) vaccination program has further contributed to preventing the insurgence of most new infections and, hence, of new precancerous lesions. However, this is not the case with endometrial carcinoma, which has demonstrated a steady increase in frequency in many industrialized countries. Primary endometrial cancer is the sixth most common neoplasm among women worldwide [1]. It was estimated that by the year 2018 more than 382,069 new cases will have been diagnosed each year throughout the world [2]. Although many strategies and devices for endometrial cell sampling have been proposed and have shown higher rates of lesion detection, until fairly recently, no effective screening program clinically comparable to cervical cytology was currently available for the detection of endometrial cancer and its precursor lesions. Nonetheless, there is no doubt that the inherent difficulties in the cytologic interpretation of endometrial cytology specimens have been the major roadblocks to date. Notwithstanding this latter, cytology is now the most common test for an initial evaluation of endometrial cancer in Japan [3] and has been encouraged as the first-level screening method for women at high risk for endometrial cancer [4].

It is thought that endometrial adenocarcinoma develops along two independent pathogenetic pathways, and these types are referred to as type I and II, respectively. Apparently, the majority of type I adenocarcinomas develop as a result of unopposed estrogen stimulation with various levels of hyperplasia serving as intermediate morphologic stages, while type II lesions occur in atrophic endometria in the post-menopausal age, are not generally proceeded by a pre-neoplastic lesion and are

T. K. Kobayashi (✉)
Division of Health Sciences, Cancer Education and Research Center, Osaka University
Graduate School of Medicine, Osaka, Japan
e-mail: tada.tkkobaya@gmail.com

© The Author(s), under exclusive license to Springer Nature Singapore Pte Ltd. 2022   1
Y. Hirai, F. Fulciniti (eds.), *The Yokohama System for Reporting Endometrial Cytology*, https://doi.org/10.1007/978-981-16-5011-6_1

related to the appearance of mutations of several oncogenes [5]. Most women are diagnosed at an early stage and have a relatively good survival rate; however, women who are diagnosed at an advanced stage or with recurrent disease have a poor prognosis [5]. Since the early detection of endometrial cancer is important for the improvement of long-term survival of patients, the adoption of endometrial cytology is a great challenge for both cytopathologists and cytotechnologists [6–15].

The primary concern of this textbook is to establish the concept that endometrial cytology has a future in gynecological practice by presenting it as a potentially worthy method of detection for malignant and premalignant lesions of the endometrium in women attending an outpatient clinic. The purpose of this textbook is also to introduce the beginner and also practicing, skilled cytopathologists and cytotechnologists to the cytopathologic interpretation of directly sampled endometrial liquid-based cytology (LBC) samples (which, up to now, was not so commonly performed worldwide), and to the clinical consequences of abnormal cytologic findings. A book with a somewhat similar title, "The Atlas of Endometrial Cytology: based on descriptive and standardized reporting system" had originally been written in Japanese for its dedicated audience and was published in 2015 [16].

1.2   Dawning of the Endometrial Cytology

The acceptance of diagnostic cytology as a current and valid discipline in medicine is largely due to the work of George Papanicolaou, the father of modern cytology. In the 1920s Papanicolaou began to publish material on the cytologic method for hormonal evaluation and, in 1928 he suggested that this method was of value in the diagnosis of cancer of the uterine cervix [17, 18]. Independently and earlier in the same period, Babes had published material on the same subject in Rumania [19]. Recognition of cytology as a valuable diagnostic tool occurred following further publications, i.e., in 1941 Papanicolaou and Traut published an article (and a book in 1943) on the value of vaginal smear cytology that demonstrated that it was also possible to diagnose endometrial cancer in asymptomatic patients [17, 18, 20]. Since that time many papers have been written delineating the accuracy of this technique. The development of a technique for sampling the endometrial mucosa was encouraged by Papanicolaou himself, who acknowledged the inadequacy of the cervical smear for detecting adenocarcinoma of the uterus. This is demonstrated by a typewritten official report of endometrial cytology signed personally by Dr. Papanicolaou during his tenure as it may be seen in Fig. 1.1, which has been taken from the book entitled "The Pap Smear; Life of George N. Papanicolaou" authored by Carmichael (Fig. 1.2) [17].

Moreover, Papanicolaou's interest in human endometrium also involved several articles concerning research on tissue culture of endometrial explants, as it can be resumed from articles he published between 1953 and 1961. Papanicolaou acknowledges carrying out a 6-year study on the behavior and the potentialities for differentiation and growth of cultured normal, benign, and malignant endometrial cells [20].

Fig. 1.1 Cover page of the book entitled "The Pap Smear: Life of George N. Papanicolaou" by Carmichael in 1973. Reprinted with permission from the publisher

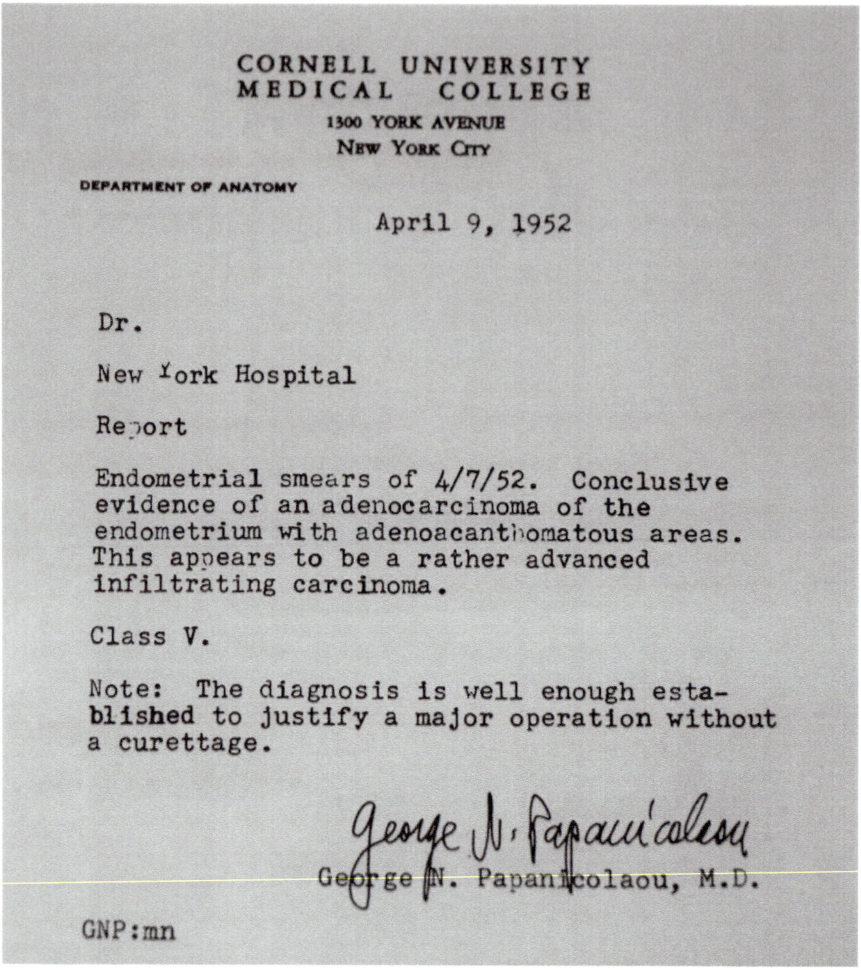

CORNELL   UNIVERSITY
MEDICAL   COLLEGE
1300 YORK AVENUE
NEW YORK CITY

DEPARTMENT OF ANATOMY

April 9, 1952

Dr.

New York Hospital

Report

Endometrial smears of 4/7/52. Conclusive
evidence of an adenocarcinoma of the
endometrium with adenoacanthomatous areas.
This appears to be a rather advanced
infiltrating carcinoma.

Class V.

Note: The diagnosis is well enough established to justify a major operation without
a curettage.

George N. Papanicolaou

George N. Papanicolaou, M.D.

GNP:mn

Fig. 1.2 Copy of the typewritten official report of endometrial cytology signed personally by Dr. Papanicolaou, which has been taken from the book by Carmichael in 1973. Reprinted with permission from the publisher

Anyhow the cytologic detection of outspoken endometrial cancer was unable to significantly reduce its occurrence like with cervical cancer.

Needless to say, modern endometrial cytology is characterized by its simple sampling procedure, and has the advantage that it can be performed in the outpatient unit without anesthesia. Koss et al. investigated whether endometrial cytology is effective as a screening method for high-risk groups for endometrial cancer in clinically asymptomatic women over 45-years old. In 1984, Koss et al. investigated 2586 asymptomatic women aged 45 years and above with the aim of evaluating the technique as a screening test for endometrial cancer in healthy individuals; 16 cases of occult carcinomas, and 21 cases of endometrial hyperplasia were detected, but, in particular, the detection rate of endometrial hyperplasia cases was so low that the value of this approach was questioned as a screening test [21], another group also reported that screening asymptomatic women for endometrial cancer is not recommended [22].

1.3 Endometrial Cytology Based on Cyto-Architecture of Tissue Fragments

Although the cellular features of endometrial hyperplasia in endometrial aspirates have been described by Morse [23], other groups report that the lesions are underdiagnosed [24, 25]. This is most likely because cytology can never provide the architectural detail needed for an accurate diagnosis of this disease. Consequently, the interpretation of endometrial smears requires special expertise apart from general histopathologic training. Many tend to think of cytology as a nucleus-centered diagnostic method that gives little consideration to tissue structure. Adopting diagnostic criteria that reflect the cytoarchitecture (i.e., the pattern of growth within cell clusters) in endometrial cytology samples has been reported to solve this problem [26–28]. Several studies have described [6, 7, 12, 14, 15, 29, 30] the cytoarchitectural diagnostic criteria for endometrial lesions by using LBC. The above-described advantages of the LBC strongly encourage the replacement of conventional preparations (direct smears) with LBCs in the routine cytologic evaluation of endometrial cellular samples.

1.4 Endometrial Sampling by Cervical Cytology

Cells exfoliated from the endometrial cavity may be present in cervical and vaginal smears, thereby presenting an opportunity to report on pathological changes in the endometrium as part of the cervical cancer screening program. Although the conventional Pap smear may be used to detect endometrial carcinoma, this specimen is not useful for endometrial cancer screening [31, 32]. As a matter of fact, the sensitivity of cervicovaginal sampling for endometrial pathology is low since significant exfoliation does not occur even in cases of endometrial cancer [12, 33]. Moreover, the likelihood of degenerative changes in spontaneously exfoliated endometrial cells including those of pregnancy-related cell changes [34] makes their interpretation difficult. Endometrial cancer detection depends on patient-related, sampling, interpretative, and screening factors. Detection correlates with shedding of neoplastic cells from the primary location into the cervix and vagina and not all adenocarcinomas readily shed tumor cells.

1.5 Development and Evaluation of Endometrial Cell Sampling Device

In 1943, Cary was the first to develop a practical technique for obtaining intrauterine sampling of endometrial cells. He used a metal cannula intended for artificial insemination, which he introduced into the uterine cavity. The endometrial mucosa was then aspirated with a 3-ml syringe [35]. From the latter half of the 1950s, the aspiration method was converted to an endometrial washing. Torres et al. [36] showed that washing the endometrial cavity with saline under positive pressure provided suitable samples for light microscopic analysis. Concern about the possible complications of this procedure such as uterine infection, dissemination of tumor through the fallopian tubes into the peritoneal cavity, and perforation of the uterus, caused gynecologists to resist this approach for several years. Subsequently, instruments were designed for washing the endometrial cavity with physiological saline under negative pressure [37] so that the risk of dissemination of tumor into the peritoneal cavity was eliminated. Unfortunately, endometrial washings were difficult to interpret [38] and the laboratory processing was a time-consuming task, so the method was not widely accepted. In 1968, Johnsson and Stormby reported the use of a cytological brushing technique to obtain cells from endometrial lesions [39].

In the 1970s, a new generation of endometrial sampling instruments was introduced. These depended on abrasion of the endometrium by a plastic helix, the Mi-Mark (Milan-Markley) [24, 40] or by a curette, the Accurette [41], and immediate transfer of the cytological material to a glass slide. A third instrument, the Isaacs endometrial sampler [42] depended on the aspiration of endometrial cavity. The Isaacs Sampler had a limited diffusion due to the risk of spreading malignant cells into the endometrial cavity [36, 43]. In 1993, Tao developed the Tao Brush endometrial cell collector that avoids cell spreading outside the uterine cavity and is easy to operate, so it was adopted by the gynecologists and is widely used [44]. In 2006, Fujihara et al. reported that the Uterobrush samples prepared by the "flicked" method have a much greater quantity of cell clusters than those using the earlier Endocyte sample [45].

1.6   Endometrial Cytopathology and Cell Block Preparation

Koss recommended that the cell block method be used complementarily to endometrial brushings because it enhances the specificity of gynecologic cytology [32]. A variety of cell block (CB) techniques have been in use for over a century [46]. CB provides a method for immunocytochemistry (ICC) that has revolutionized cytopathology by making it possible to apply panels of antibodies to multiple sequential sections. CB is prepared from residual LBC cytological specimens or from dedicated cytological samples injected in fixative/preservative fluid. Both the morphology of endometrial cells and the glandular architecture are critical to endometrial cancer diagnosis. In 1998, Yang et al. [47], reported a series of endometrial cytology samples obtained by using a Tao Brush with subsequent preparation of a cell block. This method enabled good morphological evaluation comparable to ordinary endometrial biopsy. In 2000, Diaz-Rosario and Kabawat devised a novel technique to prepare CBs that avoid disruption of glandular structure during preparation [48]. The fluid remaining after the preparation of the ThinPrep slide is subjected to sedimentation in the inverted ThinPrep filter used for slide preparation. CB's can maintain cell morphology and tissue architecture and are thus a useful complement to liquid-based smears for definitive diagnosis [49].

1.7   Endometrial Cytology by Means of Liquid-Based Cytology and Its Reporting System

Cytopathological interpretation of conventional preparations of brushing by direct smear may be problematic due to several factors such as the presence of excess obscuring blood or inflammation, excessive cellular overlapping, distortion of cells or presence of thick cell clusters, scant cellularity, and presence of fixation artifacts, so that the adoption of endometrial cytology as a diagnostic procedure has been hampered [12–15]. Other difficulties in the interpretation of endometrial cellular samples were also represented by metaplastic changes or by ovarian hormonal disorders such as endometrial glandular and stromal breakdown (EGBD); in these cases, it may be often difficult to make a correct cytological diagnosis [12, 14, 15, 50]. This seems to be the reason why endometrial cytology is only partially accepted in Japan and not generally practiced in the rest of the world.

LBC was introduced in the mid-1990s as a way to improve performance in cervical cancer screening. In a short time, LBC was applied to endometrial cytology and achieved a diagnostic sensitivity as accurate as conventional preparations, especially for its excellent cell preservation and clearance of background debris, blood, and inflammation, which decreased the number of inadequate diagnoses [51]. ThinPrep cytology combined with endometrial sampling was reported to be a useful tool for the outpatient diagnosis of endometrial lesions, and reduced the number of unnecessary curettages [6, 7]. Several studies have also achieved the same results by using SurePath LBC method [9, 10, 14, 15, 52].

Several studies have described in recent years the cytoarchitectural diagnostic criteria for endometrial lesions by using LBCs. LBC represents an opportunity to re-evaluate endometrial cytology; an increasing interest in it is demonstrated by a number of articles that report interesting results in terms of diagnostic accuracy [5–7, 14, 15]. Although a number of endometrial cytology experiences have been introduced over the years, many of these methods do not use a standardized reporting system. In 2018, Fulciniti et al. [4] published a standardized reporting system inclusive of specific diagnostic categories and cytomorphologic criteria for uniform and reliable diagnosis of endometrial malignancies on directly sampled endometrial samples. This was summarized through the serious and fruitful discussion in the symposium of the International Academy of Cytology (IAC) Meeting, 2016 at Yokohama, Japan.

1.8   Molecular-Based Approach with LBC Samples

The purpose of molecular cytopathology is to elucidate the mechanisms of disease by identifying molecular and pathway alterations. Mutation of phosphatase and tensin homolog (PTEN), beta-catenin, and p53 genes are among the most frequent molecular defects in type I and II endometrial carcinomas, respectively [12, 15, 53]. PTEN encodes a dual-specificity phosphatase with lipid phosphatase and protein tyrosine phosphatase activities that regulates both apoptosis and interactions with the extracellular matrix. In a study of 38 cases of endometrial intraepithelial neoplasia (EIN) on endometrial biopsy samples, Norimatsu et al. found that immunohistochemical loss of PTEN and positive nuclear staining of beta-catenin were frequently seen in EIN but not in normal proliferative endometrium cases. The combination of PTEN-negative/beta-catenin-positive results may become a reliable marker for detecting endometrial intraepithelial neoplasia [14, 15, 30, 53–55]. Because the overexpression of gene products of types I and II endometrial carcinomas correlates with clinicopathological factors and prognosis, it is also important to evaluate the immunocytochemical expression of endometrial malignancy markers including precursor lesions on LBC samples. It is of note that frequent mutations of p53 have been found in aggressive endometrial carcinoma including high-grade serous types and carcinosarcoma. With regard to p53, a nuclear expression threshold >4 may be useful for evaluating type II endometrial carcinoma. In 2014, Kosmas et al. [53] reported that immunocytochemical positive expression of p53 on imprint cytologic smears was correlated with surgical-pathological stage, histological grade, and lymph node metastasis.

The role of Molecular Pathology and Molecular Cytopathology of directly sampled endometrial brushings will be covered in detail in Chaps. 15 and 16 of this book.

References

1. Lortet-Tieulent J, Ferlay J, Bray F, et al. International patterns and trends in endometrial incidence, 1978-2013. JNCI J Natl Cancer Inst. 2018;110:354–61.
2. Bray F, Ferlay J, Soerjomataram I, et al. Global cancer statistics 2018; GLOBOCAN estimates of incidence and mortality worldwide for 36 cancers in 185 countries. CA Cancer J Clin. 2018;68:393–424.
3. Fujiwara H, Takahashi Y, Takano M, et al. Evaluation of endometrial cytology: cytohistological correlations in 1441 cancer patients. Oncology. 2015;88:86–64.
4. Fulciniti F, Yanoh K, Karakitsos P, et al. The Yokohama system for reporting directly sampled endometrial cytology: the quest to develop a standardized terminology. Diagn Cytopathol. 2018;46:400–12.
5. Murdock TA. Endometrial carcinoma. In: Mazur MT, Kurman RJ, editors. Diagnosis of endometrial biopsies and curettages: a practical approach. New York: Springer; 2019. p. 261–332.
6. Papaefthimou M, Symiakaki H, Mentzelopoulou P, et al. Study on the morphology and reproducibility of the diagnosis of endometrial lesions utilizing liquid-based cytology. Cancer. 2005;105:56–64.
7. Buccoliero AM, Gheri LF, Castiglione F, et al. Liquid-based endometrial cytology: cytohistological correlation in a population of 917 women. Cytopathology. 2007;18:241–9.

8. Jimenez-Ayala M, Jimenez-Ayala PB. Endometrial adenocarcinoma: prevention and early diagnosis. In: Orell SR, editor. Monographs in clinical cytology. Basel, Switzerland: Karger AG; 2008.
9. Norimatsu Y, Miyamoto M, Kobayashi TK, et al. Diagnostic utility of phosphatase and tensin homolog, beta-catenin, and p53 for endometrial carcinoma by thin-layer endometrial preparations. Cancer. 2008;114:155–64.
10. Norimatsu Y, Kouda H, Kobayashi TK, et al. Utility of liquid-based cytology in endometrial pathology: diagnosis of endometrial carcinoma. Cytopathology. 2009;20:395–402.
11. Hudson E, Blenkinsopp WK. The corpus uteri. In: Coleman DV, Chapman PA, editors. Clinical cytotechnology. London: Butterworths; 1989. p. 220–34.
12. Kobayashi TK, Norimatsu Y, Buccoliero AM. Cytology of the body of the uterus. In: Gray W, Kocjan G, editors. Diagnostic cytopathology. 3rd ed. London: Churchill Livingstone; 2010. p. 689–719.
13. Yanoh K, Hirai Y, Sakamoto A, et al. New terminology for intrauterine endometrial samples: a group study by the Japanese Society of Clinical Cytology. Acta Cytol. 2012;56:233–41.
14. Norimatsu Y, Yanoh K, Kobayashi TK. The role of liquid-based preparation in the evaluation of endometrial cytology. Acta Cytol. 2013;57:423–35.
15. Norimatsu Y, Yanoh K, Hirai Y, et al. A diagnostic approach to endometrial cytology by means of liquid-based preparations. Acta Cytol. 2020;64:195–207.
16. Hirai Y, Yanoh K, Norimatsu Y. Atlas of endometrial cytology; based on descriptive and standardized reporting system. Tokyo: Igaku Shoin; 2015. (in Japanese).
17. Carmichael DE. The Pap Smear: Life of George N. Papanicolaou Illinois. Springfield, IL: Charles C Thomas Publisher; 1973.
18. Boon ME, Chantziantoniou N. Papanicolaou revisited. Leiden: Coulomb Press Leyden; 2013. p. 19–118.
19. Babes A. Diagnostic du cancer du col uterin par les frottis. Presse Med. 1928;36:451–4. (in French).
20. Austin RM. George Papanicolaou's efforts to develop novel cytologic methods for the early diagnosis of endometrial carcinoma. Acta Cytol. 2017;61:281–98.
21. Koss LG, Schreiber K, Oberlander SG, et al. Detection of endometrial carcinoma and hyperplasia in asymptomatic women. Obstet Gynecol. 1984;64:1–11.
22. Jobo T, Arai T, Sato R, et al. Clinicopathologic relevance of asymptomatic endometrial carcinoma. Acta Cytol. 2003;47:611–5.
23. Morse AR. The value of endometrial aspiration in gynaecological practice. In: Koss LG, Coleman D, editors. Advances in clinical cytology. London: Butterworths; 1981. p. 44–63.
24. Crow J, Gordon H, Hudson E. An assessment of the Mi-Mark endometrial sampling technique. J Clin Pathol. 1980;33:72–80.
25. Ginsberg NA, Padleckas R, Javaheri G. Diagnostic reliability of Mi-Mark helix technique in endometrial neoplasia. Obstet Gynecol. 1983;62:225–30.
26. Skaarland E. New concept in diagnostic endometrial cytology: diagnostic criteria based on composition and architecture of large tissue fragments in smears. J Clin Pathol. 1986;39:36–43.
27. Byren AJ. Endocyte endometrial smears in the cytodiagnosis of endometrial carcinoma. Acta Cytol. 1990;34:373–81.
28. Ishii Y, Fujii M. Criteria for differential diagnosis of complex hyperplasia or beyond in endometrial cytology. Acta Cytol. 1997;41:1095–102.
29. Norimatsu Y, Shimizu K, Kobayashi TK, et al. Cellular features of endometrial hyperplasia and well differentiated adenocarcinoma using the endocyte sampler; diagnostic criteria based on the cytoarchitecture of tissue fragments. Cancer. 2006;108:77–85.
30. Norimatsu Y, Miyamoto M, Kobayashi TK, et al. Diagnostic utility of phosphatase and tensin homolog, beta-catenin, and p53 for endometrial carcinoma by thin-layer endometrial preparatins. Cancer. 2008;114:155–64.
31. Burk JR, Lehman HF, Wolf FS. Inadequacy of Papanicolaou smears in the detection of endometrial cancer. N Engl J Med. 1974;291:191–2.

32. Koss LG, Melamed MR. Proliferative disorders and carcinoma of the endometrium. In: Koss LG, Melamed MR, editors. Koss' diagnostic cytology and its histopathologic bases. New York: Lippincott Williams & Wilkins; 2006. p. 422–5.

33. Geldenbuys L, Murray ML. Sensitivity and specificity of the pap smear for glandular lesions of the cervix and endometrium. Acta Cytol. 2007;51:47–50.

34. Kobayashi TK, Okamoto H. Cytopathology of pregnancy-induced cell patterns in cervicovaginal smears. Am J Clin Pathol. 2000;114:S6–S20.

35. Cary WH. A method of obtaining endometrial smears for study of their cellular content. Am J Obstet Gynecol. 1943;46:422–3.

36. Torres JE, Holmquist ND, Danos ML. The endometrial irrigation smear in the detection of adenocarcinoma of the endometrium. Acta Cytol. 1969;13:163–8.

37. Dowling EA, Gravlee LC, Hutchins KE. A new technique for the detection of adenocarcinoma of the endometrium. Acta Cytol. 1969;13:496–501.

38. Afonso JF. Value of the Gravlee jet washer in the diagnosis of endometrial cancer. Obstet Gynecol. 1975;46:141–6.

39. Johnsson JE, Stormby NG. Cytological brush technique in malignant disease of the endometrium. Acta Obstet Gynecol Scand. 1968;47:38–51.

40. Milan AR, Markley RL. Endometrial cytology by a new technic. Obstet Gynecol. 1973;42:469–75.

41. Goldberg GL, Tsalacopoulos G, Davey DA. A comparison of endometrial sampling with the Accurette and Vabra aspirator and uterine curettage. S Afr Med J. 1982;61:114–6.

42. Isaacs JH, Wilhoite RW. Aspiration cytology of the endometrium: office and hospital sampling procedures. Am J Obstet Gynecol. 1974;118:679–84.

43. Anderson DG, Eaton CJ, Galinkin LJ, et al. The cytologic diagnosis of endometrial adenocarcinoma. Am J Obstet Gynecol. 1976;125:376–83.

44. Maksem JA, Robboy SJ, Bishop JW, et al. Endometrial cytology with tissue correlations. New York: Springer; 2009.

45. Fujihara A, Norimatsu Y, Kobayashi TK, et al. Direct intrauterine sampling with uterobrush: cell prepararion by the "flicked" method. Diagn Cytopathol. 2006;34:486–90.

46. Akalin A, Lu D, Woda B, et al. Rapid cell blocks improve accuracy of breast FNAs beyond that provided by conventional cell blocks regardless of immediate adequacy evaluation. Diagn Cytopathol. 2008;36:523–9.

47. Yang GC, Wan LS, Papellas J, et al. Compact cell blocks. Use for body fluids, fine needle aspirations and endometrial brush biopsies. Acta Cytol. 1998;42:703–6.

48. Diaz-Rosario LA, Kabawat SE. Cell block preparation by inverted filter sedimentation is useful in the differential diagnosis of atypical glandular cells of undetermined significance in ThinPrep specimens. Cancer. 2000;90:265–72.

49. Nathan NA, Narayan E, Smith MM, et al. Cell block cytology: improved preparation and its efficacy in diagnostic cyology. Am J Clin Pathol. 2000;114:599–606.

50. Shimizu K, Norimatsu Y, Kobayashi TK, et al. Endometrial glandular and stromal breakdown, part 1: cytological appearance. Diagn Cytopathol. 2006;34:609–13.

51. Garcia F, Barker B, Davis J, et al. Thin-layer cytology and histopathology in the evaluation of abnormal uterine bleeding. J Reprod Med. 2003;48:882–8.

52. Watanabe J, Nishimura Y, Tsunoda S, et al. Liquid-based preparation for endometrial cytology: usefulness for detecting the prognosis of endometrial carcinoma preoperatively. Cancer. 2009;117:254–63.

53. Kosmas K, Stamoulas M, Marouga A, et al. Expression of p53 in imprint smears of endometrial carcinoma. Diagn Cytopathol. 2014;42:416–22.

54. Norimatsu Y, Moriya K, Sakurai T, et al. Immunocytochemical expression of PTEN and beta-catenin for endometrial intraepithelial neoplasia in Japanese women. Ann Diagn Pathol. 2007;11:103–8.

55. Di Lorito A, Zappacosta R, Capanna S, et al. Expression of PTEN in endometrial liquid-based cytology. Acta Cytol. 2014;58:495–500.

Overview of the Yokohama System for Reporting Endometrial Cytology

2

Yasuo Hirai and Jun Watanabe

2.1 Background

The most important role of endometrial cytology is to detect malignant tumors as accurately as possible. However, endometrial malignancies cannot be diagnosed definitively by cytology alone. In principle, clinical treatment is administered after a definitive diagnosis based on tissue biopsy. The diagnostic accuracy of endometrial cytology is important because the results determine the necessity for tissue biopsy as a definitive method. It is commonly believed that the pain experienced during endometrial cytological sampling is considerably greater than that associated with collecting cells from the uterine cervix. In reality, due to the development of very thin endometrial brushes in recent years, the pain experienced by the patient is minimal.

To confirm the final diagnosis, endometrial tissue collection is required, although the process is invasive. An increase in the frequency of false positives for endometrial cytology can lead to an increase in unnecessary tissue tests; hence, it should be avoided. The Bethesda style descriptive reporting format is intended to best communicate the results of the endometrial cytology clinically and to enhance mutual communication. The Bethesda style descriptive reporting format is currently the most important form in the world, and the Yokohama System (TYS) has adopted this style. Endometrial cytology shows abnormalities in individual cells and cell architecture of cell clusters, but there are gray-zone cases where it is

Y. Hirai (✉)
Department of Obstetrics and Gynecology, Faculty of Medicine, Dokkyo Medical University, Tochigi, Japan

PCL Japan Pathology and Cytology Center, PCL Inc., Saitama, Japan
e-mail: yhirai-ind@umin.ac.jp

J. Watanabe
Department of Bioscience and Laboratory Medicine, Hirosaki University Graduate School of Health Science, Hirosaki, Japan

difficult to definitively determine whether the result is benign or malignant. In the descriptive reporting format like TYS, it is possible to separate and describe findings in the category that should be considered benign from findings suspected to be malignant.

Initially, we had two Bethesda-style reporting systems or formats for endometrial cytology: (1) the descriptive reporting format for endometrial cytology devised by the Japanese Society of Clinical Cytology (JSCC) study group [1]; and (2) the new classification scheme devised by Petros Karakitsos, et al. (Greek group) [2] (Table 2.1). While a correlation between the diagnostic categories and the risk of malignancy was found in the Greek system, the cytomorphologic criteria in the

Table 2.1 Cytological interpretention/result and histological diagnosis

Cytological result (JSCC)	Cytological result (Greece)	Histological diagnosis (WHO classification)	Descriptive terminology for cytological result	TYS grading category	Explanatory comments
Unsatisfactory	Inadequate		Unsatisfactory for evaluation	TYS 0	To be compliant with the Bethesda style
Negative for malignancy	Without evidence of malignancy	Proliferative endometrium Secretory endometrium Menstrual endometrium Atrophic endometrium Benign reactive change Endometrial polyp Endometrial metaplasia Arias-Stella reaction Endometrial glandular and stromal break down (EGBD)	Negative for malignant tumors and precursors	TYS 1	To be compliant with the Bethesda style
Endometrial hyperplasia	ACE-L	Endometrial hyperplasia without atypia	Endometrial hyperplasia without atypia	TYS 3	To be more descriptive
Atypical endometrial hyperplasia	ACE-H	Endometrial atypical hyperplasia, endometrioid intraepithelial neoplasia (EAH/EIN)	Endometrial atypical hyperplasia/ Endometrioid intraepithelial neoplasia (EAH/EIN)	TYS 5	To be more descriptive, and to be compliant with WHO histological classification, 5th edition

Table 2.1 (continued)

Cytological result (JSCC)	Cytological result (Greece)	Histological diagnosis (WHO classification)	Descriptive terminology for cytological result	TYS grading category	Explanatory comments
Malignant tumor	Malignant	All malignant tumors, including endometrioid carcinoma (G1, G2, G3, squamous differentiation), serous carcinoma, clear cell carcinoma, undifferentiated and dedifferentiated carcinoma, mixed carcinoma, other endometrial carcinomas (mesonephric carcinoma, squamous cell carcinoma, gastric or gastrointestinal type mucinous carcinoma, masonephric-like carcinoma), carcinosarcoma, leiomyosarcoma, endometrial stromal sarcoma, undifferentiated uterine sarcoma, adenosarcoma and extrauterine malignant tumors	Malignant neoplasms	TYS 6	To be more descriptive, and to be compliant with WHO histological classification, 5th edition
ATEC-US	ACE-US		Atypical endometrial cells, of undetermined significance (ATEC-US)	TYS 2	Gray zone that is compliant with the Bethesda style, shoul be less than 5% of specimens
ATEC-A	ACE-H		Atypical endometrial cells, cannot exclude EAH/EIN or malignant condition (ATEC-AE)	TYS 4	Gray zone that is compliant with the Bethesda style

various diagnostic categories were particularly emphasized in the Japanese system. We combined these two reporting systems in a new one, TYS, for reporting endometrial cytology, which took into consideration every scientific and epidemiologic evidence accumulated heretofore in each gross local area (Japan and Greece) to be used in a constructive way (Tables 2.2 and 2.3). The estimated risk of malignancy for each cytological interpretational category was deduced from literature in both geographical areas; hence, the clinical management guidelines were developed based on these findings (Table 2.4).

Table 2.2 Descriptive categories of The Yokohama System for Reporting Endometrial Cytology

1. Unsatisfactory for evaluation	TYS 0
2. Negative for malignant tumors and precursors	TYS 1
3. Atypical endometrial cells of undetermined significance (ATEC-US)	TYS 2
4. Endometrial hyperplasia without atypia	TYS 3
5. Atypical endometrial cells, cannot exclude EAH/EIN or malignant condition (ATEC-AE)	TYS 4
6. Endometrial atypical hyperplasia/Endometrioid intraepithelial neoplasia (EAH/EIN)	TYS 5
7. Malignant neoplasms	TYS 6

Table 2.3 Relative risk of the descriptive categories outlined in The Yokohama System, based on Japanese studies to date

	Risk of malignancy, %		
Categories	**Malignant tumor**	**Atypical endometrial hyperplasia**	**Management**
Unsatisfactory for evaluation:TYS0	No data, so far	No data, so far	Repeat cytology in 3 months if increased clinical suspicion
Negative for malignant tumors and precursors:TYS1	0.40%	0.20%	Clinical follow-up as needed
Atypical endometrial cells of undetermined significance (ATEC-US):TYS2	No data, so far	No data, so far	Repeat cytology in 3 months as needed
Endometrial hyperplasia without atypia:TYS3	18.20%	18.20%	More follow-up, hysteroscope, biopsy
Atypical endometrial cells, cannot exclude EAH/EIN or malignant condition (ATEC-AE):TYS4	60.00%	5.70%	More aggressive follow-up, hysteroscope, biopsy
Endometrial atypical hyperplasia/ Endometrioid intraepithelial neoplasia (EAH/EIN):TYS5	61.50%	30.80%	More aggressive follow-up, hysteroscope, biopsy
Malignant neoplasms:TYS6	94.50%	0.30%	More aggressive follow-up, hysteroscope, biopsy, staging

Table 2.4 Descriptive reporting format and diagnostic categories of The Yokohama System for Reporting Endometrial Cytology

Specimen type
Conventional method, liquid-based method
Specimen adequacy
Satisfactory, unsatisfactory (rejected specimen, fully evaluated, unsatisfactory specimen)
Result
Negative for malignant tumors and precursors:TYS1
Endometrium in proliferative phase, in secretory phase, in menstrual phase, atrophic endometrium, benign reactive change (IUD, TAM, etc.), endometrial polyp, endometrial metaplasia, arias-Stella reaction, endometrial glandular and stromal break down (EGBD)
Endometrial hyperplasia without atypia:TYS3
Endometrial hyperplasia without atypia
Endometrial atypical hyperplasia/Endometrioid intraepithelial neoplasia (EAH/EIN):TYS5
EAH, EIN, atypical polypoid adenomyoma
Malignant neoplasms:TYS6
All malignant tumors, including endometrioid carcinoma (G1, G2, G3, squamous differentiation), serous carcinoma, clear cell carcinoma, undifferentiated and dedifferentiated carcinoma, mixed carcinoma, other endometrial carcinomas (mesonephric carcinoma, squamous cell carcinoma, gastric or gastrointestinal type mucinous carcinoma, masonephric-like carcinoma), carcinosarcoma, leiomyosarcoma, endometrial stromal sarcoma, undifferentiated uterine sarcoma, adenosarcoma and extrauterine malignant tumors
Atypical endometrial cells of undetermined significance (ATEC-US):TYS2
Atypical endometrial cells, cannot exclude EAH/EIN or malignant condition (ATEC-AE):TYS4

IUD intrauterine device; *TAM* tamoxifen

2.2   The Reproducibility Assessment of the Yokohama System

The evaluation of the reproducibility of the cytological interpretation of endometrial lesions by TYS is described in detail in Chap. 5. The conclusions are summarized below.

We evaluated the reproducibility of the cytological diagnosis of endometrial lesions by TYS using BD (Beckton-Dickinson) SurePath™-LBC [3]. To confirm the reproducibility of the diagnosis and to further study the inter- and intra-observer agreement, a second review was conducted after 3 months, including three additional cytopathologists. As a result, the inter-observer agreement of the negative classes improved progressively from "good to fair" to "excellent," with values increasing from 0.70 to 0.81, respectively. Both the EGBD and malignant classes improved progressively from "good to fair" to "excellent," with values increasing from 0.62–0.63 to 0.84–0.95, respectively. As for the diagnostic agreement of ATEC in two rounds of the diagnostic review, the results were similar with a value of 0.43 (moderate) for ATEC-AE and 0.69 (good to fair) for ATEC-US, and the results of reproducibility were surprisingly good. The overall intra-observer agreement between the first and the second rounds were "good to fair" to "excellent," with values ranging from 0.73 to 0.90, and all kappa

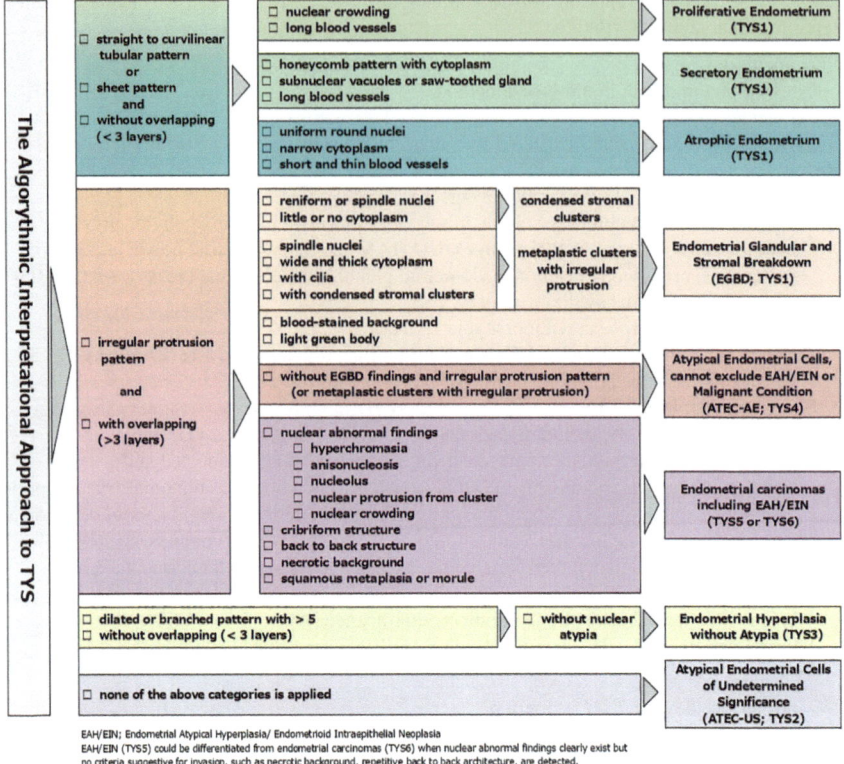

EAH/EIN; Endometrial Atypical Hyperplasia/ Endometrioid Intraepithelial Neoplasia
EAH/EIN (TYS5) could be differentiated from endometrial carcinomas (TYS6) when nuclear abnormal findings clearly exist but
no criteria suggestive for invasion, such as necrotic background, repetitive back to back architecture, are detected.

Fig. 2.1 The Algorithmic interpretational approach to TYS

improvements were significant ($p < 0.0001$). Applying the algorithmic approach to TYS (Fig. 2.1) may be a valid method to improve the precision (reproducibility) of endometrial cytology.

2.3 Clinical Handling in the Yokohama System for Reporting Endometrial Cytology

Based on the already accumulated data on the accuracy of endometrial cytology, clinical handling for each category of TYS is recommended in Japan, as shown in Table 2.5. These recommendations do not result in firm guidelines due to limited clinical data.

Table 2.5 Provisional recommendations of handling methods for each category of TYS in Japan

TYS category	Clinical handling
TYS 0 Unsatisfactory for evaluation	Re-examination
TYS 1 Negative for malignant tumors and precursors	Routine medical examination
TYS 2 Atypical endometrial cells, of undetermined significance (ATEC-US)	Re-examination using cytology or endometrial tissue biopsy
TYS 3 Endometrial hyperplasia without atypia	Endometrial tissue biopsy
TYS 4 Atypical endometrial cells, cannot exclude EAH/ EIN or malignant condition (ATEC-AE)	Endometrial tissue biopsy
TYS 5 Endometrial atypical hyperplasia/Endometrioid intraepithelial neoplasia (EAH/EIN)	Endometrial tissue biopsy
TYS 6 Malignant neoplasms	Endometrial tissue biopsy

References

1. Yanoh K, Hirai Y, Sakamoto A, et al. New terminology for intrauterine endometrial samples: a group study by the Japanese Society of Clinical Cytology. Acta Cytol. 2012;56:233–41.
2. Margari N, Pouliakis A, Anoinos D, et al. A reporting system for endometrial cytology: cyto-morphologic criteria - implied risk of malignancy. Diagn Cytopathol. 2016;44:888–901.
3. Norimatsu Y, Yamaguchi T, Taira T, et al. Inter-observer reproducibility of endometrial cytology by the Osaki Study Group method: utilizing the Becton Dickinson SurePath™ liquid-based cytology. Cytopathology. 2016;27:472–8.

Pathogenetic Bases of the Yokohama System for Reporting Endometrial Cytology

3

Yasuo Hirai and Jun Watanabe

In developing The Yokohama System(TYS) for reporting endometrial cytology, current understanding of initial/precursor lesions, which sequentially take place in the process of carcinogenesis of endometrial cancer, is of enormous importance. In this regard, the chapter on endometrial carcinoma in the Fourth Edition World Health Organization (WHO) Classification of Tumors of the uterine corpus offers timely and valuable insights [1] (Table 3.1).

When developing the TYS classification, first, we aimed to create a category for the precursor lesions of endometrial carcinoma, on which we have sufficient information due to extensive clinical research experience. According to the Fourth Edition WHO Classification of Tumors [1], the precursor lesion of endometrial carcinoma is endometrial atypical hyperplasia/endometrioid intraepithelial neoplasia (EAH/EIN). EAH is considered as a noninvasive histological lesion with biologically similar abnormalities to endometrioid carcinoma. Therefore, in TYS, we designated "Endometrial atypical hyperplasia/endometrioid intraepithelial neoplasia (EAH /EIN)" as a cytological category for this histological entity.

Second, we aimed to determine an applicable cytology category for the precursor lesion of serous carcinoma. Serous carcinoma is more aggressive and has a poorer prognosis than endometrioid carcinoma. Therefore, serous endometrial intraepithelial carcinoma (SEIC), regarded as an initial lesion of the serous carcinoma, should be considered as a malignant tumor, because of its frequent association with extrauterine spread [2]. Thus, in TYS, the cytological interpretational category for SEIC

Y. Hirai (✉)
Department of Obstetrics and Gynecology, Faculty of Medicine, Dokkyo Medical University, Tochigi, Japan

PCL Japan Pathology and Cytology Center, PCL Inc., Saitama, Japan
e-mail: yhirai-ind@umin.ac.jp

J. Watanabe
Department of Bioscience and Laboratory Medicine, Hirosaki University Graduate School of Health Science, Hirosaki, Japan

Table 3.1 Descriptive reporting format and diagnostic categories of TYS for Reporting Endometrial Cytology (Compatible version for WHO 5th edition[a])

(i) Specimen type   Conventional method, liquid-based method
(ii) Specimen adequacy   Satisfactory, unsatisfactory (rejected specimen, fully evaluated, unsatisfactory specimen)

(iii) Result
   – Negative for malignant tumors and precursors: TYS1
   Endometrium in proliferative phase, in secretory phase, in menstrual phase, atrophic endometrium, benign reactive change (IUD, TAM, etc.), endometrial polyp, endometrial metaplasia, arias-Stella reaction, endometrial glandular and stromal breakdown (EGBD)
   – Endometrial hyperplasia without atypia: TYS3
   Endometrial hyperplasia without atypia
   – Endometrial atypical hyperplasia/Endometrioid intraepithelial neoplasia (EAH/EIN): TYS5
   EAH, EIN, atypical polypoid adenomyoma
   – Malignant neoplasms(endometrial carcinomas): TYS6
   All malignant tumors, including endometrioid carcinoma (G1, G2, G3, squamous differentiation), serous carcinoma, clear cell carcinoma, undifferentiated and dedifferentiated carcinoma, mixed carcinoma, other endometrial carcinomas (mesonephric carcinoma, squamous cell carcinoma, gastric or gastrointestinal type mucinous carcinoma, mesonephric-like carcinoma), carcinosarcoma, leiomyosarcoma, endometrial stromal sarcoma, undifferentiated uterine sarcoma, adenosarcoma and extrauterine malignant tumors
   – Atypical endometrial cells of undetermined significance (ATEC-US): TYS2
   – Atypical endometrial cells, cannot exclude EAH/EIN
   **   or malignant condition (ATEC-AE): TYS4**

IUD intrauterine device; *TAM* tamoxifen
[a]Tumors of the uterine corpus, Kim K-R, Lax SF, Lazar AJ, et al. editors. In: WHO Classification of Tumor. Editorial Board, editor. WHO Classification of Tumors. Female genital tumors, 5th ed. Lyon: IARC press; 2020. p246–308

should be the "Malignant neoplasms" category instead of the "Endometrial atypical hyperplasia/endometrioid intraepithelial neoplasia (EAH/EIN)" category.

In the fifth edition of the WHO Classification of Tumors of the uterine corpus [3], the morphological classification proposed in the fourth edition (endometrial atypical hyperplasia/endometrioid intraepithelial neoplasia (EAH /EIN), endometrioid carcinoma, and serous carcinoma) remained unchanged. However, in a subset of high-grade cancers with overlapping molecular features, considerable inter-observer variability existed in the microscopic diagnosis. Therefore, four groups of carcinomas with different molecular features identified by The Cancer Genome Atlas (TCGA) have been added to the endometrial carcinoma category in the WHO 5th edition to be used integrally with conventional morphological classification. The four new subtypes include POLE-ultramutated endometrioid carcinoma (EC), mismatch repair (MMR)-deficient EC, p53-mutant EC, and non-specific molecular profile (NSMP) EC. In relation to TYS, there is no need to change each category of TYS, which is based on cytomorphological findings. However, as the molecular features of the lesions in each category become clearer, a new category could be required.

References

1. Tumours of the uterine corpus. In: Kurman RJ, Carcangiu ML, Herrigton CS, editors. WHO classification of tumours of female reproductive organs. 4th ed. Lyon: IARC press; 2014. p. 125–54.
2. Pathiraja P, Dhar S, Haldar K. Serous endometrial intraepithelial carcinoma: a case series and literature review. Cancer Manag Res. 2013;175:117–22.
3. Tumours of the uterine corpus, Kim K-R, Lax SF, Lazar AJ, et al. editors. In: WHO Classification of Tumor Editorial Board, editor. WHO Classification of Tumors. Female genital tumours, 5th ed. Lyon: IARC press; 2020. p. 246–308.

Endometrial Cell Sampling Procedure to Obtain Cytologic Specimens

4

Tetsuji Kurokawa, Toshimichi Onuma, Akiko Shinagawa, and Yoshio Yoshida

4.1 Significance of Endometrial Cytology for Women with Abnormal Uterine Bleeding at the Outpatient Clinic

Endometrial carcinoma is the most common malignancy of the female genital tract. In 2018, the total estimated number was 382,069, and the number of deaths due to endometrial carcinoma was 89,929 worldwide [1]. The incidence rates of endometrial carcinoma (EC) are generally high. If detected at an early stage, EC is a highly curable disease. However, screening for endometrial cancer has not been performed worldwide because of the lack of an acceptable test, which would reduce mortality. Routine Papanicolaou (Pap) testing is inadequate. In addition, textbooks have reported that directly sampled endometrial cytology is also too insensitive and non-specific to be useful for screening because the rate of inadequate endometrial cytology is higher than that of endometrial biopsy, in the absence of established criteria for inadequate specimens in both methods [2]. Directly sampled endometrial cytology as an acceptable screening test was dismissed after 1976 [3]. By contrast, in Japan, many gynaecologists and cytopathologists have continued to improve directly sampled endometrial cytology for a long time. As it concerns the specimen preparation technique, some manuscripts have reported methods for reducing inadequate samples [4, 5]. With regard to the diagnostic criteria for endometrial cytology, some manuscripts have reported a clear and easy diagnostic flow [6–8]. At present, almost all gynaecologists in Japan have a preferred technique to directly

The original version of the chapter has been revised. A correction to this chapter can be found at
https://doi.org/10.1007/978-981-16-5011-6_19

T. Kurokawa (✉) · T. Onuma · A. Shinagawa · Y. Yoshida
Department of Gynecology and Obstetrics, Faculty of Medical Sciences, University of Fukui, Fukui, Japan
e-mail: kurotetu@u-fukui.ac.jp

23

Table 4.1 Indications and contraindications for endometrial cytology in many Japanese institutions [9]

Indications
  Symptomatic patients
    Abnormal uterine bleeding in the last 6 months
    Abnormal perimenopausal uterine bleeding
    Postmenopausal uterine bleeding
    Abnormal menses
    (hypermenorrhoea, irregular menstrual cycle etc.)
    Abnormal vaginal discharge
    Suspected endometrial cancer
  Asymptomatic patients
    Abnormal endometrial findings on transvaginal ultrasonography
    Postmenopausal oestrogen therapy without progestins
    Hereditary nonpolyposis colorectal cancer (lynch syndrome)
    Obesity, type II diabetes mellitus, hypertension
Contraindications
  Pregnancy or suspected pregnancy

obtain endometrial cells. Japanese outpatient guidelines for gynaecological practice include the indications and the technique of directly sampled endometrial cytology [9] (Table 4.1). Endometrial cytology is the accepted first step in evaluating women with abnormal uterine bleeding at an outpatient clinic.

4.2   Indications for the Use of Directly Sampled Endometrial Cytology (Table 4.1)

The chief symptom of EC is abnormal uterine bleeding, which may be due to a variety of conditions. In particular, postmenopausal women with abnormal uterine bleeding are at a particularly high risk of EC. Patients at high risk of EC are divided into symptomatic and asymptomatic groups. The symptomatic group includes women with abnormal uterine bleeding, abnormal menses (hypermenorrhoea, irregular menstrual cycle etc.), or abnormal vaginal discharge. The asymptomatic group includes high-risk women, such as women with obesity or receiving postmenopausal oestrogen therapy without progestin; those taking tamoxifen with abnormal findings on transvaginal ultrasonography; or members of families with hereditary nonpolyposis colorectal cancer (Lynch syndrome) [2].

4.3   Collection Instruments and Evaluation of Endometrial Lesions by Endometrial Cytology

The collection instruments can be divided into endometrial aspiration devices and endometrial brushing instruments. For each technique, many types of instruments have been reported in previous manuscripts (for example, the Isaac endometrial cell sampler, Masubuchi's method, Medhosa cannula, Mi-Mark endometrial sampling kit, Endobrush®, Endopap®, Tao brush, and Endocyte). We introduce two instruments for aspiration techniques. One is the Isaac endometrial cell sampler (IECS), which has a metallic cannula and a specimen collector [10] (Fig. 4.1). The other is the instrument using Masubuchi's method [11]. The diagnostic accuracy is 77.8% for the IECS and 92.3% for Masubuchi's method [10, 11]. We introduce two instruments for brushing techniques. One is the Mi-Mark helix endometrial sampling kit, which has semi-sharp edges to abrade the endometrial lining [12] (Fig. 4.2). The other is the endobrush, which is composed of a plastic tube and a helicoidal thin wire with a nylon brush, with a plastic enlargement on the end [12] (Fig. 4.3). The sensitivity is 97% for the Mi-Mark helix and 96.4% for the endobrush [12, 13].

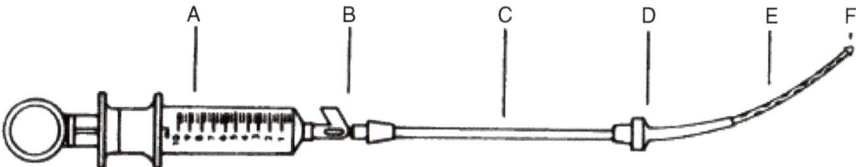

Isaacs endometrial cell sampler. (A) Syringe. (B) Adapter. (C) Shield. (D) Cervical stop. (E) Cannula with multiple perforations. (F) Cannula tip.

Fig. 4.1   The Isaac endometrial cell sampler (IECS). Reproduced with permission from [10]

Fig. 4.2   The Mi-Mark Helix endometrial sampling kit. Reproduced with permission from [12]

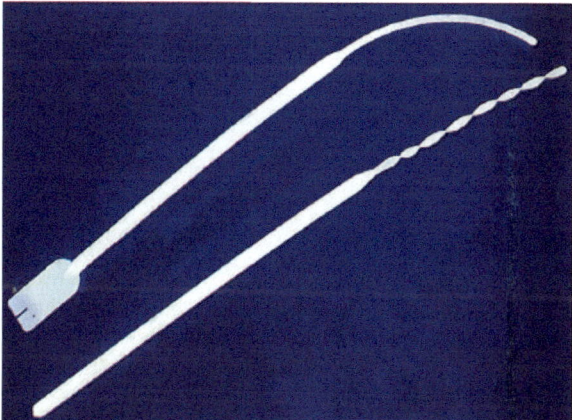

Fig. 4.3   The endobrush. Reproduced with permission from [12]

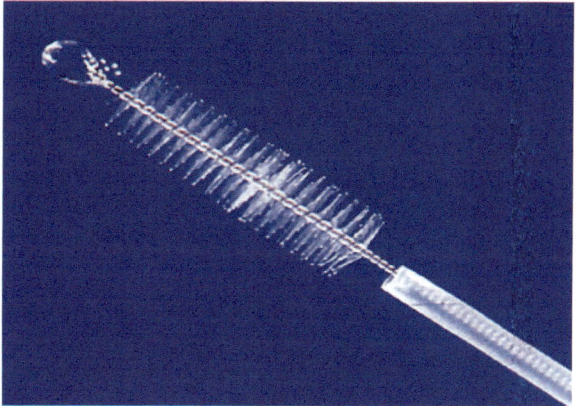

It is easier to directly obtain endometrial cytology samples using endometrial brushing techniques than endometrial aspiration techniques. The endobrush causes less pain and contamination of endocervical cells than the Mi-Mark helix. Therefore, we chose the Honest Super Brush (Honest Medical Company) as the preferred endometrial brush. Many Japanese institutions have also used the endobrush.

4.4   Techniques of Endometrial Sampling and Processing

1. The patient should have an empty bladder, as this allows the speculum to be inserted with less discomfort.
2. The patient should remove all clothing from the waist down and take the lithotomy position on the pelvic examination table.
3. A sheet should be placed under the hip to prepare for bleeding.
4. A gynaecologist should confirm the size, position, and tilt of the uterus by transvaginal ultrasonography before inserting the instrument.
5. The speculum (for example Cusco's speculum) should be gently inserted into the vagina.
6. The instrument is inside the plastic bag. Before removing the instrument from the plastic bag, the tip of the tube covering the brush should be slightly bent to match the direction from the cervical canal to the uterine cavity (Fig. 4.4).

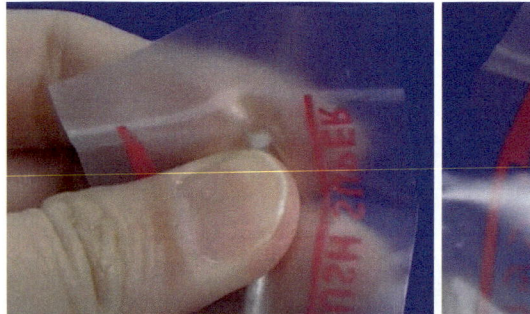
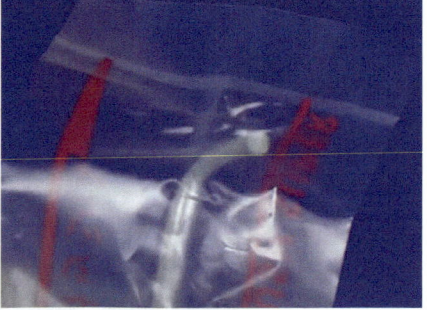
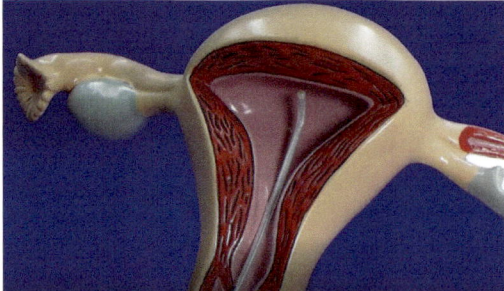

Fig. 4.4 Inserting the instrument. The tip of the tube covering the brush should be slightly bent using the thumb to match the direction from the cervical canal to the uterine cavity. The angle is 60–90 degrees. The instrument should be gently rotated or slowly moved up and down to collect endometrial cells. In this process, modulation of mechanical pressure exerted is needed to avoid perforation of the uterus

Fig. 4.5 The instrument in uterine cavity. The instrument should be rotated or moved up and down

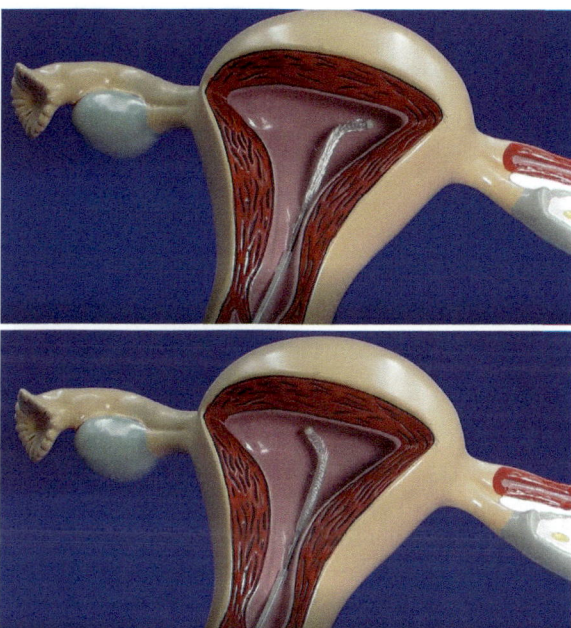

7. The instrument should be removed from the bag, and the external uterine orifice should be sterilized.
8. The instrument should be inserted into the external uterine orifice. The brush is inside the tube cover.
9. After the tip of the tube cover reaches the fundus of the uterus, the tube should be pulled back to expose the brush.
10. The instrument should be rotated or moved up and down to collect endometrial cells (Fig. 4.5).
11. The brush should be placed back inside the tube cover to prevent the contamination of endocervical cells.
12. The instrument should be removed from the uterus.
13. The brush should be immersed in the preservative fluid and gently stirred to remove endometrial cells (Fig. 4.6) (URL: https://www.youtube.com/watch?v=Q-lllraSt2w&feature=youtu.be)

4.5   Benefits and Harm of Directly Sampled Endometrial Cytology

Gynaecologists should interview patients about the possibility of pregnancy or of menstrual bleeding before the examination. Pregnant women have contraindications for directly sampled endometrial cytology. Gynaecologists should inform

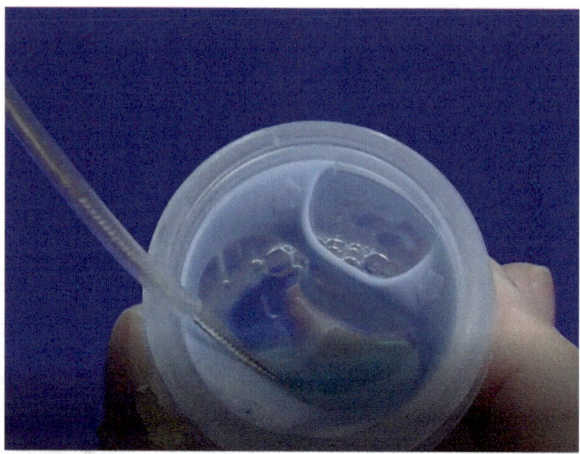

Fig. 4.6 The brush in the preservative fluid. The brush should be immersed in preservative fluid and gently stirred to remove endometrial cells

cytopathologists about the date of the last menstrual cycle because the cytologic findings in endometrial samples are strongly affected by the hormonal environment. The insertion of the instrument may be difficult in patients with fibroids or a history of uterine operation, such as caesarean section. If difficulty is expected, the gynaecologist should better confirm the size, position, and tilt of the uterus by transvaginal ultrasonography before inserting the instrument to ensure a smooth entry and minimize the patient's discomfort. Based on our direct experience, we consider the possibility of uterine perforation unlikely, unless the instrument is inserted in an extremely energetic modality. Anaesthesia was almost never needed in our practice.

The technique to directly obtain endometrial cytology samples has three major advantages for patients and gynaecologists. One is that endometrial cytology compared to endometrial biopsy reduces stress both for patients and gynecologists because endometrial brushing is less invasive than endometrial biopsy. The second advantage is that directly sampled endometrial cytology using a brush is less invasive towards the endometrial mucosa than endometrial histology using curettage. The third advantage is that the technique is extremely simple, and the time to obtain endometrial cells is very short (Fig. 4.7).

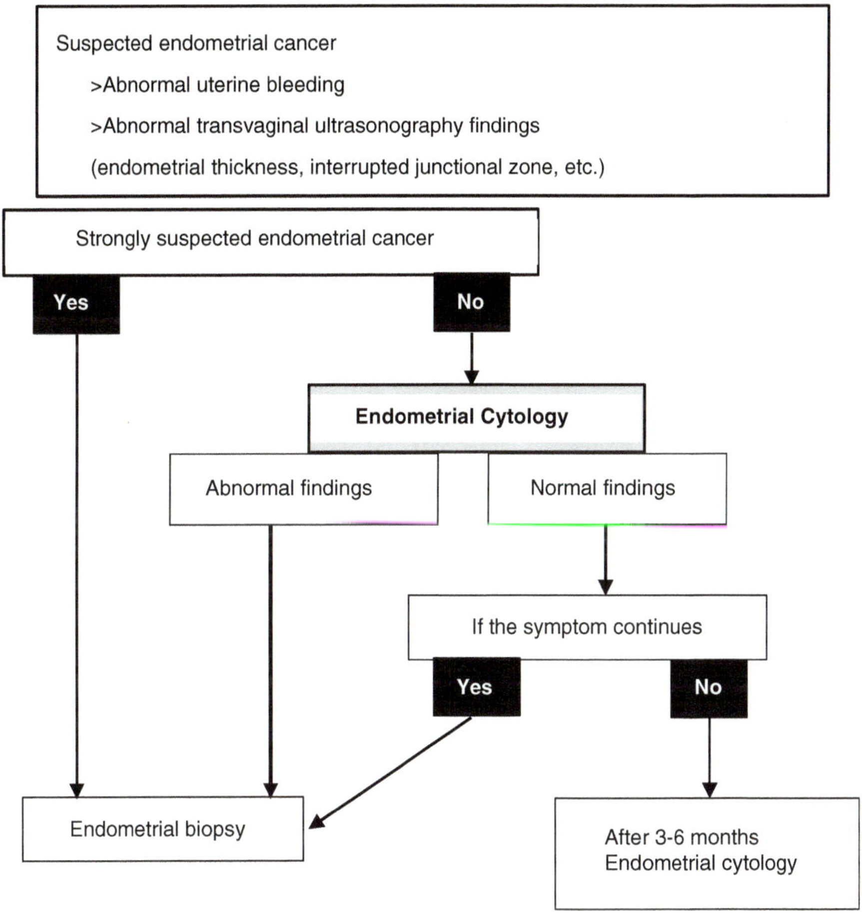

Fig. 4.7 Flowchart for patients suspected of having endometrial cancer in many Japanese institutions

References

1. Wild CP, Weiderpass E, Stewart BW, editors. World Cancer report cancer research for cancer prevention. Lyon: International Agency for Research on Cancer, World Health Organization; 2020. p. 403–10.
2. Berek JS. Berek & Novak's gynecology. 15th ed. Philadelphia: Wolters Kluwer; 2012. p. 1254–5.
3. Anderson DG, Eaton CJ, Galinkin LJ, et al. The cytologic diagnosis of endometrial adenocarcinoma. Am J Obstet Gynecol. 1976;125:376–83.
4. Fujihara A, Norimatsu Y, Kobayashi TK, et al. Direct intrauterine sampling with Uterobrush: cell preparation by the "flicked" method. Diagn Cytopathol. 2006;34:486–90.

5. Kurokawa T, Yoshida Y, Yakihara A, et al. Liquid-based endometrial cytology by preservative fluid. J Jpn Soc Clin Cytol. 2005;44:6–10.
6. Yanoh K, Norimatsu Y, Hirai Y, et al. New diagnostic reporting format for endometrial cytology based on cytoarchitectural criteria. Cytopathology. 2009;20:388–94. (in Japanese with English abstract).
7. Shinagawa A, Kurokawa T, Yamamoto M, et al. Evaluation of the benefit and use of the new terminology in endometrial cytology reporting system. Diagn Cytopathol. 2018;46:314–9.
8. Norimatsu Y, Yanoh K, Hirai Y, et al. A diagnostic approach to endometrial cytology by means of liquid-based preparations. Acta Cytol. 2020;64:195–207.
9. Kawaguchi R, et al. Guideline for gynecological practice in Japan: Japan Society of Obstetrics and Gynecology and Japan Association of Obstetricians and Gynecologists 2020 edition. Obstetrics Gynecol. 2020;47:54–7. (in Japanese).
10. Hutton JD, Morse AR, Anderson MC, et al. Endometrial assessment with Isaacs cell sampler. Br Med J. 1978;1:947–9.
11. Kuramoto H, Jobo T, Morisawa T, Kato Y, Hata K, Ono E, Imai T. Endometrial aspiration cytology clinic. J Jpn Soc Clin Cytol. 1982;21:527–34. (in Japanese with English abstract).
12. Jimenez-Ayala M, Jimenez-Ayala PB. Endometrial adenocarcinoma prevention and early diagnosis. In: Orell SR, editor. Monographs in clinical cytology. Basel: Karger AG; 2008. pp.13–20.
13. Yanoh K, Norimatsu Y, Munakata S, et al. Evaluation of endometrial cytology prepared with the Becton Dickinson SurePath™ method: a pilot study by the Osaki Study Group. Acta Cytol. 2014;58:153–61.

Algorithmic Interpretational and Diagnostic Approach to Endometrial Cytology for the Yokohama System

5

Yoshiaki Norimatsu, Tadao K. Kobayashi, Yasuo Hirai, and Franco Fulciniti

5.1 Definition and Criteria

Figure 5.1 shows the "Algorithmic Interpretational approach to endometrial cytology to the Yokohama System (Algorithmic approach to TYS)." This diagnostic scheme was developed for TYS [1–6] by modifying the algorithmic diagnostic scheme of the descriptive reporting format for endometrial cytology already devised by the JSCC study group [2, 3, 6]. This "algorithmic approach to TYS" is designed to be easily applicable in practice by following three diagnostic steps, progressively evaluating microscopic fields of a given sample from low to high power magnification, with special attention to the architectural and cytopathological features.

Y. Norimatsu (✉)
Department of Medical Technology, Faculty of Health Sciences, Ehime Prefectural University of Health Sciences, Tobe-cho, Iyo-gun, Ehime, Japan
e-mail: ynorimatsu@epu.ac.jp

T. K. Kobayashi
Division of Health Sciences, Cancer Education and Research Center, Osaka University Graduate School of Medicine, Osaka, Japan

Y. Hirai
Department of Obstetrics and Gynecology, Faculty of Medicine, Dokkyo Medical University, Tochigi, Japan

PCL Japan Pathology and Cytology Center, PCL Inc., Saitama, Japan

F. Fulciniti
Clinical Cytology Service, Istituto Cantonale dì Patologia, Ente Ospedaliero Cantonale, Locarno, Switzerland

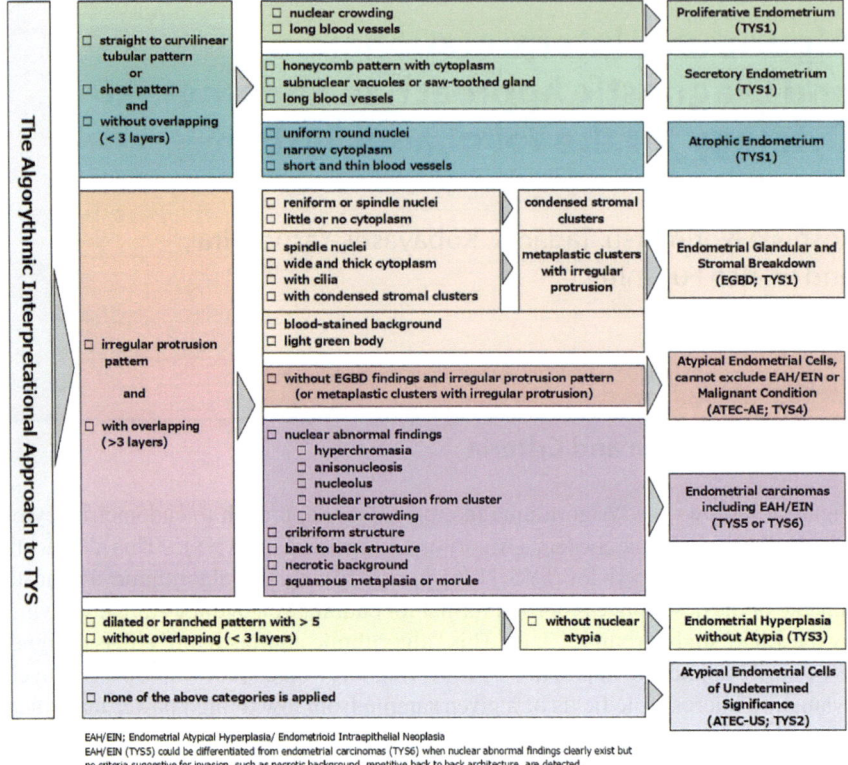

Fig. 5.1 Algorithmic interpretational approach to endometrial cytology to TYS

The first step is to determine whether the morphology of the endometrial cell clusters is regular or irregular[1] and whether nuclear overlapping of three or more layers exists or not.[2] Second, the cytomorphological features at medium and high

---

[1] "The morphology of the endometrial cell clusters is regular" refers to the presence of "straight to curvilinear tubular pattern." The width of the tubular-shaped gland is approximately uniform, and cohesion of the endometrial stromal cells to the margins of the gland is characteristic.

When the diction, "the morphology of the endometrial cell clusters is irregular" is used, there are several possibilities, all including the finding of "irregular protrusions patterns." Some irregular small projections are noted on the margin of cell clusters. The margin of the cytoplasm of those small projections is clearly observed. Cohesion of the endometrial stromal cells is not noted in margins of the gland.

In malignancy, a papillary growth pattern with irregular branching is also noted and the absence of endometrial stromal cells cohesion to the glandular margins is characteristic (Fig. 5.2). Occasionally, a fibrovascular core may be observed in the epithelial papillary clusters. When the papillary structure becomes complex and confluent, back to back and a cribriform pattern can be easily recognized (Figs. 5.3 and 5.4).

[2] "Nuclear overlapping" is defined by the presence of two or more layers of cell nuclei in the sheets, and this can clearly be seen under the microscope by fine focusing. In this way, the presence of two layers (Fig. 5.5a,b) or of three (or more) layers can easily be established (Fig. 5.5c,d).

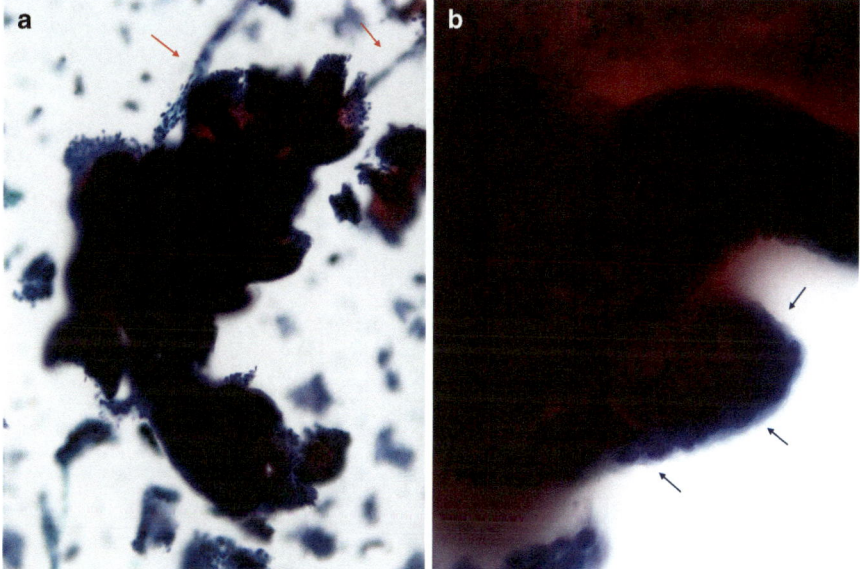

Fig. 5.2 Endometrial endometrioid carcinoma, Grade 1 (TYS6). In malignancy, a papillary growth pattern with irregular branching is also noted. Absence of adherent endometrial stromal cells (blue arrows) to the glandular margins is characteristic. Occasionally, a fibrovascular core (red arrows) may be observed within the epithelial papillary clusters (Papanicolaou stain, original magnification **a**: 10×, **b**: 40×)

power microscopic magnification, in various situations are considered prior to proceeding to the third step of cytological interpretation. The diagnostic procedure by using this "algorithmic approach to TYS" is described below (See also Fig. 5.1) [6, 7].

5.1.1 Inadequate Sample for Evaluation (TYS 0)

This condition includes every case in which the obtained sample cannot be evaluated for diagnosis for a variable series of causes (e.g., excess obscuring blood or inflammation, excess vaginal contamination, no endometrial cells present, etc. Please refer to Figs. 7.3–7.9 in Chap. 7). In order to assess the adequacy of endometrial LBC specimens, the evaluation of specimen cellularity (numerical criterion)[3] is among the most important issues.

---

[3] In a very recent study, Nimura et al. [8] proposed that the presence of ≥10 clusters with ≥30 endometrial cells per cluster could be used as a specimen adequacy criterion for endometrial LBC in non-menopausal patients, while the presence of 5 or more cell clusters was considered satisfactory in postmenopausal patients.

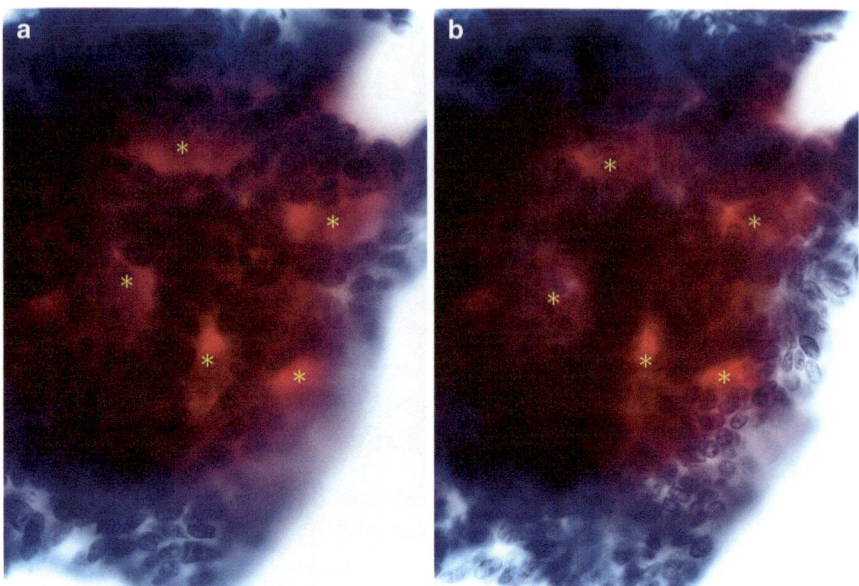

Fig. 5.3 Endometrial endometrioid carcinoma, Grade 1 (TYS6). A cribriform pattern with the formation of glandular cavities is observed. The case may hence be qualified as: "endometrial carcinomas including EAH/EIN." (Papanicolaou stain, original magnification **a, b**: 40×)

5.1.2   Negative for Malignant Tumors and Precursors (TYS1)

When all cell clusters in the preparation display physiological "tubular or sheet patterns" accompanied by stromal cells and "nuclear overlapping amounts to less than three layers," the interpretation is "Negative for Malignant Tumors and Precursors (TYS1)." Proliferative endometrium (PE), secretory endometrium (SE), and atrophic endometrium (AE) are included as normal (physiological) endometrium.

1. **Proliferative endometrium**

   Microscopic evaluation is performed in three steps: if, at low magnification "straight to curvilinear tubular pattern (Fig. 5.6)," a "sheet pattern (Fig. 5.7)" with "nuclear overlapping with less than 3 layers (Fig. 5.8)" are recognized, then, at intermediate magnification, features like "nuclear crowding,"and/or, "long blood vessels" are looked for (Fig. 5.9), and a presumptive diagnosis of proliferative endometrium is finally made.

2. **Secretory endometrium**

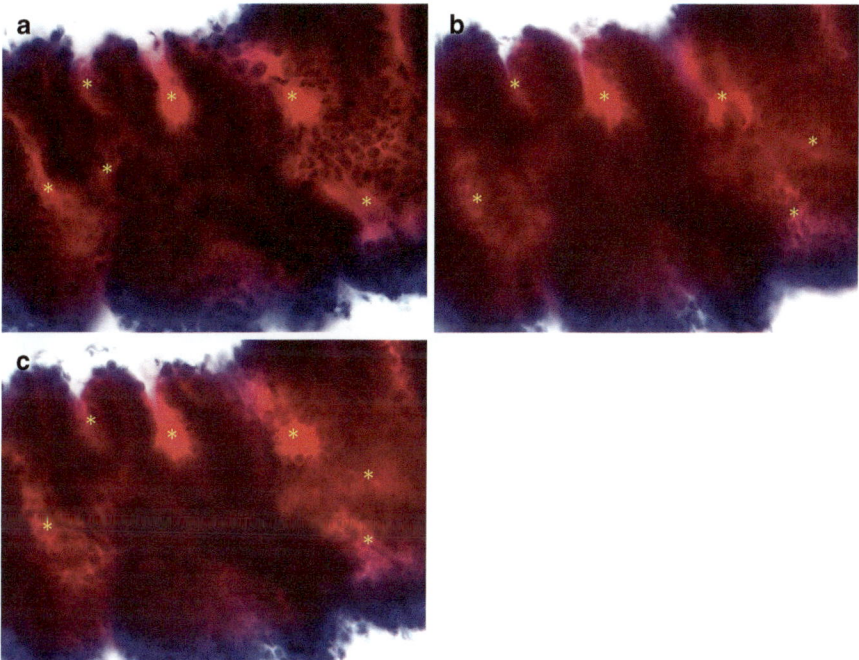

Fig. 5.4 Endometrial endometrioid carcinoma, Grade 1 (TYS6). In this case, a prominent "back to back" architecture can be observed (Papanicolaou stain, original magnification **a–c**: 40×)

The same first three cytological criteria: "straight to curvilinear tubular pattern (Fig. 5.10a,b)" or "sheet pattern (Fig. 5.10c)" and "nuclear overlapping with less than 3 layers" apply, then, if an "honeycomb pattern with cytoplasmic subnuclear vacuoles (Fig. 5.11a) or saw-toothed glands (Fig. 5.11b,c)," and/or, "long blood vessels" are identified, a presumptive diagnosis of secretory endometrium is finally performed.

3. **Atrophic endometrium**

   Similarly, at intermediate microscopic magnification, if "uniform round nuclei with scarce cytoplasm (Fig. 5.12)," and/or, thin and "short blood vessels (Fig. 5.13)" are observed, a presumptive diagnosis of atrophic endometrium is finally performed.

4. **Endometrial glandular and stromal breakdown**

   Endometrial glandular and stromal breakdown (EGBD), a non-pathological endometrium, must be classified as TYS1. The cytohistological findings of EGBD will be described in detail in Chap. 12.

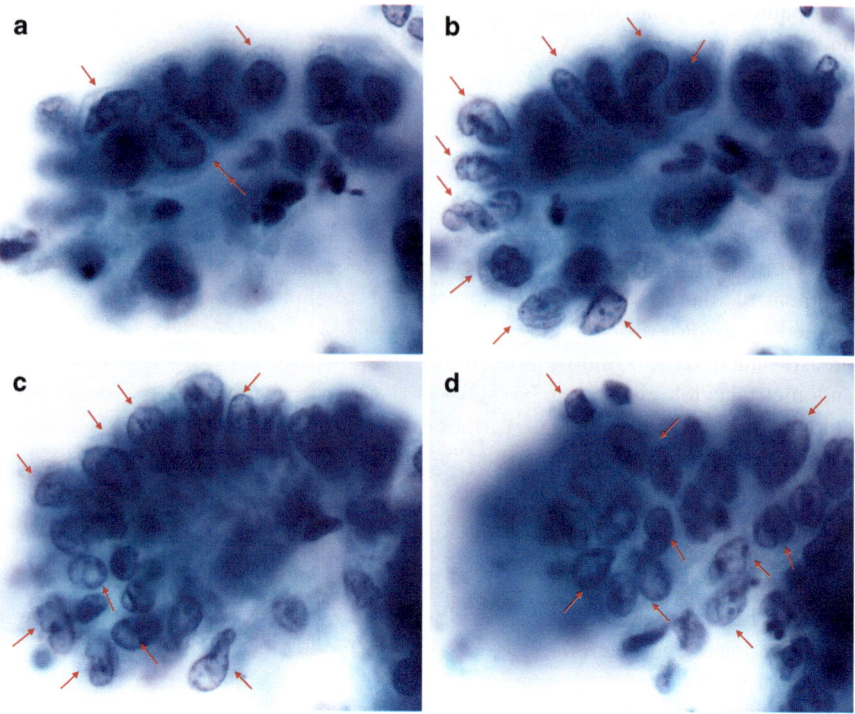

Fig. 5.5 The "nuclear overlapping". These pictures are taken while finely focusing within the same microscopic field. Four distinct nuclear layers (red arrows) can be easily appreciated in the progression from **a** to **d** (Papanicolaou stain, original magnification **a–d**: 100×)

5.1.3   Atypical Endometrial Cells, Cannot Exclude EAH/EIN or Malignant Condition (ATEC-AE) (TYS4)

The term "ATEC" is employed when atypical endometrial cells are observed; "ATEC-AE" is used when the possibility of EAH/EIN or a malignant tumor is suggested but the findings are insufficient to interpret them as "Malignant Neoplasms (TYS6)" because the number of atypical cells is limited or the atypia may be caused by inflammation, metaplastic changes or iatrogenic influences. In such cases, endometrial biopsy is recommended as the subsequent triage method.

In cases of ATEC-AE, both the "irregular protrusion pattern" and the "nuclear overlapping with more than 3 layers" pattern should be initially recognized during the cytological preparation. Subsequently, if "cell clusters with irregular protrusion patterns including MCIP" are observed despite the absence of typical EGBD findings, the case should be interpreted as ATEC-AE (Figs. 5.14 and 5.15).

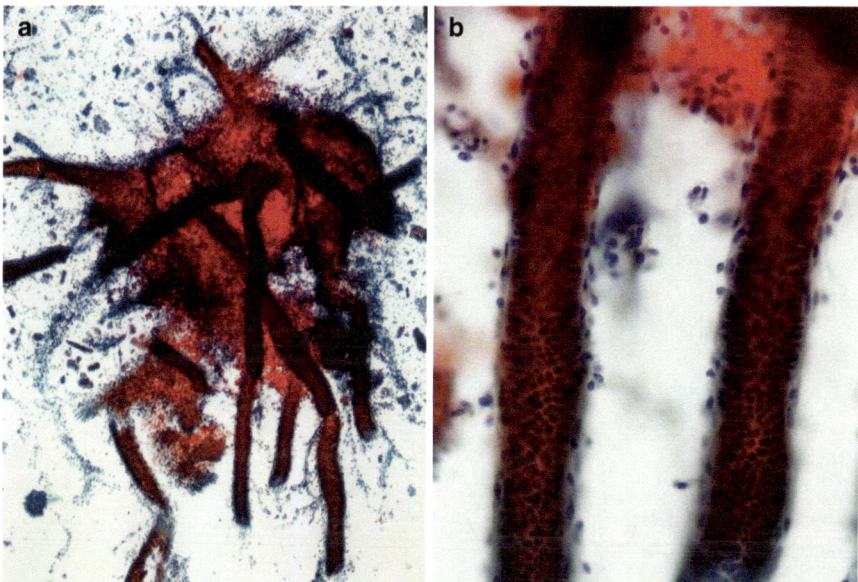

Fig. 5.6 Proliferative endometrium (TYS1). At low power microscopic magnification, a "straight to curvilinear tubular pattern" is recognized (Papanicolaou stain, original magnification **a**: 4×, **b**: 20×)

5.1.4 Malignant Neoplasms (Endometrial Carcinomas) (TYS6) or Endometrial Atypical Hyperplasia/Endometrioid Intraepithelial Neoplasia (EAH/EIN) (TYS5)

In the original "TYS reporting format," a cytological category named TYS5 was designated to indicate EAH/EIN.[4] Furthermore, it was judged as rationally justified to consider the TYS5 and "Malignant Neoplasms (Endometrial Carcinomas) (TYS6)" categories jointly as "Endometrial Carcinomas including EAH/EIN (TYS5 or TYS6)."[5].

---

[4] In the fourth edition of the WHO classification [9], EH and EAH/EIN were categorized as precursors to lesions of endometrioid carcinoma. The EAH/EIN category qualified as a noninvasive lesion with biologically similar abnormalities to endometrioid carcinoma. Nuclear features in EIN are different from those in EAH, and the cases with mild nuclear atypia or without atypia can be included in this diagnostic group. Because the definition of ATEC-AE is focused on the degree of nuclear overlapping and the patterns of cell clusters, this category seems to be possibly applicable to EIN with mild nuclear atypia, too.

[5] EAH/EIN (TYS5) category might be differentiated from endometrial carcinomas (TYS6), when, in presence of well-defined nuclear abnormalities, additional findings suggesting invasion, such as an obvious necrotic background, the presence of isolated malignant cells, marked back to back arrangement, or cribriform structure, are lacking. However, as far as the reproducibility of the

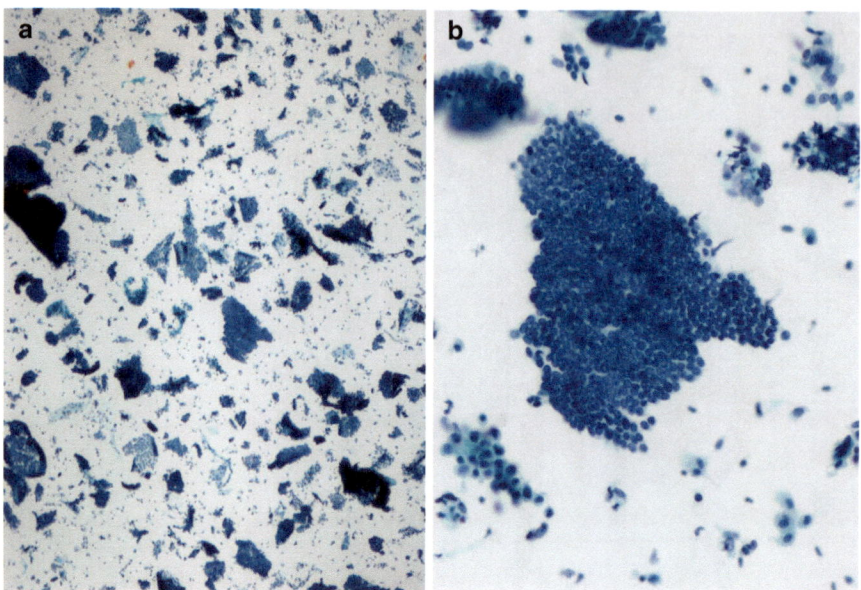

Fig. 5.7 Proliferative endometrium (TYS1). At low and intermediate microscopic evaluation, a "sheet pattern" is recognized (Papanicolaou stain, original magnification **a**: 4×, **b**: 20×)

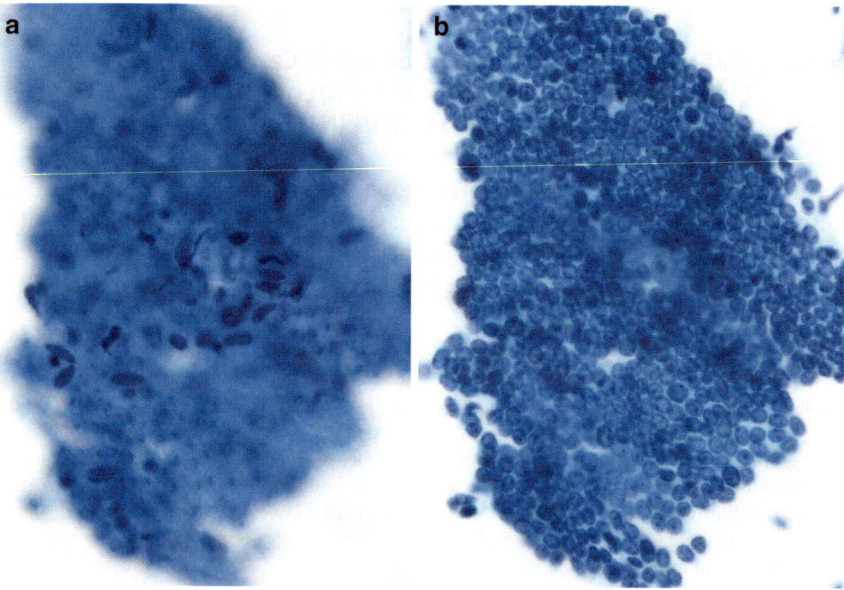

Fig. 5.8 Proliferative endometrium (TYS1). "Nuclear overlapping" amounts to less than 3 layers, by using fine focusing (Papanicolaou stain, original magnification **a, b**: 40×)

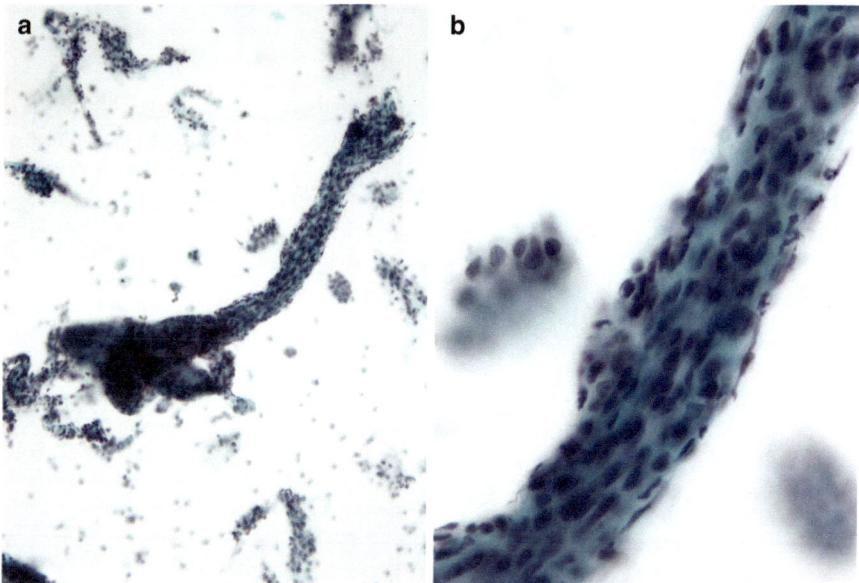

Fig. 5.9 Proliferative endometrium (TYS1). In addition, also long vessels are shown, with a diagnosis of proliferative endometrium (Papanicolaou stain, original magnification **a**: 10×, **b**: 40×)

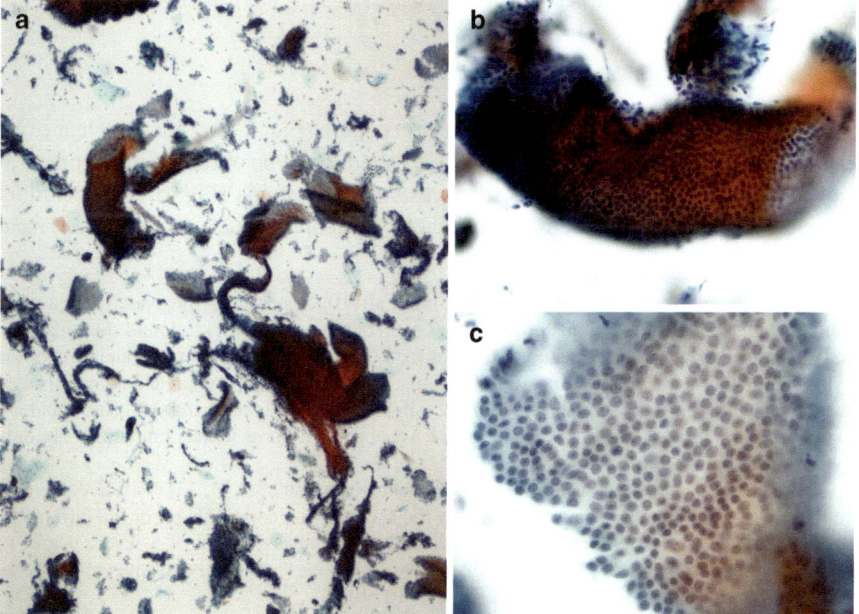

Fig. 5.10 Secretory endometrium (TYS1). At low microscopic evaluation, straight to curvilinear tubular glands (**a**) or sheets (**b**) are recognized. In addition "nuclear overlapping amounts to less than 3 layers" (**c**) (Papanicolaou stain, original magnification **a**: 4×, **b**: 20×, **c**: 40×)

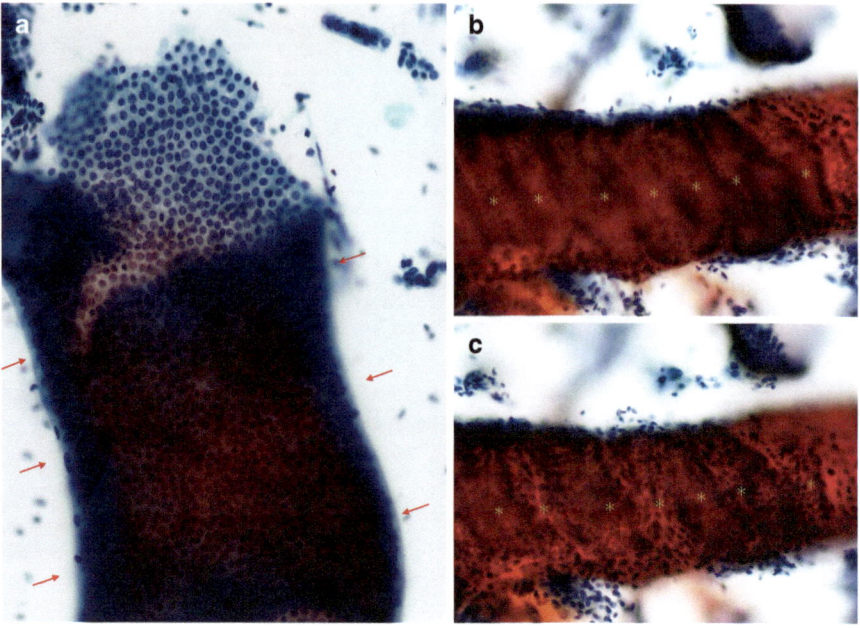

Fig. 5.11 Secretory endometrium (TYS1). This latter is characterized by an "honeycomb pattern with cytoplasm subnuclear vacuoles (**a**) or saw-toothed glands (**b**, **c**)" (Papanicolaou stain, original magnification **a–c**: 20×)

Microscopically, the "irregular protrusion pattern (Fig. 5.16)" and "nuclear overlapping with more than 3 layers (Fig. 5.17)" should first be observed in the preparation. Then, if further diagnostic criteria as: "nuclear atypia (Fig. 5.18)" including hyperchromasia, anisonucleosis, prominent nucleolus,[6] nuclear protrusion from

---

histological diagnosis of EAH is concerned, Zaino et al. [10] reported that the overall kappa value by three gynecological pathologists was 0.28 (fair), Izadi-Mood et al. [11] found that the mean interobserver kappa value by 5 pathologists was 0.34 (fair). In the study on reproducibility of the cytologic diagnosis of EAH by Papaefthimiou et al. [12], the interobserver agreement in EAH over the three review rounds was, respectively, 0.22 (slight), 0.41 (moderate), and 0.33 (fair), and not significantly improved. A correct diagnosis of EAH appears to have serious limitations not only cytologically but also histologically.

Based on the above-mentioned limitations, Yanoh, Norimatsu et al. [3, 13] considered that the role of cytological diagnosis was not primarily the accurate diagnosis of EAH or the differential diagnosis between EAH and endometrial carcinomas, but the assessment of a severe degree of atypia. Hence, rationally, those cytologic diagnostic categories were jointly considered as "endometrial carcinomas including EAH/EIN (TYS5 or TYS6)."

[6] Recently, Norimatsu et al. [14] investigated whether an objective evaluation of nuclear findings in endometrial cytology could be more accurate than the simple cytomorphological evaluation. Nuclear image morphometry for geometric and texture features (area, gray value, aspect ratio, internuclear distance, nucleolar diameter) was performed using the ImageJ software (version 1.51) by the U.S. National Institutes of Health (http://rsbweb.nih.gov/ij/). This study demonstrated that

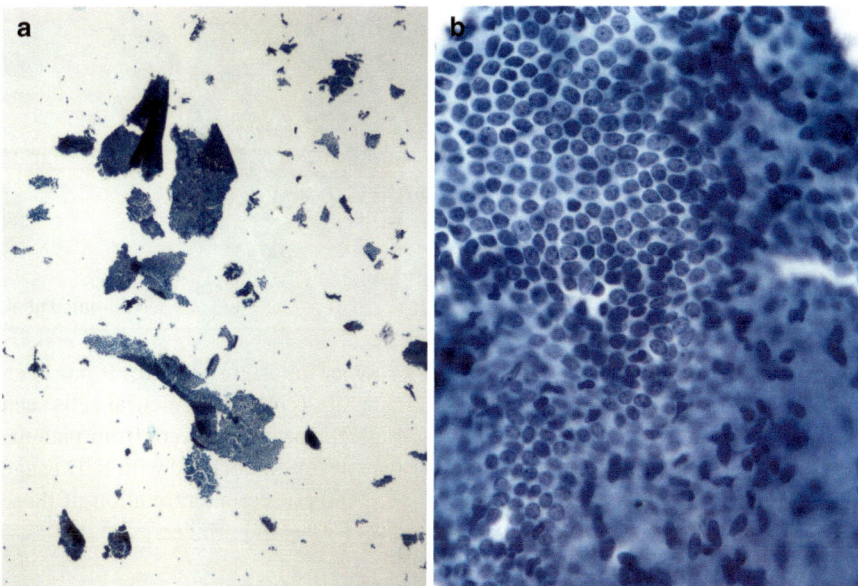

Fig. 5.12 Atrophic endometrium (TYS1). A sheet pattern (**a**) with nuclear overlapping with less than 3 layers (**b**) is recognized. Atrophic endometrial mucosa shows uniform round nuclei with narrow cytoplasm (Papanicolaou stain, original magnification **a**: 4×, **b**: 40×)

cluster and nuclear crowding, and/or, cytoarchitecturally abnormal cell clusters including "cribriform structure (Fig. 5.2)," "back to back structure (Fig. 5.3)," and/ or, "necrotic background (Fig. 5.21)" are observed, the case can be finally defined as "Endometrial Carcinomas" including EAH/EIN.

Along with the algorithmic interpretational approach mentioned above, the finding of atypical mitoses has a definite diagnostic value and should also be mentioned in the report.

---

one single morphometric parameter (nucleolar diameter as nucleolar size; Fig. 5.19) distinguished endometrial endometrioid carcinomas (EECs) and endometrial serous carcinomas (ESCs) from surface papillary syncytial change with EGBD (EGBD-SPSC), with a marked increase in nucleolar size (≥2.0 μm) in the former two entities and, again, a marked increase in nucleolar size in ESCs vs. Grade3 (G3)-EECs (≥3.0 μm).

Consequently, a nucleolar size of ≥2.0 μm, as determined by image morphometry, was considered an useful and objective parameter for diagnosing Grade1 (G1) or higher grade EECs, and to further differentiate G1-EECs from EGBD-SPSC (Fig. 5.20). In addition, a nucleolar size of ≥3.0 μm was deemed as an useful parameter to distinguish ESCs from G3-EECs. In this way, it became obvious that image morphometry is more useful than simple microscopic evaluation to score nuclear atypia in LBC of directly sampled endometrial mucosa.

Fig. 5.13 Atrophic endometrium (TYS1). Atrophic endometrium also shows short and thin blood vessels (Papanicolaou stain, original magnification **c**: 40×)

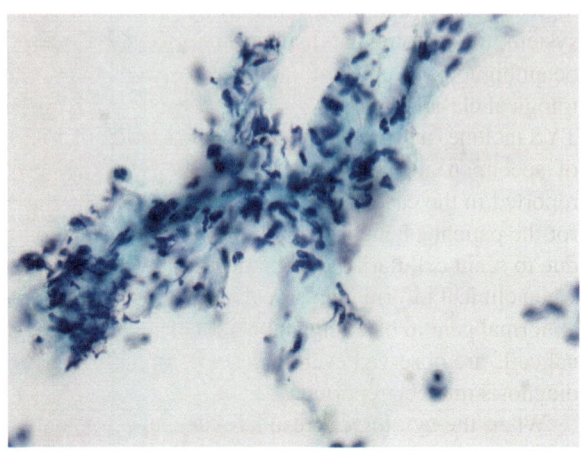

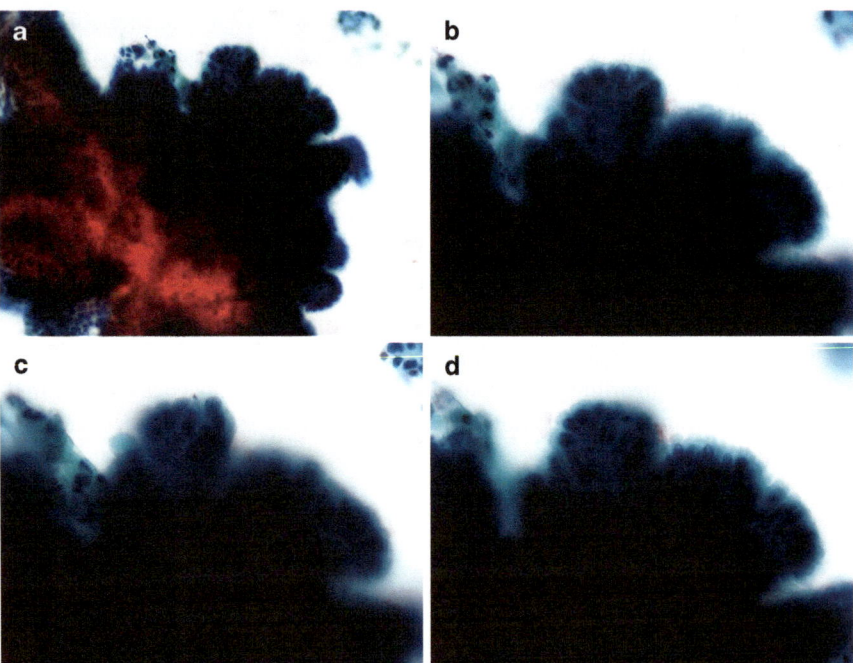

Fig. 5.14 ATEC-AE (TYS4). If an irregular protrusion pattern with nuclear overlapping with more than 3 layers is observed on the preparation, in spite of the absence of typical EGBD findings, it should be interpreted as ATEC-AE. The case was a 23-year-old woman diagnosed as EAH. The endometrial cytologic findings after progesterone treatment are insufficient to interpret the picture as "malignant neoplasms (TYS6)." (Papanicolaou stain, original magnification **a**: 20×, **b–d**: 40×)

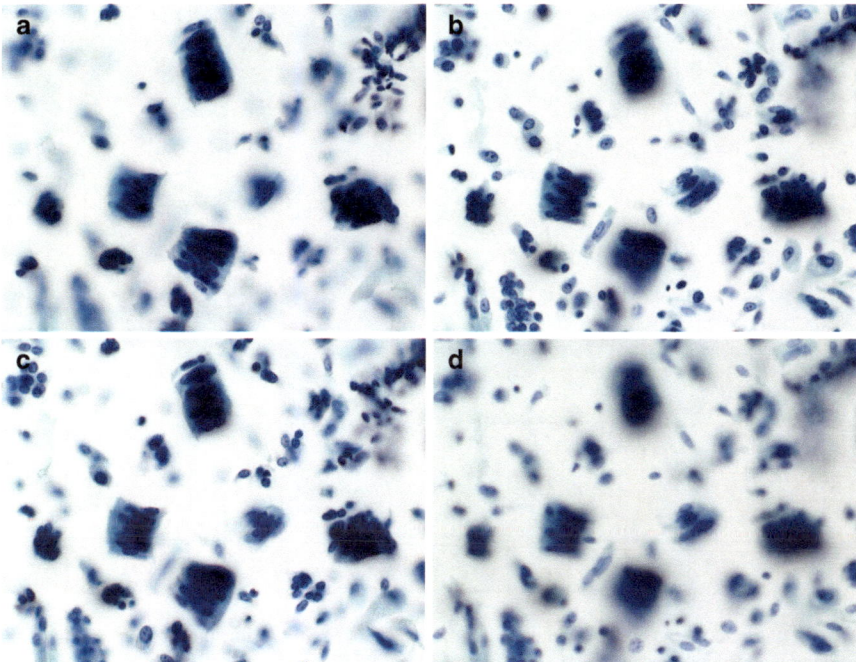

Fig. 5.15 ATEC-AE (TYS4). If an irregular protrusion pattern with nuclear overlapping with more than 3 layers is observed in the preparation, in spite of the absence of typical EGBD findings, it should be interpreted as ATEC-AE. The case was a 56-year-old woman histologically diagnosed with pseudomyxoma peritonei. The endometrial cytologic findings are insufficient to interpret as "malignant neoplasms (TYS6)." (Papanicolaou stain, original magnification **a–d**: 40×)

5.1.5   Endometrial Hyperplasia Without Atypia (TYS3)

At low power microscopic observation, 5 or more cell clusters with "dilated or branched gland pattern (Fig. 5.22)[7]" should be seen, with nuclear overlapping not more than three layers. Nuclear atypia should be absent to make the diagnosis of "endometrial hyperplasia without atypia (EH)". An additional diagnostic comment could be written, such as "disordered endometrial maturation cannot be excluded."

---

[7] A "Dilated or branched gland pattern" is noted in the tubular-shaped glands. The maximum width of a gland is more than twice that of its minimum width and the cohesion of the endometrial stromal cells to the margins of the gland are characteristic.

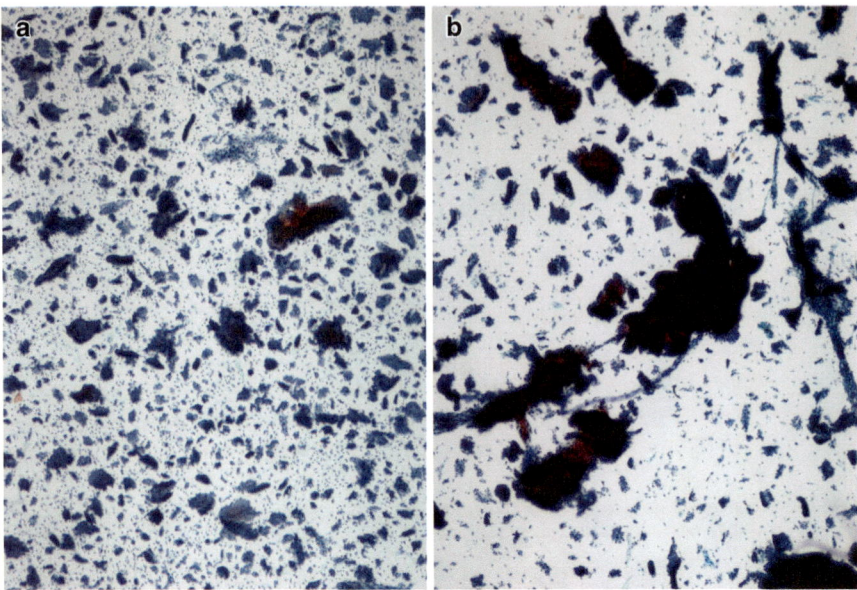

Fig. 5.16 Endometrial endometrioid carcinoma, grade 1 (TYS6). At low power microscopic magnification an irregular protrusion pattern can be observed (Papanicolaou stain, original magnification **a**, **b**: 4×)

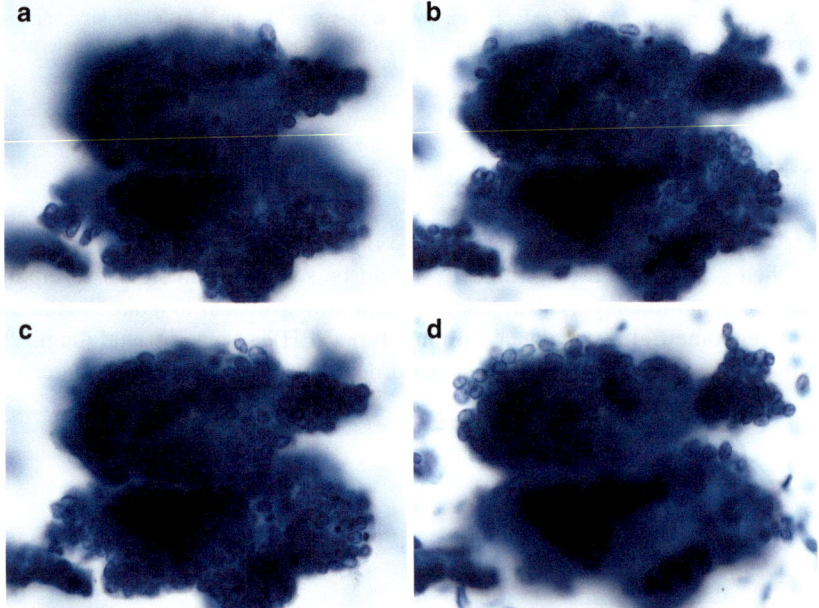

Fig. 5.17 Endometrial endometrioid carcinoma, Grade 1 (TYS6). At intermediate microscopic magnification with fine focusing, nuclear overlapping with more than 3 layers may easily be observed (Papanicolaou stain, original magnification **a**–**d**: 40×)

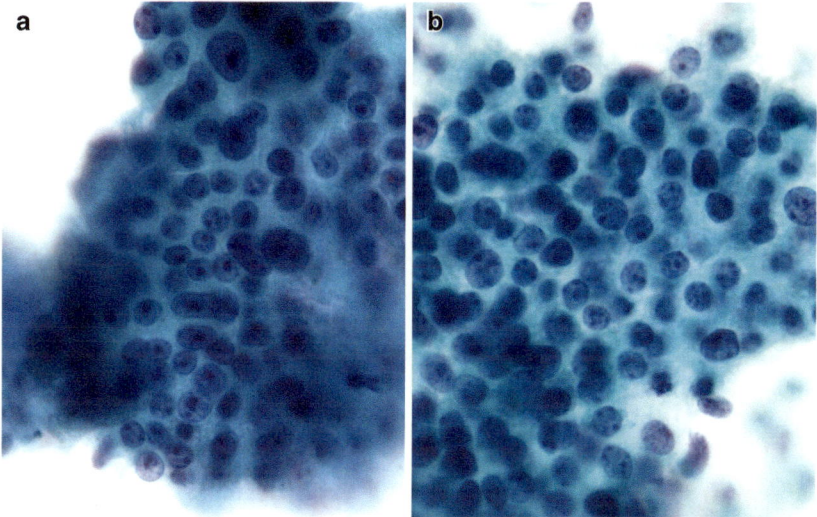

Fig. 5.18 Endometrial endometrioid carcinoma Grade 1 (TYS6). High power microscopic magnification shows several features of nuclear and architectural atypia, as hyperchromasia, anisonucleosis, prominent nucleolus, nuclear protrusion from cluster and nuclear crowding. The case can now be defined as "endometrial carcinomas including EAH/EIN." (Papanicolaou stain, original magnification **a**, **b**: 60×)

Fig. 5.19 Image morphometry by ESC case of LBC preparation. Nuclear image morphometry for nucleolar diameter was performed using ImageJ software. A line on nucleolar diameter (yellow line) was drawn by the use of the mouse for manual nuclei selection (Papanicolaou stain, original magnification 60×)

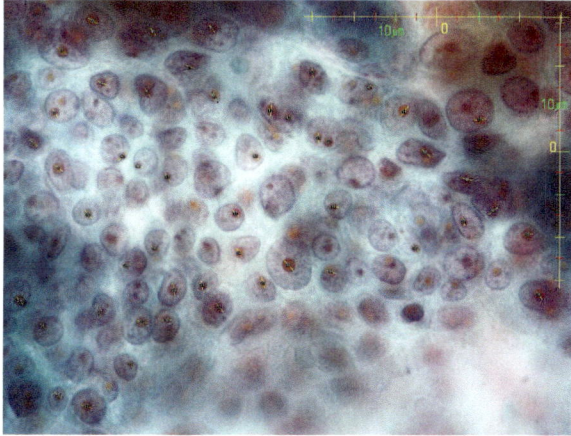

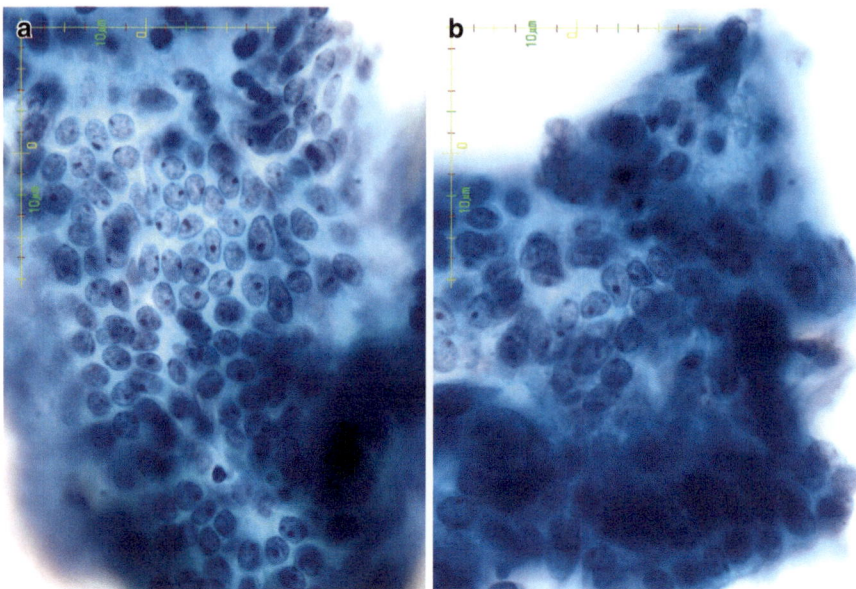

Fig. 5.20 In some cases, EGBD-SPSC (**a**) has a similar degree of nucleolar size compared to G1-EEC (**b**), so it is difficult to distinguish these cases by visual inspection in the cytologic diagnosis (Papanicolaou stain, original magnification **a**, **b**: 60×)

5.1.6 Atypical Endometrial Cells of Undetermined Significance (ATEC-US) (TYS2)

"ATEC-US (TYS2)" is selected when atypical endometrial cells are observed but their significance cannot be determined because of interfering inflammatory, metaplastic, or iatrogenic changes. For cases with cytological findings insufficient to classify the case into any other of the diagnostic categories mentioned above, "ATEC-US" should be selected (Fig. 5.23). In such cases, subsequent endometrial biopsy is not always necessary unless the change persists repeatedly.[8]

---

[8] Shinagawa et al. [15] studied the cytohistologic correlation in cases diagnosed according to the TYS. As to the result, cases in ATEC-US category did not show atypical endometrial hyperplasia lesions or malignancy after 3 months. These data suggest that patients with ATEC-US results can be followed up for at least 3 months, and the introduction of the TYS decreased the number of unnecessary endometrial biopsies.

In this way, when the concept of ATEC will have further been exploited and its subsequent management established, diffuse adoption of the TYS could help avoid unnecessary endometrial biopsies.

Fig. 5.21 Endometrial endometrioid carcinoma, Grade 1 (TYS6). This microscopic field shows necrotic debris admixed with atypical epithelial cells (Papanicolaou stain, original magnification 40×)

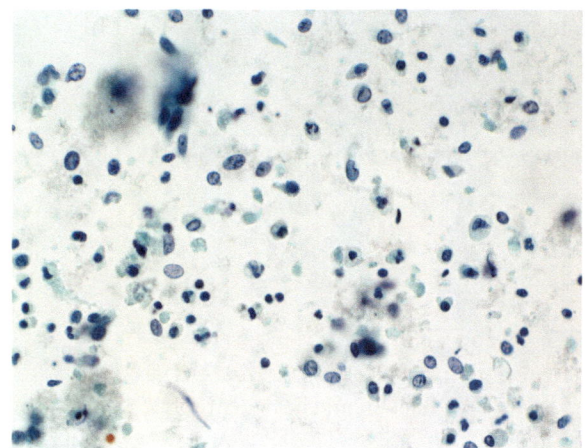

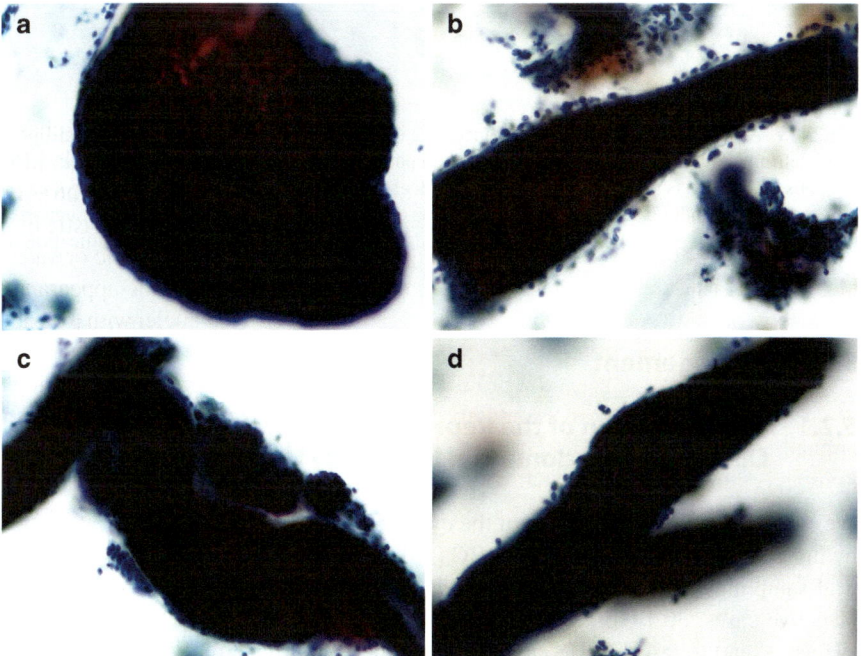

Fig. 5.22 Endometrial Hyperplasia without Atypia (TYS3). At intermediate microscopic magnification, 5 or more cell clusters with "dilated or branched gland pattern" should be seen, with nuclear overlapping not more than three layers. In addition, nuclear atypia is absent (Papanicolaou stain, original magnification **a–d**: 20×)

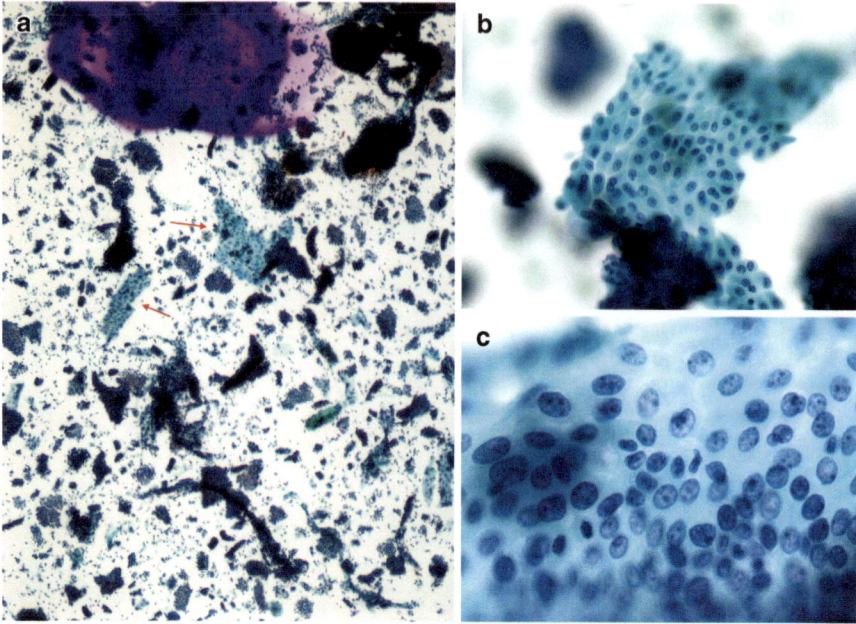

Fig. 5.23 ATEC-US (TYS2). The case concerns a (56-year-old) woman diagnosed with atrophic endometrium. Although some irregular protrusion pattern with mild nuclear atypia (nuclear swelling, anisonucleosis) is seen, nuclear overlapping does not exceed more than three layers. For cases with cytological findings insufficient to classify the case into any other of the diagnostic categories mentioned above, "ATEC-US" should be selected (Papanicolaou stain, original magnification **a**: 4×, **b**: 20×, **c**: 60×)

5.2   Management

5.2.1   The Evaluation of the Reproducibility of the Cytological Diagnosis of Endometrial Lesions by the TYS

We evaluated the reproducibility of the cytological diagnosis of endometrial lesions by the TYS using BD SurePath™-LBC [13]. For this study, a total of 244 endometrial samplings were classified by two academic cytopathologists as follows: 147 Negative cases, 36 EGBD cases, 47 Malignant cases, eight ATEC-AE cases, two EH cases, and four ATEC-US cases. To confirm the reproducibility of the diagnosis and to further study the inter- and intra-observer agreement, a second review round followed at 3-month intervals, which included three additional cytopathologists.

As a result, the inter-observer agreement of Negative classes improved progressively from "good to fair" to "excellent," with values increasing from 0.70 to 0.81. Both EGBD and Malignancy classes improved progressively from "good to fair" to "excellent," with values increasing from 0.62–0.63 to 0.84–0.95, respectively. Because ATEC represents an uncertain diagnostic category, and is expected to include a spectrum ranging from a benign endometrium to neoplastic changes, we believe that, in analogy to the ASC-US/ASC-H categories of the Bethesda classification for gynecological cytology, the ROM (Risk Of Malignancy) of the applied diagnostic criteria may vary according to the entity of cytologic atypia and of the diagnostic experience of the cytopathologist.

As for the diagnostic agreement of ATEC in the two rounds of the diagnostic review, the results were the same with a value of 0.43 (moderate) for ATEC-AE and 0.69 (good to fair) for ATEC-US, respectively, and the results of reproducibility were surprisingly good. The overall intra-observer agreement between the first and the second rounds in three cytopathologists was "good to fair" to "excellent," with values changing from 0.73–0.90, and all kappa improvements were significant ($P < 0.0001$). Applying the algorithmic approach to TYS may be a valid method to improve the precision (reproducibility) of endometrial cytology.

In a very recent study, Hirai et al. [16] reported that the clinical ability of the TYS using LBC for detecting endometrial malignancies was almost identical to the ability of suction endometrial tissue biopsy. This result indicated cytology was not inferior to suction endometrial tissue biopsy for the detection of endometrial malignancies.

In addition, Wang et al. [17] performed a meta-analysis for assessment of the endometrial cytological method in endometrial cancer diagnosis on about 4179 patients with various endometrial lesions with cytohistopathological correlation and found that endometrial cytology is an efficient diagnostic method and is therefore applicable to the diagnosis of endometrial disorders. We believe that the TYS using LBC preparations is the most promising tool for the initial pathological evaluation and cancer surveillance of women with highly suspicious signs suggestive of endometrial malignancies such as Lynch syndrome.

References

1. Yanoh K, Norimatsu Y, Hirai Y, et al. New diagnostic reporting format for endometrial cytology based on cytoarchitectural criteria. Cytopathology. 2009;20:388–94.
2. Yanoh K, Hirai Y, Sakamoto A, et al. New terminology for intrauterine endometrial samples: a group study by the Japanese Society of Clinical Cytology. Acta Cytol. 2012;56:233–41.
3. Yanoh K, Norimatsu Y, Munakata S, et al. Evaluation of endometrial cytology prepared with the Becton Dickinson SurePath™ method: a pilot study by the Osaki Study Group. Acta Cytol. 2014;58:153–61.
4. Margari N, Pouliakis A, Anoinos D, et al. A reporting system for endometrial cytology: cytomorphologic criteria-implied risk of malignancy. Diagn Cytopathol. 2016;4:888–901.
5. Margari N, Pouliakis A, Aninos D, et al. Internal quality control in an academic cytopathology laboratory for the introduction of a new reporting system for endometrial cytology. Diagn Cytopathol. 2017;45:883–8.
6. Fulciniti F, Yanoh K, Karakitsos P, et al. The Yokohama system for reporting directly sampled endometrial cytology: the quest to develop a standardized terminology. Diagn Cytopathol. 2018;46:400–12.
7. Norimatsu Y, Yanoh K, Hirai Y, et al. A diagnostic approach to endometrial cytology by means of liquid-based preparations. Acta Cytol. 2020;64:195–207.

8. Nimura A, Ishitani K, Norimatsu Y, et al. Evaluation of cellular adequacy in endometrial liquid-based cytology. Cytopathology. 2019;30:526–31.
9. Zaino R, Carinelli SG, Ellenson LH, et al. Tumours of the uterine corpus, epithelial tumours and precursors. In: Kurman RJ, Carcangiu ML, Herrigton CS, Young RJ, editors. World Health Organization classification of tumours of female reproductive organs. Lyon, France: IARC Press; 2014. p. 125–6.
10. Zaino RJ, Kauderer J, Trimble CL, et al. Reproducibility of the diagnosis of atypical endometrial hyperplasia: a Gynecologic Oncology Group study. Cancer. 2006;106:804–11.
11. Izadi-Mood N, Yarmohammadi M, Ahmadi SA, et al. Reproducibility determination of WHO classification of endometrial hyperplasia/well differentiated adenocarcinoma and comparison with computerized morphometric data in curettage specimens in Iran. Diagn Pathol. 2009;25:4–10.
12. Papaefthimiou M, Symiakaki H, Mentzelopoulou P, et al. Study on the morphology and reproducibility of the diagnosis of endometrial lesions utilizing liquid-based cytology. Cancer. 2005;105:56–64.
13. Norimatsu Y, Yamaguchi T, Taira T, et al. Inter-observer reproducibility of endometrial cytology by the Osaki Study Group method: utilising the Becton Dickinson SurePath™ liquid-based cytology. Cytopathology. 2016;27:472–8.
14. Norimatsu Y, Irino S, Maeda Y, et al. Nuclear morphometry as an adjunct to cytopathologic examination of endometrial brushings on LBC samples: a prospective approach to combined evaluation in endometrial neoplasms and look alikes. Cytopathology. 2021;32:65–74.
15. Shinagawa A, Kurokawa T, Yamamoto M, et al. Evaluation of the benefit and use of the new terminology in endometrial cytology reporting system. Diagn Cytopathol. 2018;46:314–9.
16. Hirai Y, Sakamoto K, Fujiwara H, et al. Liquid-based endometrial cytology using SurePath™ is not inferior to suction endometrial tissue biopsy for detecting endometrial malignancies: midterm report of a multicentre study advocated by Japan Association of Obstetricians and Gynecologists. Cytopathology. 2019;30:223–8.
17. Wang Q, Wang Q, Zhao L, et al. Endometrial cytology as a method to improve the accuracy of diagnosis of endometrial cancer: case report and meta-analysis. Front Oncol. 2019;9:256.

Evaluation of Sample Adequacy: Cytologic Criteria

6

Yasuo Hirai, Kenji Yanoh, Yoshiaki Norimatsu, and Maki Kihara

6.1 Background

The main objects of microscopical evaluation in endometrial cytology are endometrial glandular epithelial cells and endometrial stromal cells. Both these cells and the structure of the cell-clusters exhibit morphological changes under the influence of sex hormones. Therefore, during a cytological examination, the examiner must be informed of the hormonal state of the examinee. Some drugs are known to affect the endometrial cyto-histomorphology; thus, it is also important to have examinee's drug history. Generally, in the cytologic diagnostic evaluation of glandular cells, cell-cluster findings are as important as individual cell findings. Consequently, sufficient cell quantity and quality are required for observing the structure of the cell-cluster in endometrial cytology, with special attention to the evaluation of glandular epithelial cells. When preparing specimens for cytological analysis, it is important

Y. Hirai (✉)
Department of Obstetrics and Gynecology, Faculty of Medicine, Dokkyo Medical University, Tochigi, Japan

PCL Japan Pathology and Cytology Center, PCL Inc., Saitama, Japan
e-mail: yhirai-ind@umin.ac.jp

K. Yanoh
Department of Obstetrics and Gynecology, Suzuka Chuo General Hospital, Suzuka, Mie, Japan

Y. Norimatsu
Department of Medical Technology, Faculty of Health Sciences, Ehime Prefectural University of Health Sciences, Tobe-cho, Iyo-gun, Ehime, Japan

M. Kihara
Department of Obstetrics and Gynecology, International University of Health and Welfare Narita Hospital, Chiba, Japan

51

to have enough cells and to preserve cell structure. Specifically, it is important that cell-clusters remain intact. In addition, blood and inflammatory cells should be removed or prevented as much as possible from covering the cell-clusters, as this hampers their microscopical evaluation. Therefore, it is necessary to establish a quality standard of the sample before cytologic diagnosis. TYS included criteria for assessing the adequacy of cytology samples.

Examples of variant types of inadequate specimens described in this chapter can be found in Chap. 7 (Figs. 7.3–7.9). Please refer to these for your reference.

6.2   Definition

The specimen is rejected if the label is peeled off, the glass is broken, no slip is attached, or if the slip is not suitable for cytological examination (Table 6.1). However, if the whole specimen is fully examined and judged inadequate for cytologic reporting, the specimen is considered unsatisfactory.

In this specific case, the reason for inadequacy must be clearly stated. Specific reasons for judging a specimen as inappropriate include insufficient cell volume, poor specimen fixation and drying, impaired microscopical evaluation due to inflammation and hemorrhagic background, cell damage or distortion during preparation, lack of clinical information on menstruation and bleeding, and the use of drugs and intra-uterine contraceptives (Tables 6.1 and 6.2).

Nevertheless, if atypical cells are found, they should be reported even if the appropriate adequacy criteria for the sample are not sufficiently met. In this case, the specimen should be reported as adequate, but limited by one or more of the above described factors. The clinician receiving the report should be aware that the report may have a limited diagnostic value and may hence consider the possibility to repeat sampling or to take a diagnostic biopsy or, instead, to shorten the follow-up interval according to the reason for inadequacy.

Table 6.1 Specimen adequacy

Satisfactory
Rejected specimen
Peeled or defective labels
Glass damage
No slip attached or mismatch between the slip and the label
Unsatisfactory
Poorly fixed samples, poor staining, poor preservation
Sample drying, marked inflammatory change
Significant hemorrhagic background
Distortion of cells or cell-clusters
Lack of clinical information (Table 6.2)
Insufficient cell volume (refer to explanatory note)

Table 6.2 Clinical information to be attached to cytology samples

Age
Last menstruation
Pre-menopausal and post-menopausal categories
(in the case of the post-menopausal period, the age of menopause is described)
Presence or absence of uterine bleeding
Drug history:
Intra-uterine device usage for contraception

6.3   Explanatory Note

The following provisional criteria have been established for the number of collected cells.

- **Satisfactory criteria for the number of cells collected in LBC samples:** Thirty or more endometrial epithelial cells forming a cluster are defined as "a cell-cluster," and, if the number of "cell-clusters" exceeds 10, the sample is defined as "satisfactory". However, when the patient is 60 years old or older, it is defined as "satisfactory" if the number of "cellclusters" exceeds five [1].
- **Satisfactory criteria for the number of cells collected in a conventional smear:** Fifty to one-hundred (or more) endometrial epithelial cells forming a cluster are defined as "a cell-cluster," and the sample is defined as satisfactory if 10 or more of these "cell-clusters" are recognized or if one large tissue-like cell-cluster is observed [2, 3].

In determining the adequacy of an endometrial cytology specimen, the number of cells collected from the endometrium is particularly important. The above-mentioned criteria are tentatively proposed as quantitative criteria for the number of cells collected on the basis of several pieces of evidence. The above satisfactory criteria for the LBC sample have been evaluated in the BD SurePath samples and not yet on ThinPrep samples. Furthermore, sufficient scientific evidence has not yet been obtained for the above-mentioned satisfactory criteria with conventional smears. It is hence necessary to acquire sufficient scientific evidence in the future.

Reference

1. Nimura A, Ishitani K, Norimatsu Y, et al. Evaluation of cellular adequacy in endometrial liquid-based cytology. Cytopathology. 2019;30:526–31.
2. Kobayashi TK, Norimatsu Y, Buccoliero AM. Cytology of the body of the uterus. In: Gray W, Kocjan G, editors. Diagnostic cytopathology. 3rd ed. London: Churchill Livingstone; 2010. p. 689–719.
3. Yanoh K, Norimatsu Y, Hirai Y et al. New diagnostic reporting format for endometrial cytology based on cytoarchitectural criteria. Cytopathology. 2009;20;388–94.

Processing Methodology of Endometrial Cytology Samples

7

Yoshiaki Norimatsu, Takeshi Nishikawa,
Tadao K. Kobayashi, and Franco Fulciniti

7.1 Background

7.1.1 Processing Methodology of Endometrial Conventional Cytology Samples

Soon after the appearance of the first models of endometrial brushing devices, like the Endocyte device (Laboratoire CCD, Paris, France) [1, 2] and the Uterobrush device (CooperSurgical, Trumbull, CT, USA) [3], etc. (Fig. 7.1), the obtained material was directly smeared on glass slides by gently rolling the brush tip of the sampler. One of the most successful procedures to obtain richly cellular preparations was the so-called "flicked method" [3], in which the brush was progressively rotated gently on several contiguous areas of the glass slide, and deprived of its cells by the help of a stick or a forceps (Fig. 7.2a–c) before alcohol fixation and Papanicolaou stain (Fig. 7.2d).

However, the cytopathological interpretation of conventional preparations (CP) by direct smear may be problematic due to several confounding factors like the presence of excess obscuring blood (Fig. 7.3) or inflammation (Fig. 7.4), excessive

Y. Norimatsu (✉)
Department of Medical Technology, Faculty of Health Sciences, Ehime Prefectural University of Health Sciences, Tobe-cho, Iyo-gun, Ehime, Japan
e-mail: ynorimatsu@epu.ac.jp

T. Nishikawa
Department of Diagnostic Pathology, Nara Medical University Hospital, Kashihara, Nara, Japan

T. K. Kobayashi
Division of Health Sciences, Cancer Education and Research Center, Osaka University Graduate School of Medicine, Osaka, Japan

F. Fulciniti
Clinical Cytology Service, Istituto Cantonale dì Patologia, Ente Ospedaliero Cantonale, Locarno, Switzerland

55

Fig. 7.1 Endometrial sampling device. Endometrial mucosa is collected using various brushing devices (**a** Endocyte, **b** Uterobrush)

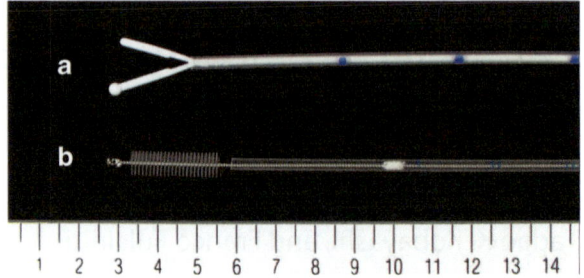

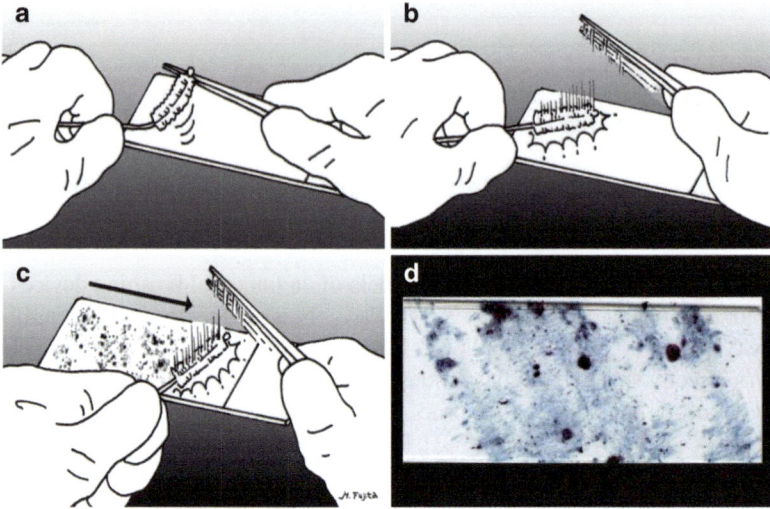

Fig. 7.2 The conventional preparation by direct smear. The procedure for the "flicked" method is as follows: (1) the stem of the brush is held in one hand, and a forceps (or a thin stick) in the other (**a**). (2) the tip of the brush is placed on a glass slide and sharply flicked by forceps or stick (**b**). (3) the brush is moved little by little along the length of the slide (**c**), repeating step 2 several times (**d**). [Modified from 3]

Fig. 7.3 Excess blood contamination is frequent if the obtained material is directly smeared onto a glass slide and interferes with or massively hampers cytologic diagnosis (Papanicolaou stain, original magnification 4×)

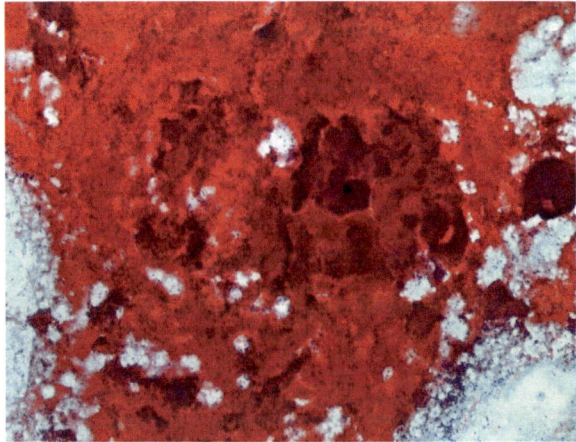

Fig. 7.4 An excess of inflammatory cells interferes with the cytological interpretation of the smear (Papanicolaou stain, original magnification 4×)

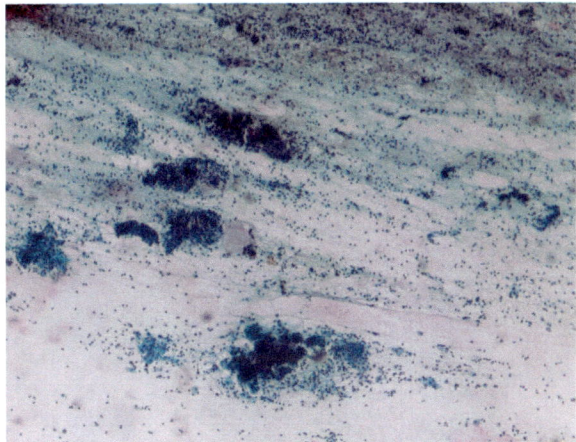

cellular overlapping, cell distortion (Fig. 7.5) or of thick cell clusters, scant cellularity and air-drying artifact (Fig. 7.6). These factors have hampered the routinary adoption of endometrial cytology for a long period [4]. This seems to be the reason why endometrial cytology is only accepted in Japan and not generally practiced in the rest of the world.

7.1.2   Processing Methodology of Endometrial Liquid-Based Cytology (LBC) Samples

The Liquid-Based Cytology (LBC) technique was originally developed for gynecologic cervical smears but progressively gained widespread consensus after having demonstrated its usefulness both in non-gynecologic and fine-needle aspiration cytology samples [5, 6]. The main advantages of this methodology are the reduction in confounding factors, the even distribution of cells on a thin layer, and the possibility to obtain more slides from the same sample. The above-described advantages of LBC strongly encourage the replacement of CP with LBC in the routine cytologic evaluation of endometrial cellular samples. Currently, LBC has become the main technique for cervical cytology in Western countries since several investigations concerning cervical cancer screening have suggested that LBC may achieve a superior or at least equivalent diagnostic sensitivity to that found in CP [7, 8].

The use of LBC in endometrial cytology has eliminated the confounding factors described in conventional endometrial cytology samples, i.e., the presence of excess obscuring blood (Fig. 7.7) or inflammation (Fig. 7.8), cell distortion, and air-drying (Fig. 7.9). Hence, the transition to LBC preparation has provided an opportunity to re-evaluate the role of cytology in endometrial pathology [9–13]. The main LBC technologies currently in use include ThinPrep® (TP; Hologic Inc., Marlborough, MA, USA) and BD SurePath™ (SP; BD Diagnostics, Burlington, NC, USA) and

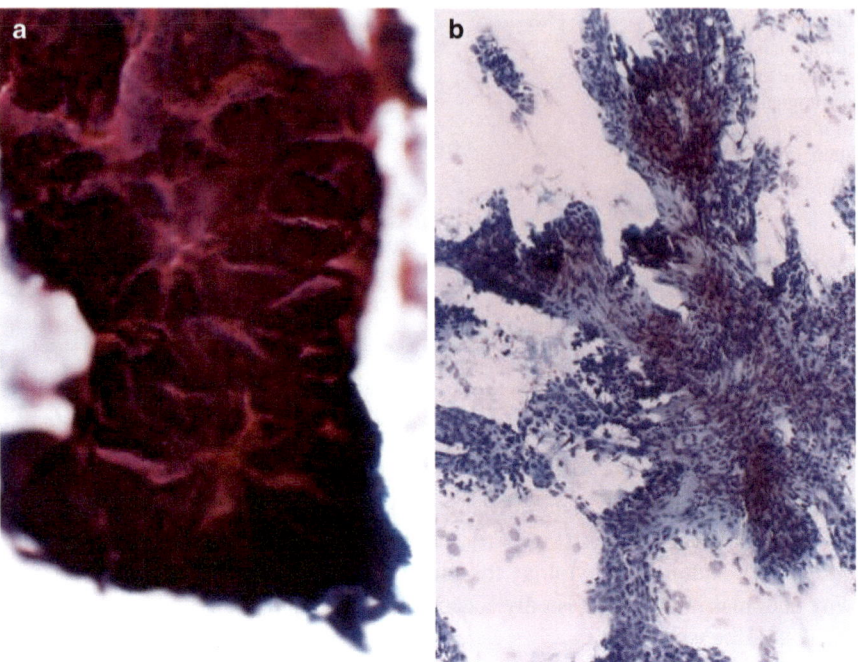

Fig. 7.5 Endometrial cell clusters with papillo-tubular structure are affected by smearing artifact, resulting, respectively, in thick three-dimensional groups (**a**) or distorted thin aggregates with broken cellular edges (**b**) with consequent architectural and cytologic distortion (Papanicolaou stain, original magnification **a**, **b**: 10×)

Fig. 7.6 Endometrial cells with air-drying artifact, which prevents detailed observation of the interior of the cell clusters (Papanicolaou stain, original magnification 10×)

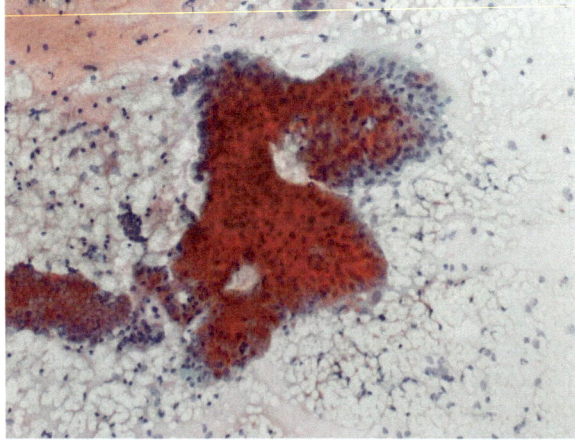

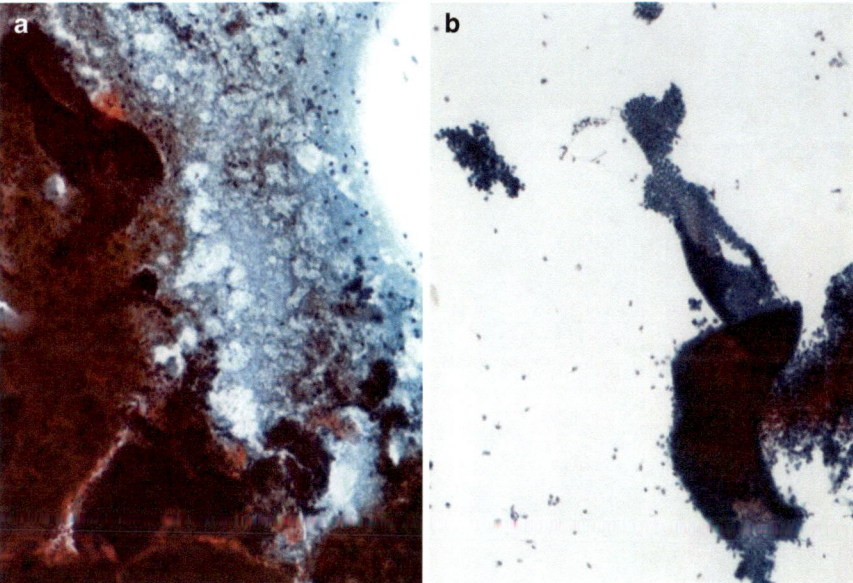

Fig. 7.7 Improvement of excessive blood due to the use of LBC. EGBD cells in a CP are contaminated by excess blood, which interferes with the observation of cellular findings (**a**). In contrast, a cleaner background in the LBC preparation makes observation of EGBD cells much easier (**b**) (EGBD case, Papanicolaou stain, original magnification **a**, **b**: 4×)

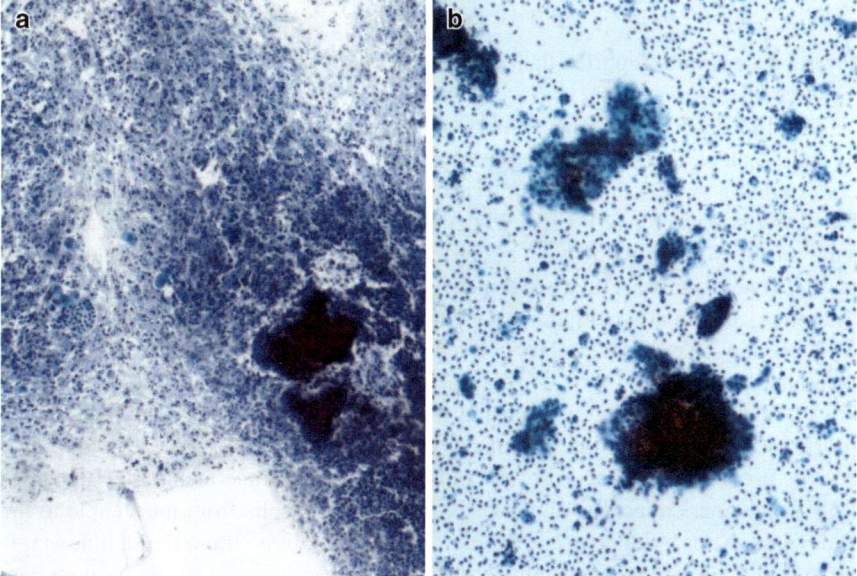

Fig. 7.8 The reduction of excess inflammatory cells due to the use of LBC. Cancer cells are contaminated by excess inflammatory cells which interfere with the observation of cellular findings in a CP (**a**). In contrast, the background of the LBC preparation is much cleaner, and observation of cancer cells is easier (**b**). (G1-EEC case, Papanicolaou stain, original magnification **a**, **b**: 4×)

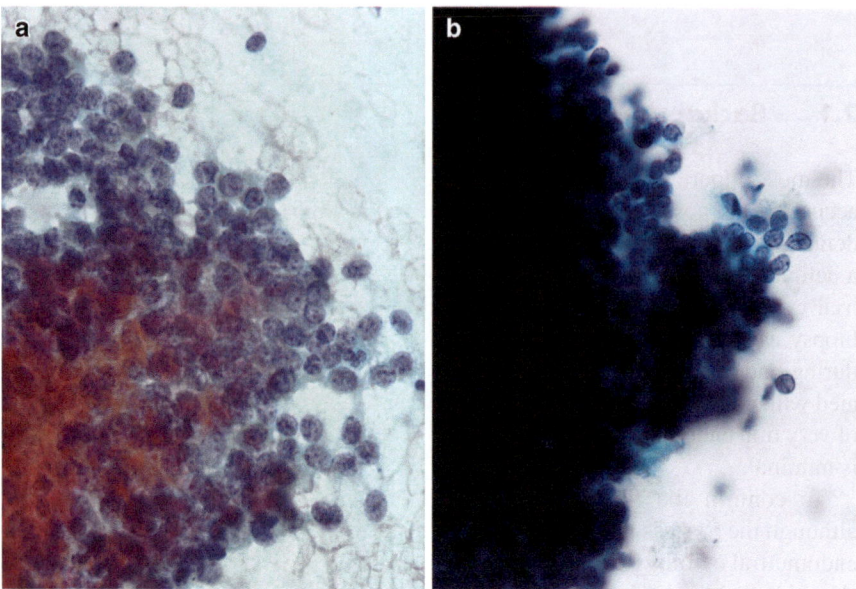

Fig. 7.9 Improvement of air-drying artifact cells due to the use of LBC. In CP, not only is it difficult to recognize the cell clusters as CSC but it is also difficult to differentiate the cell clusters from cancer cells due to the air-drying effect on cells with swollen and smudged structureless nuclei (**a**). In contrast, the LBC preparation does not show any drying artifact of CSC, so they are easy to recognize (**b**) (CSC of EGBD case, Papanicolaou stain, original magnification **a**, **b**: 40×)

several groups have reported their experience by either using TP [9, 10] or SP [11–13] methods.

TP-LBC and SP-LBC have subtle differences that reflect different sampling devices, collection media, and processing techniques [14, 15]. TP-LBC use a vacuum filtration system through an electrically charged membrane using a methanol-based fixative, whereas SP-LBC, which uses an ethanol-based fixative containing a small quantity of formaldehyde, are processed through a density gradient centrifugation system before gravity sedimentation of the specimen onto a glass slide. Therefore, the relative differences in the technical processing of TP-LBC and SP-LBC may be expected to produce some morphologic differences in endometrial glandular lesions.

All TP samples are prepared using a ThinPrep2000 (or 5000) automated slide processor® according to the manufacturer's recommendations. The ThinPrep processors use a computerized process and patented membrane technology that controls dispersion, collection, and transfer of diagnostic cells from the sample to the slide glass. After fixing the sample (Fig. 7.10a), a rotation within the sample vial is produced by an air jet to separate debris and disperse mucus; a gentle vacuum then collects cells on the exterior surface of the electrically charged filter membrane

Membrane Filtration

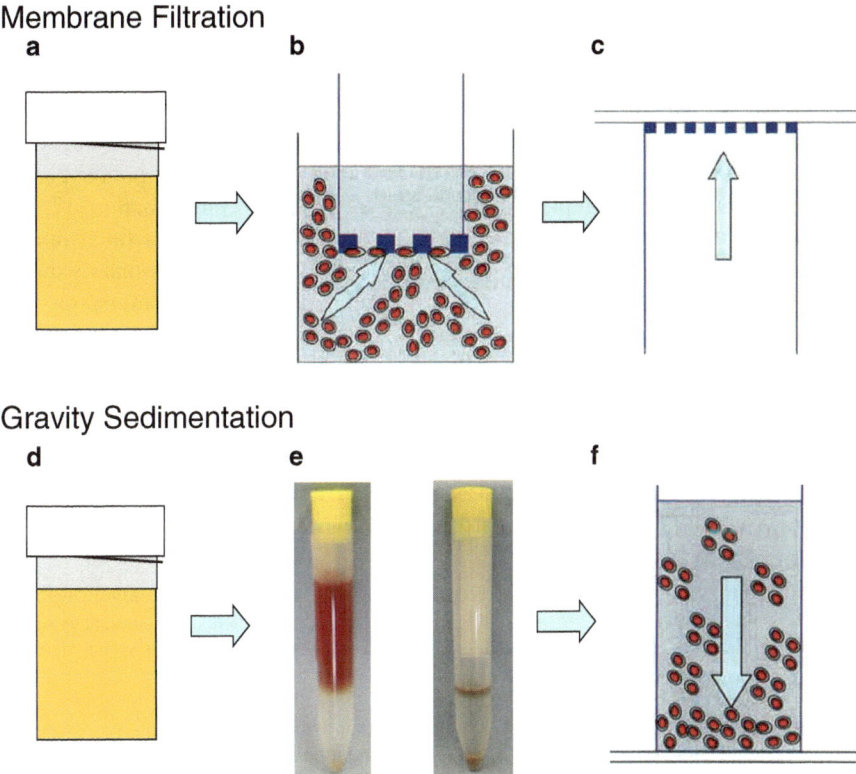

Gravity Sedimentation

Fig. 7.10 The different processing techniques in TP-LBC and SP-LBC samples. TP-LBC uses a vacuum-based electrostatically charged membrane filtration system with a methanol-based fixative (**a–c**), whereas SP-LBC using an ethanol-based fixative containing also a little formalin are processed through a density gradient centrifugation system before gravity sedimentation of the specimen onto a glass slide (**d–f**). [Modified from 16]

(Fig. 7.10b). The cells are then transferred by gently pressing the membrane against the microscope slide glass (Fig. 7.10c) [16]. The filter is automatically removed and the obtained TP preparation is immediately fixed in 95% ethanol and stained by the Papanicolaou method.

SP samples are processed according to the manufacturer's recommendations. The cell solution is transferred automatically onto a BD density reagent that uses the BD PrepMate™ System. This system uses centrifuge sedimentation through a density gradient to partially separate cell debris and inflammatory cells from the sample. After fixing the sample (Fig. 7.10d), and centrifugation through density gradient (Fig. 7.10e), the cell pellets are resuspended and mixed, subsequently transferred to a BD Settling Chamber mounted on a BD SurePath PreCoat slides (Fig. 7.10f) [16]. The cells are sedimented by gravity and then, stained on the BD Totalys ™ SlidePrep, using a modified Papanicolaou procedure.

Additionally, the remaining sample in the vial after cytologic diagnosis can be used for ancillary techniques such as immunocytochemistry, flow cytometry, and molecular biology.

Ensuing investigation of technical differences in LBC methodologies may yield additional data concerning the further improvement of the quality of the diagnostic samples. As far as it concerns past experiences, TP-LBC showed poorer performance than SP-LBC in samples with heavy blood or mucus contamination [17, 18]. To explain the differences in the quality of TP versus SP samples starting from the same original cytologic specimen, it was speculated that these were mainly due to the different abilities of the two systems to remove the blood contamination. The removal of blood components is necessary to improve the quality of the LBC preparations.

Endometrial samples are obtained by intrauterine brushes that permit to adequately sample the endometrial cavity without anesthesia and cervical dilation. As for the SP-LBC, the obtained brushing samples are immediately rinsed into a preservative-fixative solution (BD SP™ preservative fluid) and fixed[1]. On the other hand, as for the TP-LBC, CytoLyt® fluid (CyL)[2] is recommended by Hologic for

---

[1] In LBC, the brush of the device is immersed in a vial of preservative-fixative solution where it is vigorously rotated several times to ensure the release of the cells collected. The abundant cytologic material obtained is deposited as cell clusters of various sizes on the slide. The device is removed from the vial, then the sample is ready for processing after about 30 min in fixative solution and is stable for several weeks thereafter.

When endometrial cells are fixed, cell clusters maintain their three-dimensional pattern resembling, essentially, microbiopsies; this allows a direct correlation between histological and cytological architectures. Hence, the adoption of cytoarchitectural criteria reflecting the histological growth pattern appears to be more useful for the cytologic assessment of endometrial lesions. Moreover, it should be noted that endometrial cells are subjected to mechanical trauma during specimen preparation to obtain an LBC sample, such as that caused by mixing the vial with a vortex mixer.

According to a recent investigation, it was concluded that the endometrial cell clusters are not yet completely fixed in the interval between 30 min and 1 h of fixation, so the mechanical trauma during the SP-LBC process may destroy at least part of them and the three-dimensional structure of the cell clusters may be partially or totally lost (Fig. 7.11). Instead, numerous single or several endometrial cells were observed in the background of the specimen. Nishikawa et al. [19] reported that as far it concerns the number of endometrial cell clusters with the major axis of 200 μm or more, a fixation time <6 h produced an average number of 9.3 (range: 0–22), while a fixation time ≥ 18 h produced a significantly higher average number of 71.3 (range: 3–313).

On the other hand, as for the number of single or small dispersed clusters of endometrial cells, their average number with a fixation time <6 h was 132.2 (range: 29–508), while a fixation time ≥18 h showed a significantly lower average number of 35.7 (range: 5–593). This result suggests that fixation time plays a major role in the protection of endometrial cell clusters from physical impact during the LBC preparation process. A fixation time of 18 h or more is essential for the endometrial SP-LBC preparation (Fig. 7.12).

[2] CyL is a 25% methanol buffer solution; it is used for hemolysis, prevention of protein precipitation, dissolution of mucus, and morphologic maintenance in samples. If a cytologic sample from the uterine cervix is inadequate for massive blood contamination in a TP preparation, a mixture of glacial acetic acid (GAA) additive and CyL recommended by Hologic is used as hemolytic treatment, and the TP preparation is remade. However, it has been reported that this treatment adversely affects the morphologic appearance of cervical squamous cells and glandular cells due to the influence of GAA [20, 21].

Fig. 7.11 Effects of a <6 h fixation time in endometrial SP-LBC sample. In this short fixation time, endometrial glands are destroyed by the mechanical trauma received during the LBC preparation process, and most of the three-dimensional structure of the gland has been lost (Papanicolaou stain, original magnification **a**: 4×, **b**: 20×)

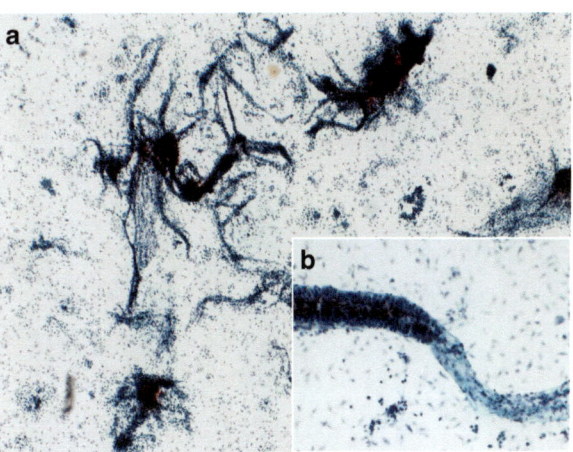

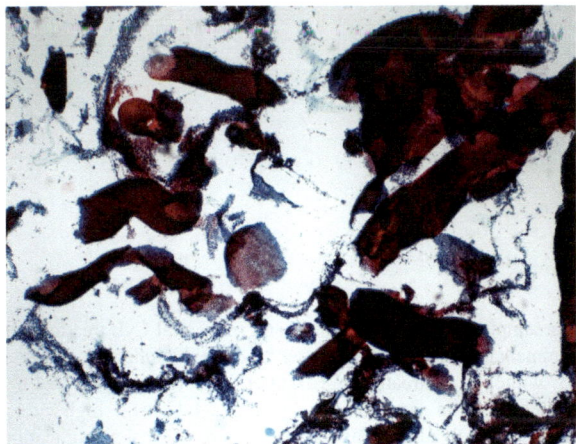

Fig. 7.12 Effects of an adequate fixation time ≥18 h in an endometrial SP-LBC sample. In contrast to a short fixation time, the three-dimensional structure of many endometrial glands is well preserved. The fixation time plays a major role in the protection of endometrial cell clusters from mechanical trauma during the LBC preparation process (Papanicolaou stain, original magnification 4×)

pre-fixation of cytologic samples and for rinsing the brush tips to obtain the endometrial cell sample [22]. The sample is left for up to 1 h[3], thereafter it is centrifuged,

---

[3] Cell suspensions were prepared by adding some blood to the cellular samples, treated with a hemolytic agent, i.e., CyL, then they were left for up to 1, 6, and 24 h, respectively. After different time intervals of hemolysis, TP preparations were made and stained with the Papanicolaou method. All of the obtained preparations were examined for cellularity, nuclear area, and brightness level of nuclear chromatin.

As a result, the cellularity "without CyL treatment" was 27.8 (Fig. 7.13a), "immediately after" was 195.4 (Fig. 7.13b), "after 1 h" was 266.2 (Fig. 7.13c), "after 6 h" was 148.6 (Fig. 7.13d), and

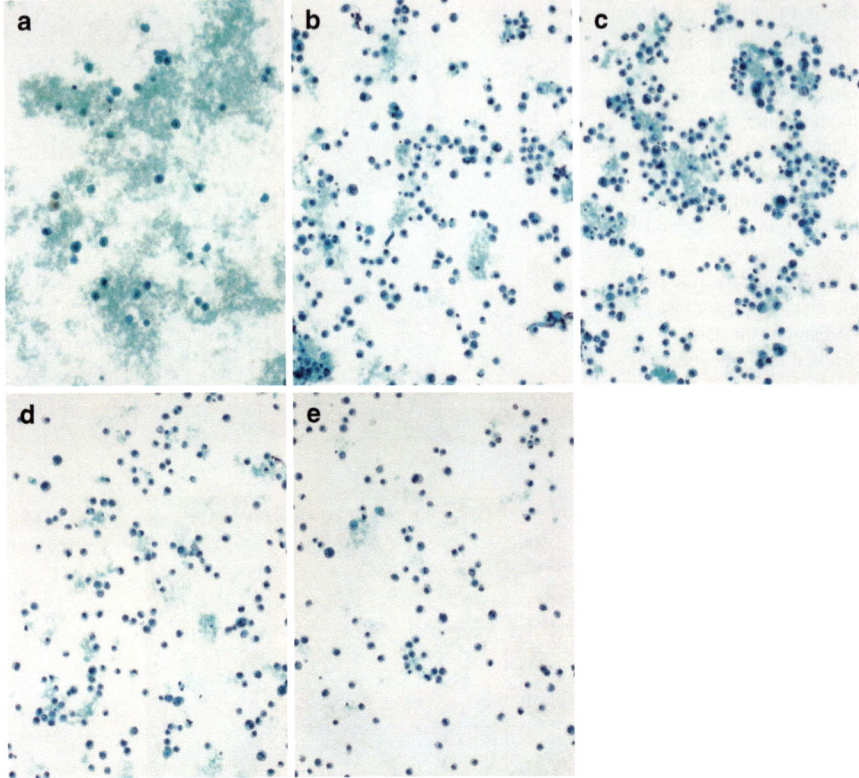

Fig. 7.13 Cellular samples with blood addition and sequential CyL treatment in TP-LBC. After CyL treatment; "Immediately after" (**b**) and after "1 h" (**c**) significantly increased the number of cells compared to "without CyL treatment" (**a**), after "6 h" (**d**) and after "24 h" (**e**) respectively. The cellularity values found in samples obtained after "1 h" were higher than those found in samples obtained "immediately after" (Papanicolaou stain, original magnification **a–e**: 20×)

then the obtained cell pellet is resuspended in PreservCyt® as a preservative-fixative solution and fixed. Then, LBC is then processed according to the chosen methodology.

In the previous study, the cytoarchitectural features of endometrial adenocarcinoma were compared by using TP-LBC and SP-LBC preparations [23]. The observed technical differences in the cytological preparations obtained with both LBC methodologies were the following: a tendency for nuclear swelling and for

---

"after 24 h" was 99.4 (Fig. 7.13e). The cellularity values found in samples obtained after "1 h" were higher than those found in samples obtained "immediately after." The cellularity in samples obtained at "24 h" was significantly lower than that found in samples obtained "immediately after," "1 h," and "6 h." Moreover "24 h" of CyL treatment in samples without adding blood caused reductions in cellularity, nuclear area, and tendency toward the brightness of nuclear chromatin.

If CyL treatment is delayed until the next day in routine practice, this may lead to a general decrease in cellularity and to an increase in degenerative changes in the cell suspension. Hence, it was estimated that a hemolytic treatment of "1 h" represented the optimal choice.

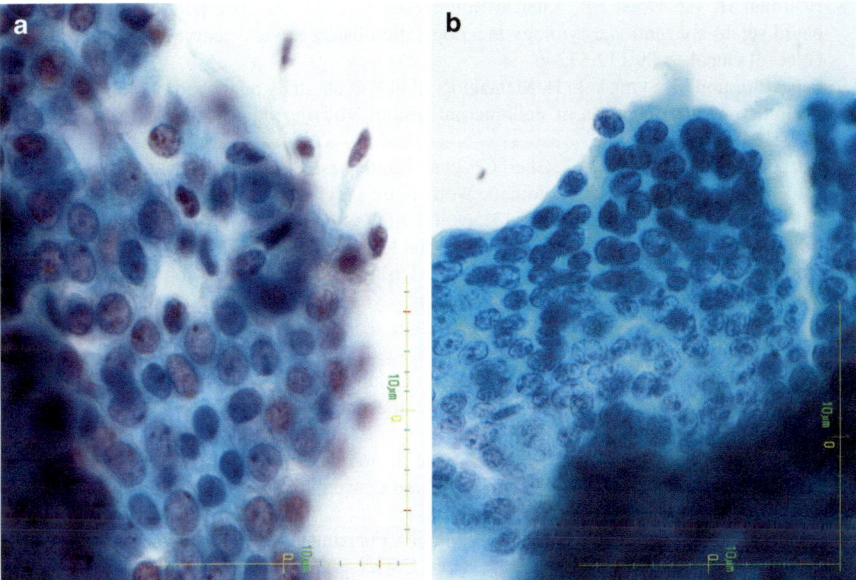

Fig. 7.14 Differences in the cytological findings caused by technical differences in LBC methodologies. (**a**, **b**) are the same case of G1-EEC. A tendency for nuclear swelling and for planar cell clusters is noted in TP-LBC (**a**), while a tendency to form larger three-dimensional cell clusters is recognized in SP-LBC (**b**). (Papanicolaou stain, original magnification **a**, **b**: 60×)

planar cell clusters was noted in TP-LBC (Fig. 7.14a), while a tendency to form larger three-dimensional cell clusters was recognized in SP-LBC (Fig. 7.14b). Generally speaking, both methods have very good diagnostic accuracy and a lower rate of unsatisfactory diagnosis [24].

References

1. Byren AJ. Endocyte endometrial smears in the cytodiagnosis of endometrial carcinoma. Acta Cytol. 1990;34:373–81.
2. Norimatsu Y, Shimizu K, Kobayashi TK, et al. Cellular features of endometrial hyperplasia and well differentiated adenocarcinoma using the endocyte sampler. Diagnostic criteria based on the cytoarchitecture of tissue fragments. Cancer. 2006;108:77–85.
3. Fujihara A, Norimatsu Y, Kobayashi TK, et al. Direct intrauterine sampling with Uterobrush: cell preparation by the flicked method. Diagn Cytopathol. 2006;34:486–90.
4. Yanoh K, Hirai Y, Sakamoto A, et al. New terminology for intrauterine endometrial samples: a group study by the Japanese Society of Clinical Cytology. Acta Cytol. 2012;56:233–41.
5. Hutchinson ML, Agarwal P, Denault T, et al. A new look at cervical cytology. ThinPrep multicenter trial results. Acta Cytol. 1992;36:499–504.
6. Fischler DF, Toddy SM. Nongynecologic cytology utilizing the ThinPrep processor. Acta Cytol. 1996;40:669–75.
7. Fremont-Smith M, Marino J, Griffin B, et al. Comparison of the SurePath liquid-based Papanicolaou smear with the conventional Papanicolaou smear in a multisite direct-to-vial study. Cancer. 2004;102:269–79.

8. Beerman H, van Dorst EB, Kuenen-Boumeester V, et al. Superior performance of liquid-based versus conventional cytology in a population-based cervical cancer screening program. Gynecol Oncol. 2009;112:572–6.

9. Papaefthimiou M, Symiakaki H, Mentzelopoulou P, et al. Study on the morphology and reproducibility of the diagnosis of endometrial lesions utilizing liquid-based cytology. Cancer. 2005;105:56–64.

10. Buccoliero AM, Castiglione F, Gheri CF, et al. Liquid based endometrial cytology: its possible value in postmenopausal asymptomatic women. Int J Gynecol Cancer. 2007;17:182–7.

11. Norimatsu Y, Kouda H, Kobayashi TK, et al. Utility of thin layer preparations in the endometrial cytology: evaluation of benign endometrial lesions. Ann Diagn Pathol. 2008;12:103–11.

12. Norimatsu Y, Kouda H, Kobayashi TK, et al. Utility of liquid-based cytology in endometrial pathology: diagnosis of endometrial carcinoma. Cytopathology. 2009;20:395–402.

13. Nishimura Y, Watanabe J, Jobo T, et al. Cytologic scoring of endometrioid adenocarcinoma of the endometrium. Cancer. 2005;105:8–12.

14. Michael CW, McConnel J, Pecott J, et al. Comparison of ThinPrep and TriPath PREP liquid-based preparations in nongynecologic specimens: a pilot study. Diagn Cytopathol. 2001;25:177–84.

15. Belsley NA, Tambouret RH, Misdraji J, et al. Cytologic features of endocervical glandular lesions: comparison of SurePath, ThinPrep, and conventional smear specimen preparations. Diagn Cytopathol. 2008;36:232–7.

16. Evered A, Shambayati B. Preparation techniques. Fundamentals of biomedical science cytopathology. 2nd In: Shambayati B., Eds.; Oxford University Press, New York; 2011: pp. 26–27.

17. Sweeney BJ, Haq Z, Happel JF, et al. Comparison of the effectiveness of two liquid-based Papanicolaou systems in the handling of adverse limiting factors, such as excessive blood. Cancer. 2006;108:27–31.

18. Kenyon S, Sweeney BJ, Happel J, et al. Comparison of BD Surepath and ThinPrep Pap systems in the processing of mucus-rich specimens. Cancer Cytopathol. 2010;118:244–9.

19. Nishikawa T, Norimatsu Y, et al. Examination of fixation time of endometrial cytology samples by BD SurePath-LBC method. J Jpn Soc Clin Cytol. 2018;57:368. (in Japanese with English summary).

20. Dalton P, MacDonald S, Boerner S. Acetic acid recovery of gynecologic liquid-based samples of apparent low squamous cellularity. Acta Cytol. 2006;50:136–40.

21. Cohen D, Shorie J, Biscotti C. Glacial acetic acid treatment and atypical endocervical glandular cells: an analysis of 92 cases. Am J Clin Pathol. 2010;133:799–801.

22. Norimatsu Y, Ohsaki H, Masuno H, et al. Efficacy of CytoLyt® hemolytic action on ThinPrep® LBC using cultured osteosarcoma cell line LM8. Acta Cytol. 2014;58:76–82.

23. Norimatsu Y, Sakamoto S, Ohsaki H, et al. Cytologic features of the endometrial adenocarcinoma: comparison of ThinPrep and BD SurePath preparations. Diagn Cytopathol. 2013;41:673–81.

24. Garcia F, Barker B, Davis J, et al. Thin-layer cytology and histopathology in the evaluation of abnormal uterine bleeding. J Reprod Med. 2003;48:882–8.

Negative for Malignant Tumors and Precursors: TYS1

8

Yoshiaki Norimatsu, Takeshi Nishikawa,
Tadao K. Kobayashi, Akihiko Kawahara, Jun Akiba,
and Franco Fulciniti

8.1 Normal Endometrium [1]

The endometrial cycle in women follows a series of morphologic and physiologic events of differentiation and regeneration of uterine mucosa which is characterized by proliferation, secretion, and shedding (menstruation). During the reproductive years, normal endometrium comprises glands, stroma, and vascular elements that synchronously proliferate, differentiate, and then disintegrate at roughly 28-day intervals. The epithelium lining the glands, stroma, and vasculature of the functional layer of the endometrium all undergo morphological changes by cyclically released ovarian estradiol and progesterone during a menstrual cycle. At the same time, the glands and stroma of the basal layer of the endometrial corpus show no significant morphological changes.

Y. Norimatsu (✉)
Department of Medical Technology, Faculty of Health Sciences, Ehime Prefectural University of Health Sciences, Tobe-cho, Iyo-gun, Ehime, Japan
e-mail: ynorimatsu@epu.ac.jp

T. Nishikawa
Department of Diagnostic Pathology, Nara Medical University Hospital, Kashihara, Nara, Japan

T. K. Kobayashi
Division of Health Sciences, Cancer Education and Research Center, Osaka University Graduate School of Medicine, Osaka, Japan

A. Kawahara
Department of Diagnostic Cytopathology, Kurume University Hospital, Fukuoka, Japan

J. Akiba
Department of Diagnostic Pathology, Kurume University Hospital, Kurume, Japan

F. Fulciniti
Clinical Cytology Service, Istituto Cantonale dì Patologia, Ente Ospedaliero Cantonale, Locarno, Switzerland

The stromal compartment decidualizes in the mid-secretory phase of the menstrual cycle, independent of pregnancy. Stromal cells, which undergo predecidual change in the upper endometrium of the functional layer, acquire abundant well-defined, dense cytoplasm and centrally located round or oval, vesicular nuclei. The coiled endometrial spiral arteries originate from the myometrial arcuate arteries and connect with the superficial capillary network, which in turn is drained by dilated veins. There is a gradual increase in the arborization and coiling of spiral arteries during the preovulatory-ovulatory and postovulatory periods up to cycle days 23–25.

8.2   Proliferative Endometrium (PE) [1]

8.2.1   Background

During its proliferative phase, the endometrium responds to increasing estrogen levels by the synchronous proliferation of glands, stroma, and blood vessels. Hence, the spiral arteries grow, the functional layer shows proliferation and thickening. Based on an average 28-day menstrual cycle, proliferative endometrial changes may be divided into early (days 4–7), mid (days 8–10), and late (days 11–13) intervals.

In the early proliferative phase, the glands are straight, narrow, and round in cross-section (Fig. 8.1). Their epithelium is cubic to columnar with oval or rounded nuclei in which the chromatin is evenly dispersed, with little epithelial stratification (Fig. 8.2). Occasional mitoses are seen in the glandular epithelium. In the mid-proliferative phase (Fig. 8.3), the glandular epithelium becomes taller, stratification of the epithelium is more obvious (Fig. 8.4), and mitotic figures are frequently seen in both the glands and stroma.

In the late proliferative phase, rows of tortuous glands arranged at regular intervals characterize the pre-ovulatory endometrium, and grow in parallel fashion no matter how oblique the plane of section (Fig. 8.5). The glandular epithelium has pseudostratified nuclei (Fig. 8.6), but mitoses decline in number in both the glands and stroma.

Fig. 8.1 Proliferative endometrium (Early proliferative phase). The glands are straight, narrow and round in cross-section. (H&E stain, original magnification 4×)

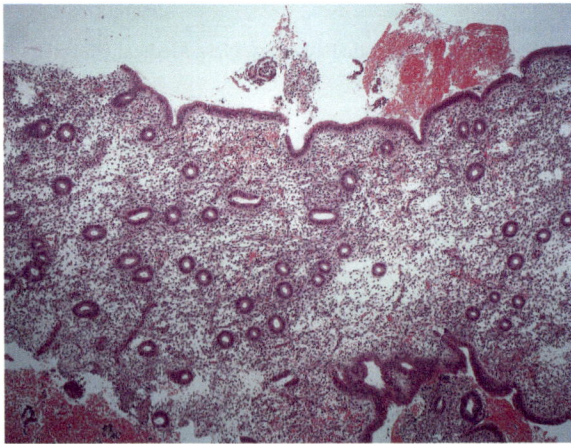

Fig. 8.2 Proliferative endometrium (Early proliferative phase). Glandular epithelium is cubic to columnar with oval or rounded nuclei with evenly dispersed chromatin and little epithelial stratification. (H&E stain, original magnification 40×)

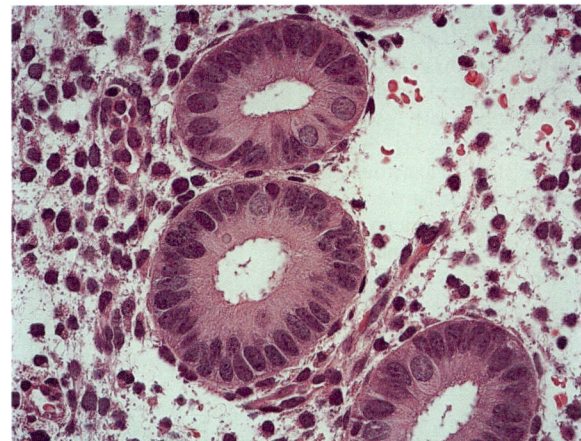

Fig. 8.3 Proliferative endometrium (Mid proliferative phase). (H&E stain, original magnification 4×)

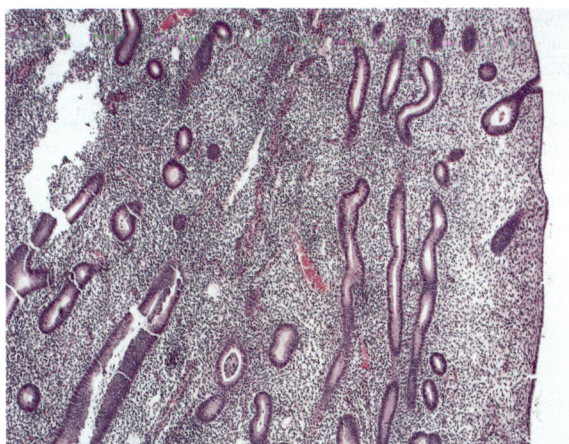

Fig. 8.4 Proliferative endometrium (Mid proliferative phase). By this phase the glandular epithelium is taller, stratification of the epithelium is more obvious. (H&E stain, original magnification 40×)

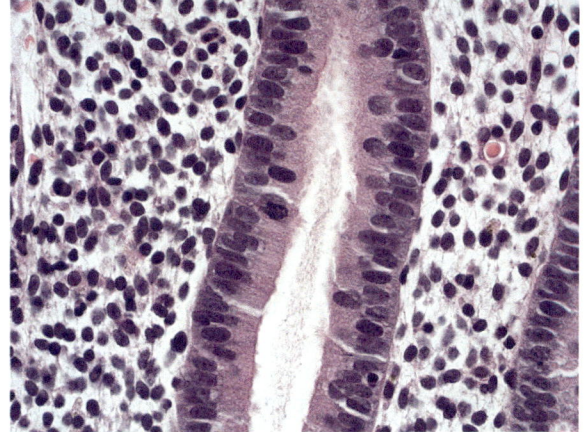

Fig. 8.5 Proliferative endometrium (Late proliferative phase). Rows of tortuous glands arranged at regular intervals characterize the preovulatory endometrium, and they tend to have a parallel arrangement no matter how oblique the plane of section. (H&E stain, original magnification 4×)

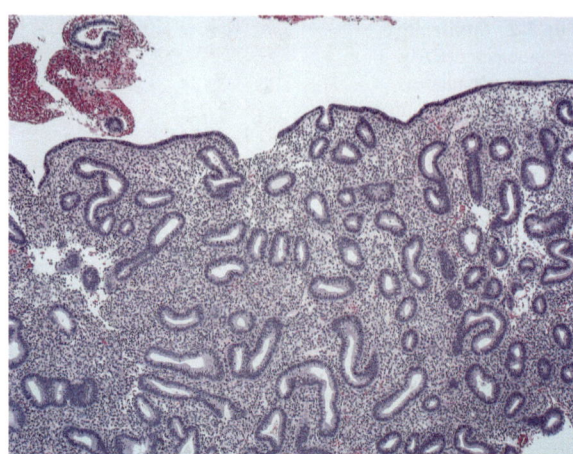

Fig. 8.6 Proliferative endometrium (Late proliferative phase). The glandular epithelium has pseudostratified nuclei. Mitotic figures are seen in both the glands and stroma (arrows). (H&E stain, original magnification 40×)

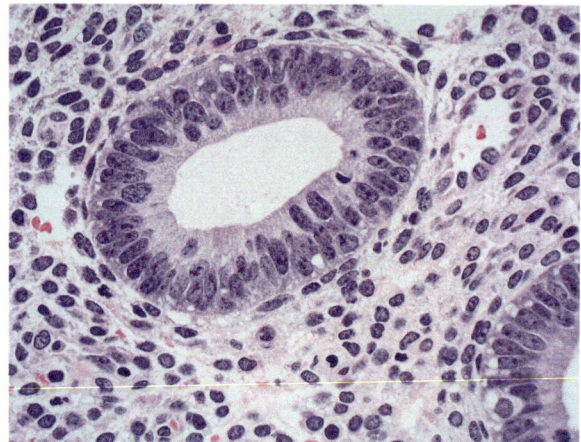

During the proliferative phase, the endometrial stroma is usually densely cellular, and the stromal cells are small and oval with hyperchromatic nuclei and indistinct cytoplasm and cell borders (Figs. 8.2, 8.4 and 8.6).

8.2.2 Definition

The cytological characteristics of the fragments that reflect the histological architecture of PE are shown below [1–3]. Endometrial glands appear as straight to curvilinear, of tubular-shape (Fig. 8.7). The width of the tubular-shaped gland is approximately uniform, and cohesion of the endometrial stromal cells to the margins of the gland is noted (Fig. 8.8). When the tubular-shaped glands are disrupted and/or opened out, they show a flat epithelial surface and adhesion of the stromal cells to the epithelium is observed (Fig. 8.9).

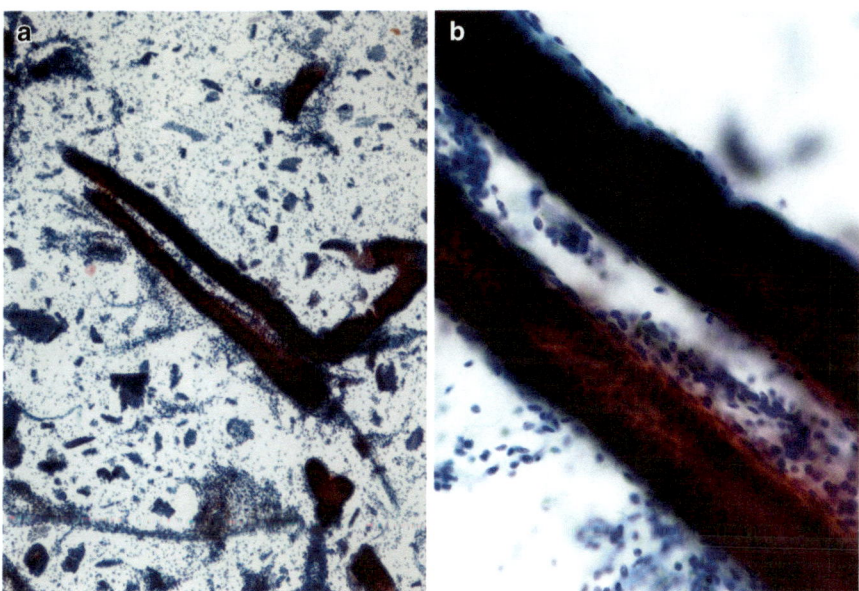

Fig. 8.7 Proliferative endometrium. Endometrial glands appear as straight to curvilinear tubular-shaped clusters. (Papanicolaou stain, original magnification **a**:4×, **b**:20×)

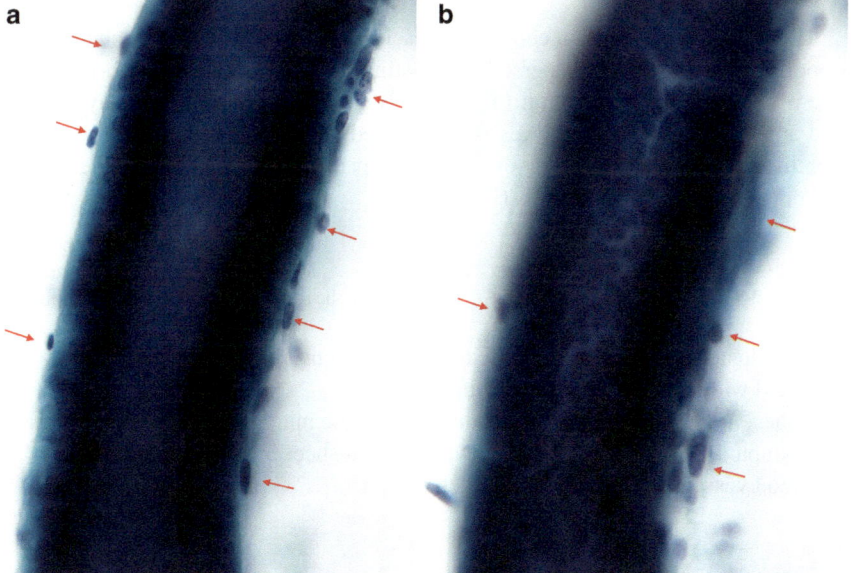

Fig. 8.8 Proliferative endometrium. The width of the tubular-shaped gland is approximately uniform, and cohesion of the endometrial stromal cells to the margins of the gland (arrows) is noted. (Papanicolaou stain, original magnification **a** and **b**:40×)

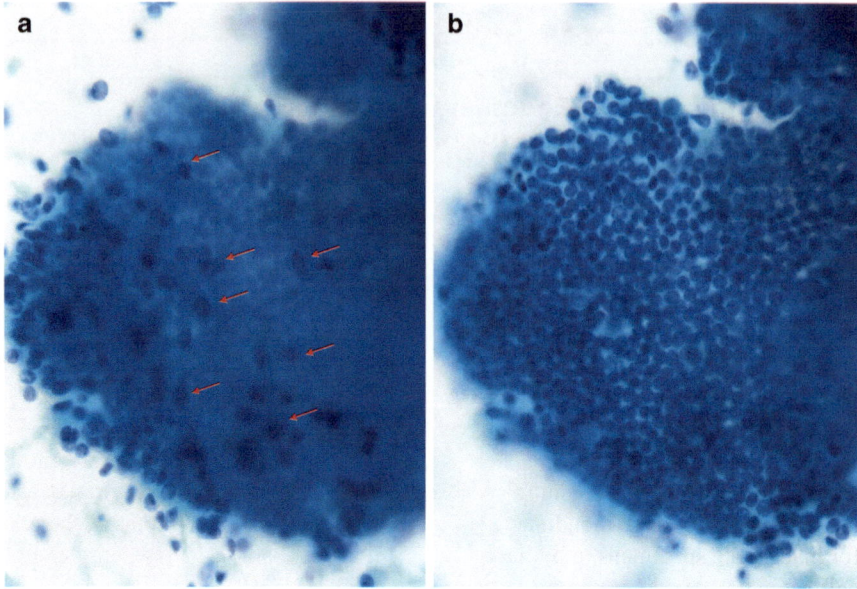

Fig. 8.9 Proliferative endometrium. When the tubular-shaped glands are disrupted and opened out, they have a flat surface and adhesion of the endometrial stromal cells (arrows) is observed. The nuclei of epithelial cells are closely packed, the nuclear-to-cytoplasmic ratio is relatively high. (Papanicolaou stain, original magnification **a** and **b**:40×)

The nuclei of epithelial cells are closely packed, oval to cigar-shaped with smooth contours, evenly dispersed chromatin, and small-sized nucleoli (Fig. 8.9). The nuclear-to-cytoplasmic ratio is relatively high. Normal mitoses may be seen. The appearance of spiral arteries is recognized in the background. These can be identified as elongated bundles of spindle-shaped cells including endothelial cells and/or perivascular smooth cells (Fig. 8.10).

8.2.3 Criteria

- Cell clusters of epithelial cells with straight to curvilinear tubular pattern.
- Gland width is approximately uniform.
- Cohesion of the endometrial stromal cells to the margins of the gland is present.
- Disrupted tubular glands open out and lie in a sheet pattern.
- Blood vessels are present in the background.

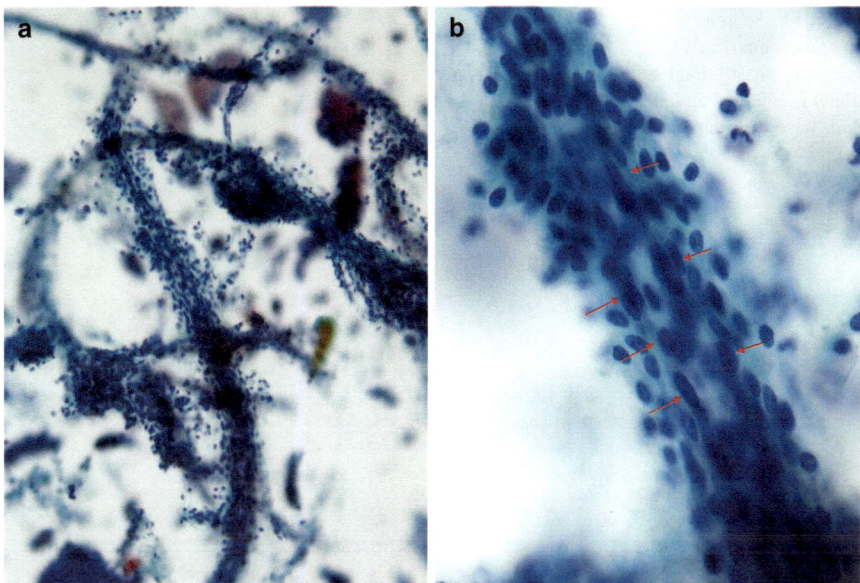

Fig. 8.10 Proliferative endometrium. The appearance of spiral arteries is recognized in the background. These are identified by an elongated bundle of spindle-shaped cells including endothelial cells and/or perivascular epithelioid cells (arrows). (Papanicolaou stain, original magnification **a**:10×, **b**:40×)

8.3   Secretory Endometrium (SE) [1]

8.3.1   Background

Once ovulation occurs, in addition to estrogen, the ovary also produces progesterone. This changes the proliferative pattern of the endometrium to a secretory pattern. During an average 28-day menstrual cycle, secretory changes are divided into ovulatory—or interval (days 14 and 15, the interval phase), early—(days 16–19, the vacuole phase), mid—(days 20–22, the secretory phase), and late (days 23 and 24 and days 25–28, i.e.: the early predecidual and late predecidual/exhaustive phase) phases, respectively.

Glandular and stromal cells in the ovulatory or interval period resemble those of the late proliferative phase. Anyhow interval glandular cells show incomplete perinuclear and subnuclear clearing which becomes especially prominent and generalized on postovulatory day 3. In the early secretory phase, each gland is enlarged and most endometrial cells show a well-developed subnuclear vacuole (Fig. 8.11), hence the nuclei are pushed towards to the center of the cytoplasm, producing nuclear palisading (Fig. 8.12). The stroma is relatively edematous.

Fig. 8.11 Secretory
Endometrium (Early
secretory phase). Each
gland is enlarged and gland
cells have well-developed
subnuclear vacuoles. (H&E
stain, original
magnification 4×)

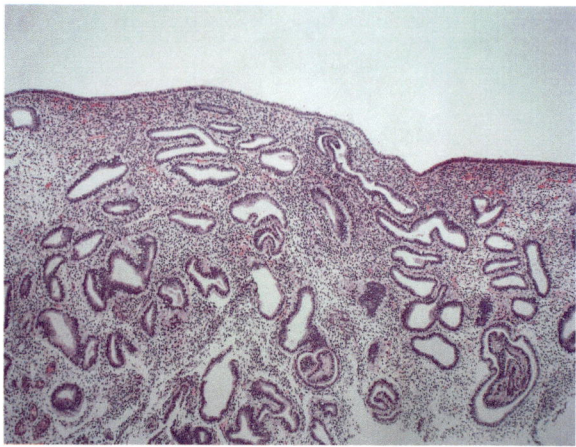

Fig. 8.12 Secretory
Endometrium (Early
secretory phase). The
nuclei are pushed up to the
center of the cytoplasms,
producing nuclear
palisading. The stroma is
relatively edematous.
(H&E stain, original
magnification 40×)

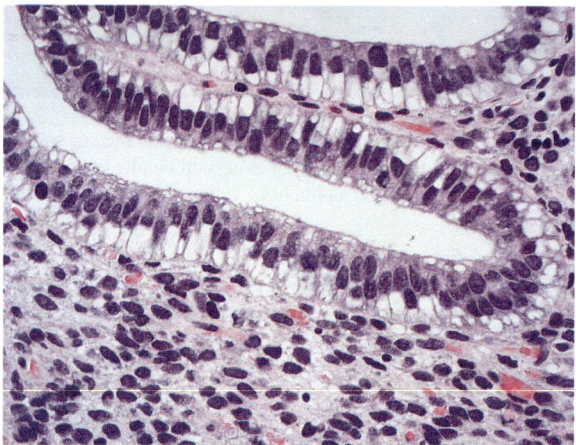

In the mid-secretory phase, the glands become more dilated, with an undulated profile (Fig. 8.13). As the glandular secretion increases progressively, cytoplasmic vacuoles became supranuclear and secretion accumulates within glandular lumina (Fig. 8.14). Stromal edema progressively increases (Fig. 8.15). Stromal cells begin to display oval to plump nuclei, with a vesicular nucleoplasm. In the late secretory phase, the glands demonstrate coiling and serrated (or saw-toothed) appearance (Fig. 8.16). Typically glandular secretory activity decreases, the upper two-thirds of the functionalis becomes predecidualized (Fig. 8.17).

Predecidual stromal change increases, initially being most evident in the cells surrounding the spiral arteries, and then further extending with formation of the so-called compact layer (stratum compactum) beneath the surface epithelium (Fig. 8.17). In the immediate premenstrual days, apoptotic activity is seen with glands, fibrin thrombi appear in small blood vessels, and there is extravasation of blood cells into the stroma (Fig. 8.18).

Fig. 8.13 Secretory Endometrium (Mid secretory phase). The glands become more dilated and tortuous. Stromal edema progressively increases. (H&E stain, original magnification 4×)

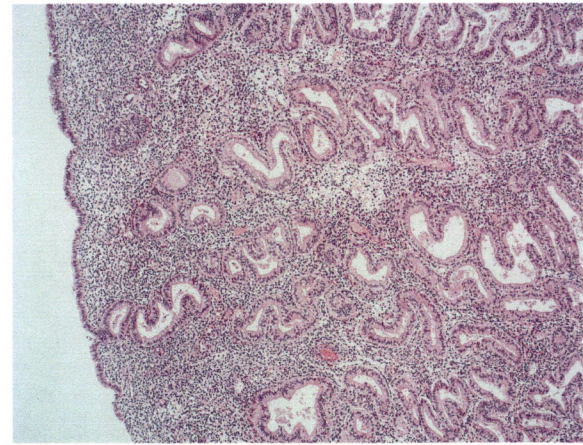

Fig. 8.14 Secretory Endometrium (Mid secretory phase). The degree of glandular secretion increases, cytoplasmic vacuoles became supranuclear and secretions are seen within glandular lumina. (H&E stain, original magnification 40×)

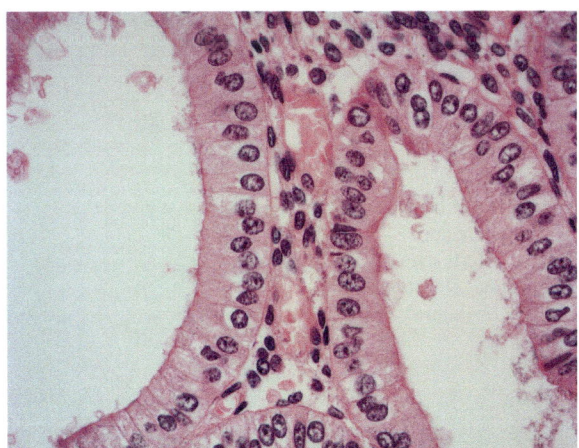

Fig. 8.15 Secretory Endometrium (Mid secretory phase). Stromal edema progressively increases. (H&E stain, original magnification 20×)

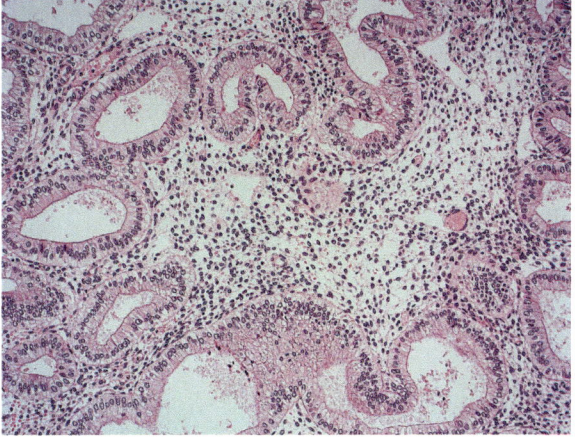

Fig. 8.16 Secretory Endometrium (Late secretory phase). The glands demonstrate coiling and serrated (or saw toothed) appearance. Reduction of glandular secretory activity is typically noticed, the upper two-thirds of the functionalis becomes predecidualized. (H&E stain, original magnification 4×)

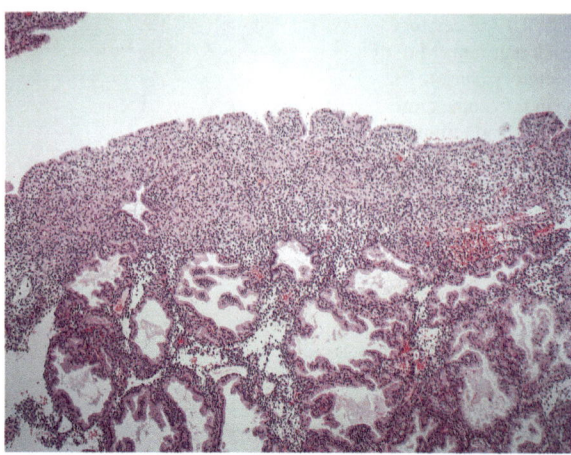

Fig. 8.17 Secretory Endometrium (Late secretory phase). Predecidual stromal change increases surrounding the spiral arteries, the so-called compact layer beneath the surface epithelium is formed. (H&E stain, original magnification 10×)

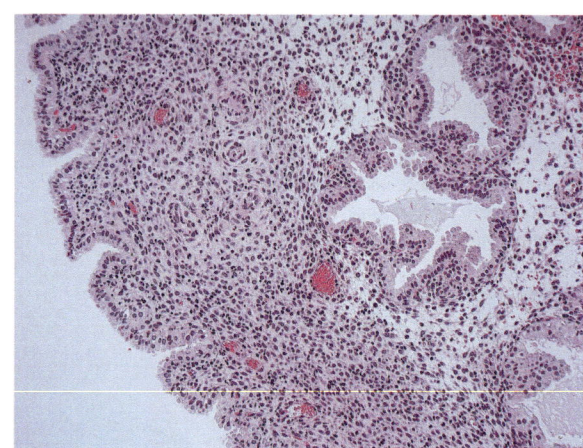

Fig. 8.18 Secretory Endometrium (Immediate premenstrual days). In the immediate premenstrual days, fibrin thrombi appear in small blood vessel, and there is extravasation of blood cells into the stroma. (H&E stain, original magnification 4×)

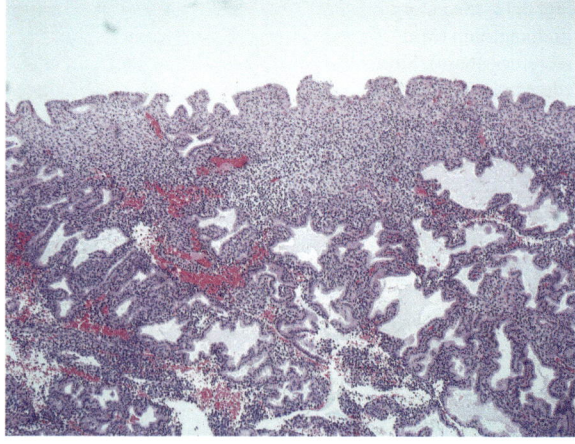

8.3.2   Definition

The cytological features of the detached endometrial fragments that reflect the histological architecture of SE are shown below [2–4]. Although early SE is similar to PE (Fig. 8.19), cells have a lower nuclear/cytoplasmic ratio, smaller inconspicuous nucleoli, and absent mitoses. The subnuclear cytoplasmic vacuoles are observed as large empty perinuclear spaces (Fig. 8.20). The cells within glands and cell sheets begin to take a "honeycomb" pattern (Fig. 8.21).

In the mid-secretory phase, the glands become more dilated, with an undulate profile (Fig. 8.22), and, towards the late secretory phase the endometrial glands display the typical three-dimensional accordion-pleated pattern which is equivalent to serrated (or saw-toothed) glands (Fig. 8.23) observed histologically.

Cells are larger and produced secretory substances show a honeycomb pattern with increased cytoplasm compared to that of early SE (Fig. 8.24). Nuclei are larger than those of PE, rounded and vesicular, and display a fine chromatin pattern.

The appearance of spiral arteries is recognized in the background. These are identified by elongated bundles of spindle-shaped cells including endothelial cells and/or perivascular epithelioid cells.

Fig. 8.19 Early secretory endometrium. Early secretory endometrium is similar to Proliferative endometrium. (Papanicolaou stain, original magnification **a**:4×, **b**:10×)

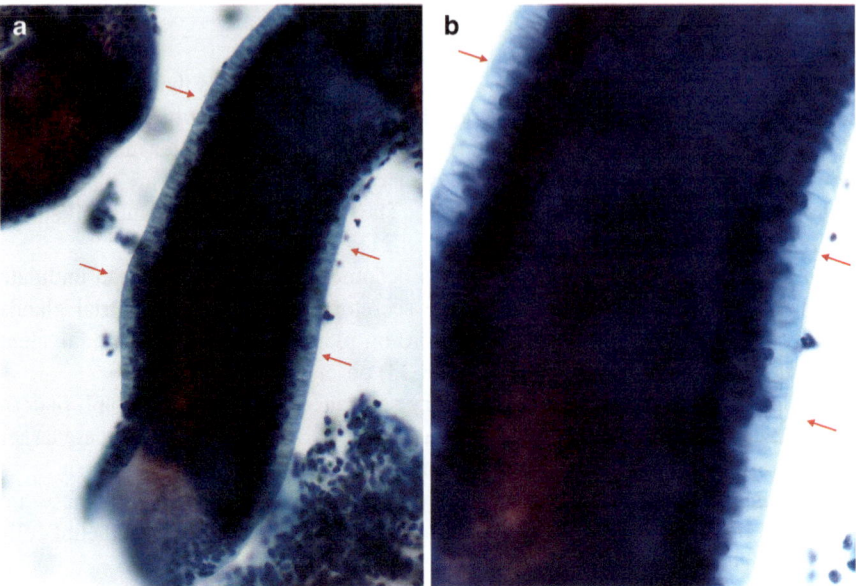

Fig. 8.20 Early secretory endometrium. Subnuclear cytoplasmic vacuoles (arrows) are observed as larger empty perinuclear spaces. (Papanicolaou stain, original magnification **a**:20×, **b**:40×)

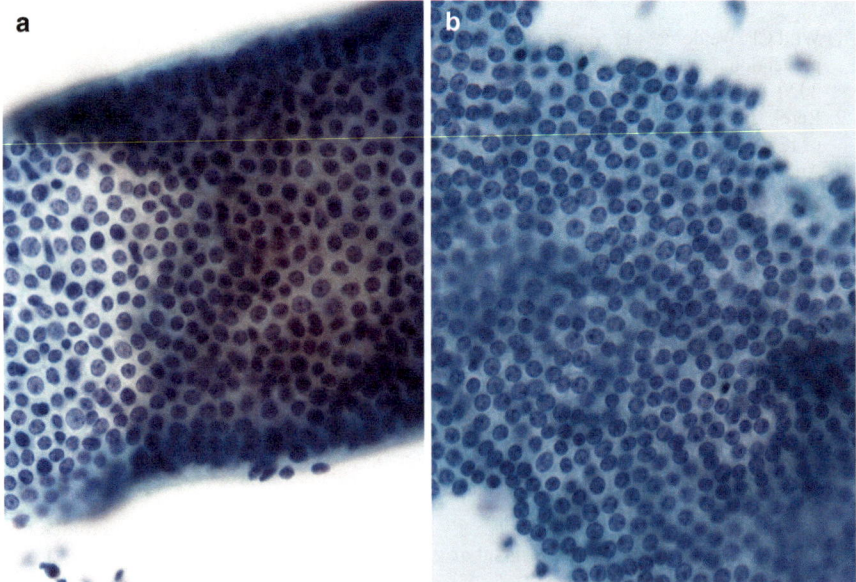

Fig. 8.21 Early secretory endometrium. The cells within glands and cell sheets start taking a honeycomb pattern. (Papanicolaou stain, original magnification **a** and **b**:40×)

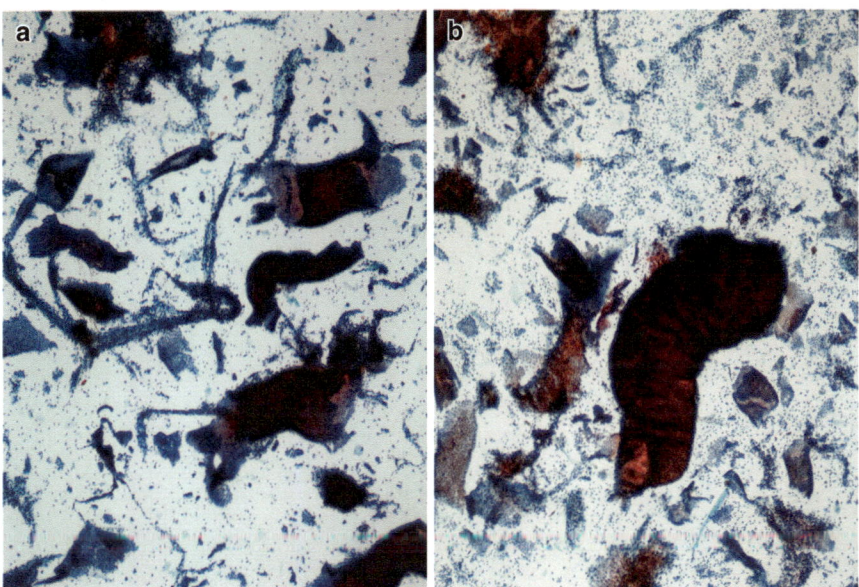

Fig. 8.22 Mid secretory endometrium. The glands become more dilated, with an undulated profile. (Papanicolaou stain, original magnification **a** and **b**:4×)

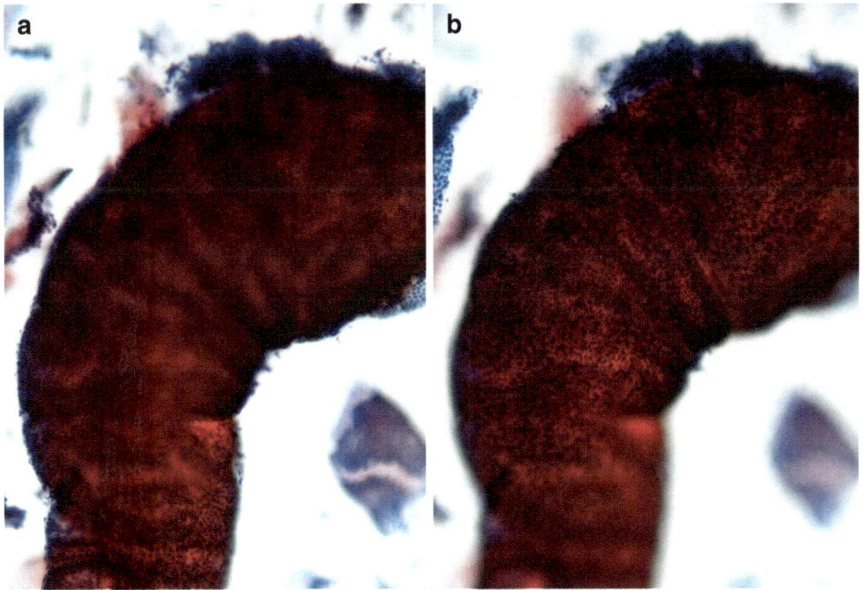

Fig. 8.23 Mid secretory endometrium. The clusters of endometrial glands display an inner three-dimensional accordion-pleated pattern. (Papanicolaou stain, original magnification **a** and **b**:10×)

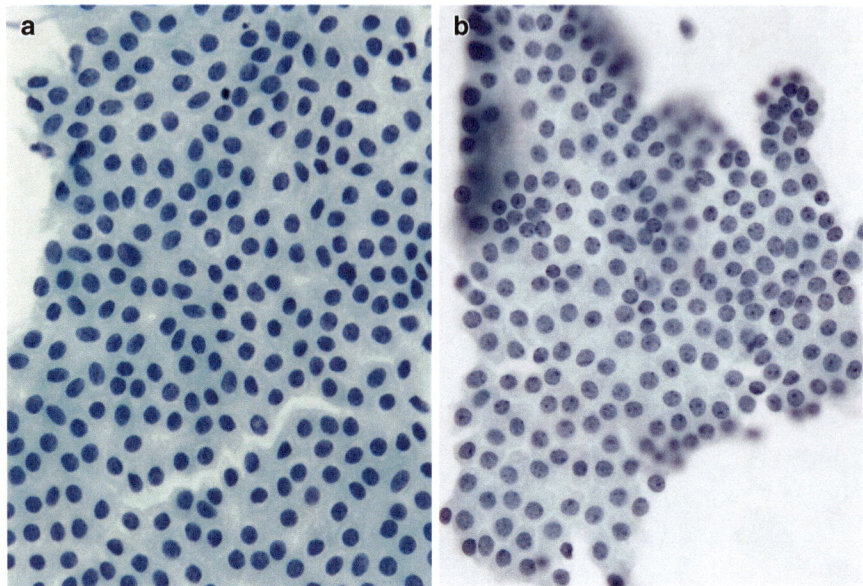

Fig. 8.24 Mid secretory endometrium. Nuclei are larger than those of PE. Cells show an honeycomb pattern with increased cytoplasm, larger than that of early SE; with initial production of secretory substances. (Papanicolaou stain, original magnification **a** and **b**:40×)

8.3.3 Criteria

- Cell clusters of straight to tubular pattern with more dilated lumen than PE.
- Gland width is approximately uniform.
- Cohesion of the endometrial stromal cells to the margins of the gland.
- The cell sheets show honeycomb pattern.
- Blood vessels in background.

8.4   Menstrual Endometrium [1]

8.4.1   Background

If conception does not occur, the spiral artery contract, causing ischemic changes following the abrupt decrease of estrogen and progesterone. As a result, ischemic tissue necrosis is promoted in the functional layer, then the exhausted mucosa exfoliates and sheds into the endometrial cavity. Overt menstruation is found primarily in humans and close evolutionary relatives species such as primates. The relatively copious bleeding of humans relates to the relatively large size of the uterine cavity in adult females and to the design of the uterine microvasculature, hence external bleeding appears to be the consequence of endometrial and stromal shedding preceding a new menstrual cycle.

Menstruation is characterized by glandular and stromal breakdown (Fig. 8.25). The endometrial glands collapse, some of the glands remain vacuolated and of secretory appearance (Fig. 8.26). The stromal cells lose their predecidual appearance take on a condensed and collapsed appearance. They aggregate into tightly packed balls with hyperchromatic nuclei (stromal blue balls) (Fig. 8.27) and separate from the glands. Other features include the presence of necrotic debris, neutrophil infiltration, interstitial hemorrhage, and fibrin deposition (Fig. 8.28). Apoptotic bodies are identified within both the glands and the stroma.

8.4.2 Definition

The endometrial tissue fragments show collapsed benign glands and "dissolved" or "ball-like" stromal aggregates (Fig. 8.29a). Some of glands remain vacuolated and

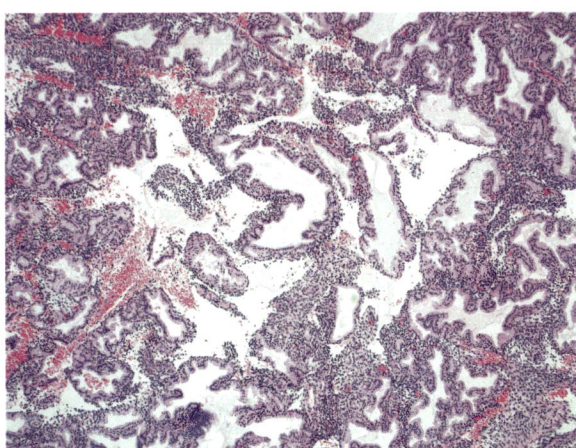

Fig. 8.25 Menstrual endometrium. Menstruation is characterized by glandular and stromal breakdown. (H&E stain, original magnification 4×)

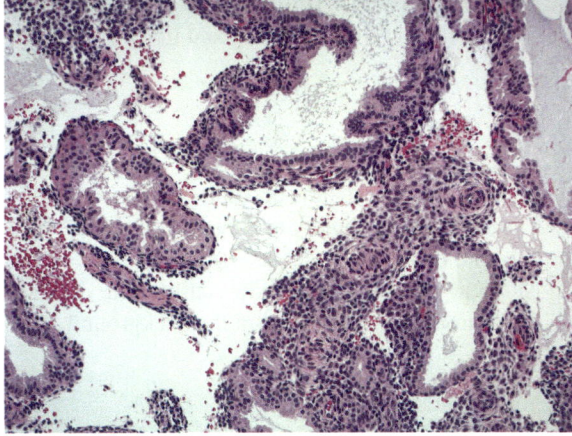

Fig. 8.26 Menstrual endometrium. The endometrial glands are collapsed, some of glands remain vacuolated and of secretory appearance. (H&E stain, original magnification 10×)

Fig. 8.27 Menstrual endometrium. The stromal cells collapse and aggregate into tightly packed cell balls with hyperchromatic nuclei (stromal blue balls). (H&E stain, original magnification 20×)

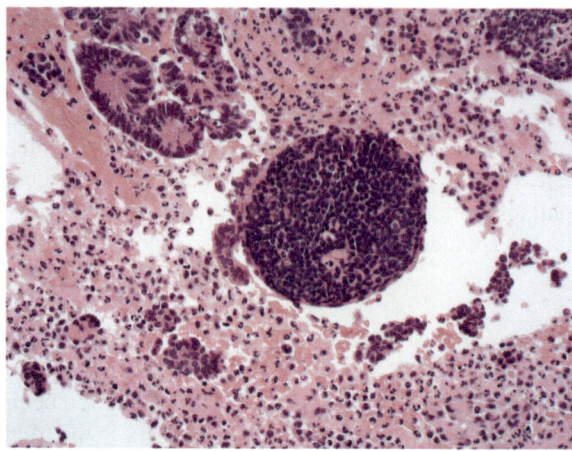

Fig. 8.28 Menstrual endometrium. Other features include the presence of necrotic debris, neutrophilic infiltration, interstitial hemorrhage, and fibrin deposition. (H&E stain, original magnification 40×)

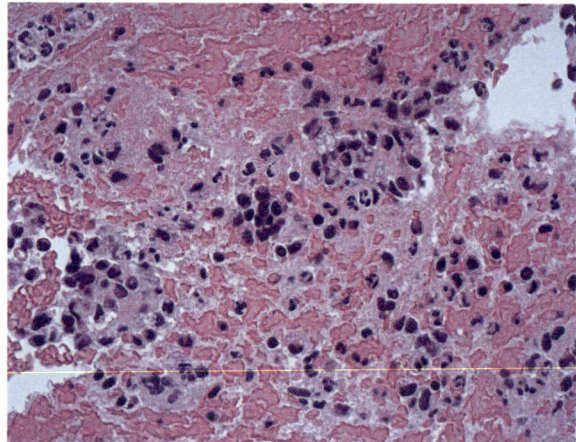

of secretory appearance (Fig. 8.29b). Other features include the presence of neutrophilic granulocytes, nuclear dust, and fibrin strands in the background.

8.4.3 Criteria

- Some of glands remain vacuolated and of secretory appearance.
- The stromal cells tightly aggregate into packed balls with hyperchromatic nuclei (stromal blue balls).
- Necrotic debris, neutrophilic infiltration, interstitial hemorrhage, and fibrin deposition can be observed in the background.

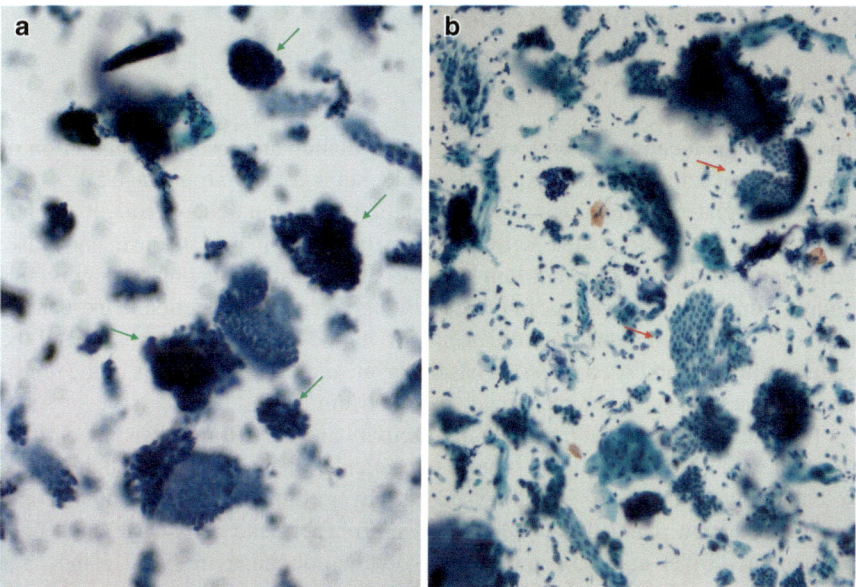

Fig. 8.29 Menstrual endometrium. The endometrial tissue fragments show collapsed benign glands and "dissolved" or "ball-like" stromal aggregates (green arrows). Some of glands remain vacuolated and of secretory appearance (red arrows). (Papanicolaou stain, original magnification **a** and **b**:4×)

8.4.4   Differential Diagnosis

Menstrual endometrium shows the greatest possible structural polymorphism seen in any of the phases of cycling endometrium, meaning that it may unwittingly be overdiagnosed as endometrial cancer. Although, the absence of stroma between the glandular elements of menstrual endometrium may give the impression of confluent glands or, in some cases, of solid epithelial growth, the menstrual endometrium lacks the coarse chromatin and nuclear pleomorphism of malignancy and a genuine cribriform pattern.

The distinction between cyclical menstrual endometrium and endometrial glandular and stromal breakdown (EGBD) is difficult. The glands of menstrual endometrium regularly exhibit some residual secretory change, and a careful search will identify typical areas of stromal fragmentation as opposed to areas of coagulative necrosis that are seen with bleeding of EGBD. Furthermore, a predecidual component is not part of irregular shedding resulting from EGBD.

8.5   Atrophic Endometrium [1]

8.5.1   Background

The age of the menopause, with cessation of ovulation and resultant diminution of hormone production by the ovaries is variable, but is usually around age 50.

An age-related morphological transformation continuum takes years, it extends from weak proliferation in the perimenopause to atrophy in the late postmenopause. Postmenopausally, the endometrium becomes thin and atrophic, unless there is continuing estrogenic drive, either in the form of endogenous production or exogenous hormone use. When there is no estrogenic drive, the functionalis is absent and the endometrium is composed only of a basalis layer, similar to the basalis of the reproductive years and of the premenarchal endometrium.

The glands do not exhibit proliferative activity and vary from consisting entirely of small widely spaced atrophic tubules to cystically dilated glands throughout (the so-called cystic atrophy or senile cystic atrophy); a mixture of small tubules and cystically dilated glands may occur (Fig. 8.30). The gland cells have small dark regular nuclei that may be round, ovoid, or low columnar; in cystic glands, the nuclei are often compressed and attenuated (Fig. 8.31).

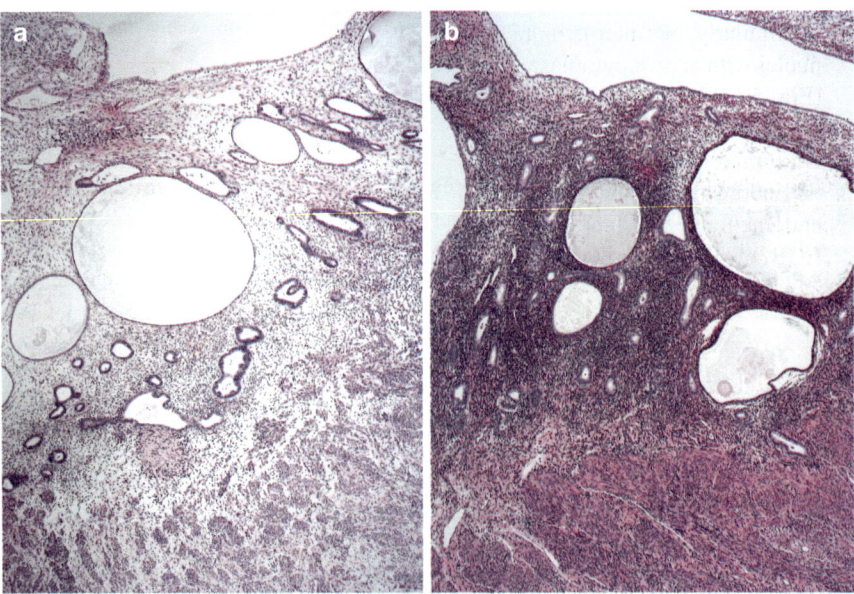

Fig. 8.30 Atrophic endometrium. The glands do not exhibit proliferative activity and vary from consisting entirely of small widely spaced atrophic tubules to cystically dilated glands throughout. (H&E stain, original magnification **a** and **b**; 4×)

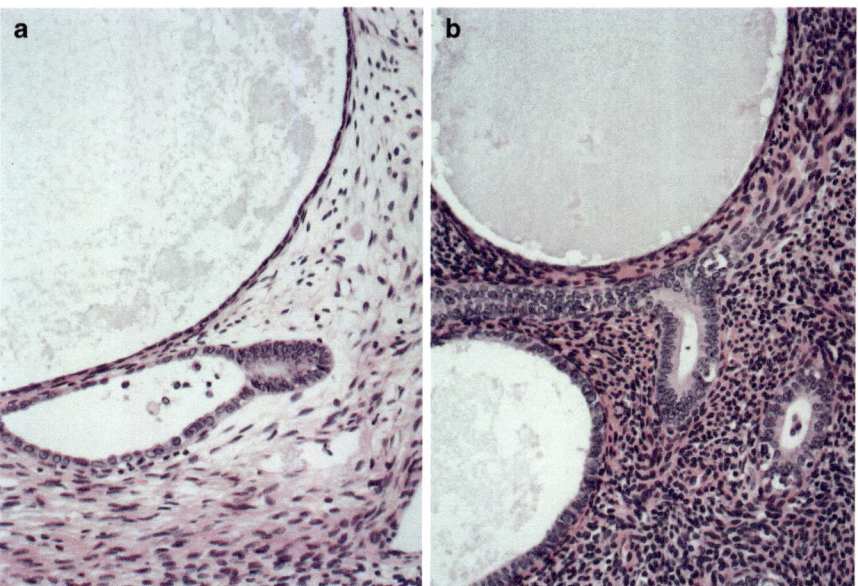

Fig. 8.31 Atrophic endometrium. The gland cells have small dark regular nuclei that may be round, ovoid, or low columnar; in cystic glands, the nuclei are often compressed and attenuated. The stroma has a more fibrous appearance (**a**) or is densely cellular and composed of ovoid to spindle-shaped cells with scant cytoplasm (**b**). (H&E stain, original magnification **a** and **b**; 20×)

The stroma in postmenopausal endometria may be densely cellular and composed of ovoid to spindle-shaped cells with scant cytoplasm (Fig. 8.31a) or have a more fibrous appearance (Fig. 8.31b) than in the premenopausal endometrium. The stroma tends to become more hypocellular and fibrous with advancing age.

8.5.2 Definition

Atrophic endometrium [2–4] may resemble the early phase of cycling proliferation but with less epithelial pseudostratification, shorter apical-base epithelial cell axes, and greatly reduced to absent mitoses. The uniform round nuclei are generally arranged in small monolayered sheets and present distinct cytoplasmic boundaries, with decreased or absent mitoses (Fig. 8.32). Because the stromal cells of atrophic endometrium are often fibroblast like, similar to the fibrous stroma of the lower uterine segment, also less abundant cytologic material is obtained and represented in the preparations. These stromal cells may occur in loose groupings; crushed blue stromal debris appears "torn up" or "mottled" (Fig. 8.33).

Unlike stromal cells in the atrophic endometrium, those in the PE are usually densely cellular and have indistinct cytoplasm and cell borders. The nuclear shape varies from oval to spindle shaped or reniform with fine chromatin (Fig. 8.34).

Fig. 8.32 Atrophic endometrium. The uniform round nuclei are generally arranged in small monolayered sheets and present distinct cytoplasmic boundaries. (Papanicolaou stain, original magnification 40×)

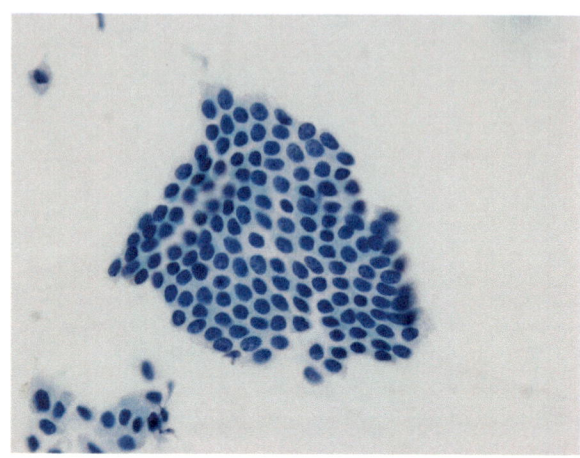

Fig. 8.33 Stromal cells in atrophic endometrium. The stromal cells may occur in loose groupings; crushed blue stromal debris appears "torn up" or "mottled". (Papanicolaou stain, original magnification 40×)

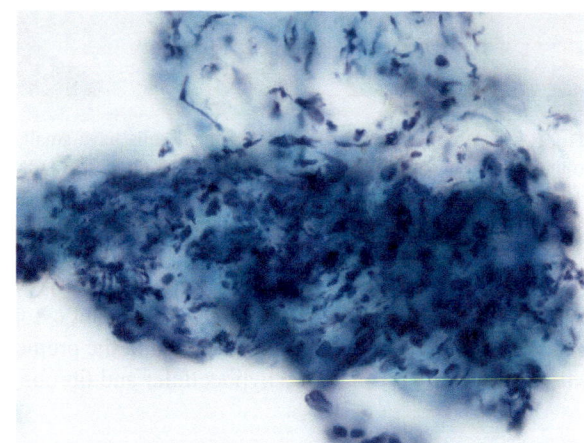

Fig. 8.34 Stromal cells in proliferative endometrium. Unlike stromal cells in the atrophic endometrium, those in the proliferative endometrium are usually densely cellular and have indistinct cytoplasmic borders. The nuclear shape is oval, spindled or reniform with fine chromatin. (Papanicolaou stain, original magnification 40×)

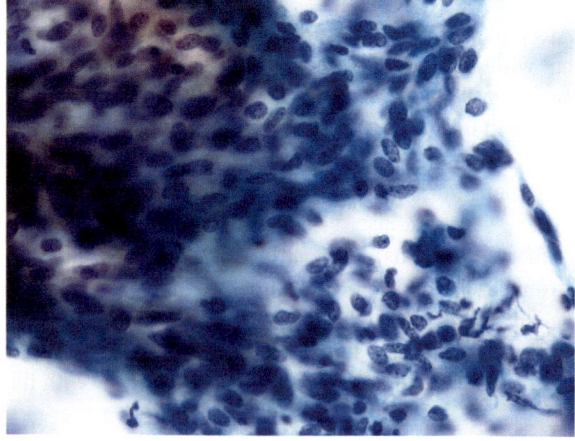

8.5.3   Criteria

- Nuclear crowding and overlapping is not as striking as in PE.
- Small monolayer epithelial sheets with uniform round nuclei.
- Distinct cytoplasmic boundaries.
- Decreased or absent mitoses.

8.6   Benign Reactive Changes: IUD (Intrauterine Contraceptive Devices)

8.6.1   Background

The IUD, also known in some places as the intrauterine contraceptive device or IUCD, is one of the world's most widely used family planning methods. It is the second most commonly used form of contraception, ranking second only to female sterilization and also the most commonly used form of reversible contraception. Globally, 14.3% of women of reproductive age use IUD, but the distribution of IUD users is strikingly non-uniform. In some countries, the percentage of women using IUD is <2%, whereas in other countries, it is >40%. The reasons for this large variation are not fully understood. Among IUD users, 92 million (60%) are concentrated in China [5]. However, IUD acceptance is also growing in other parts of the world [6].

8.6.2   Mechanism of Action of IUD

The major effect of all IUD is to induce a local inflammatory reaction in the endometrial mucosa. The cellular and humoral components of this inflammatory response permeate endometrial blood and lymph vessels and are released into the uterine cavity. Although their effect on contraception is not entirely known, it seems to depend largely from the type of device [7] In the case of inert and copper IUDs, the endometrium is the target site of the contraceptive effect with a prevailing anti-implantation effect that is secondary to various histofunctional damages of the mucosa. The first IUD-induced changes are of mechanical type and include flattening, erosion, crushing, and atrophy of the endometrium, depending basically from the shape and, especially, the size of the IUD. The second important change is the appearance of a nonspecific inflammatory response (akin to a foreign body reaction). This response consists of a focal inflammatory infiltrate [8] containing neutrophilic granulocytes, lymphocytes, eosinophils, plasma cells, macrophages, mast cells [9, 10], and, infrequently, foreign-body type multinucleated giant cells [11, 12]. However while in non-IUD users, the concentration of inflammatory cells in the uterine fluid is much lower in the luteal phase than in the proliferative, midcycle or premenstrual phases, in IUD-users it tends to be elevated in all phases of the menstrual cycle [13]. The inflammatory reaction is especially marked where the endometrium is in contact with the IUD. The IUD also causes vascular changes: altered permeability and diffuse microthrombosis of the capillaries in the endometrial stroma in contact with the IUD. Many authors have suggested that this aseptic inflammatory response is primarily responsible for the contraceptive effect of IUDs.

8.6.3 Occurrence of IUD Associated Cellular Change

It is well known that the string attached to the IUD can lead to chronic irritation of the endometrium and the endocervix (Fig. 8.35). Prolonged use of an IUD is often associated with the exfoliation of endometrial cells. These cells have been found in 60% of smears from IUD users [14, 15] throughout the menstrual cycle, while, in non-IUD users, apparently normal endometrial cells may be found in the following conditions: during the immediate postpartum period, in impending or early abortion [16], in abnormal bleeding related to the endometrium. Introduction of an IUD into the uterine cavity results in a local histiocytic and inflammatory cell response [8, 16]. Further cytological findings in IUD users both in cervical and endometrial samples include: amorphous calcified bodies with foreign-body-type giant cells (Fig. 8.36) as well as pleomorphic reactive-reparative cells (Fig. 8.37) [17, 18]. In cervical smears of IUD users, atypical endometrial cells, with marked nuclear enlargement may be observed that, unless an extreme care is taken, may be mistaken for carcinomatous glandular cells [19]. Atypical endometrial cell clusters, similar to those found in cervical smears, could also be seen in uterine fluid aspirates [20]. A study of the DNA content of those atypical cells [21] indicated that these latter had a diploid to octaploid DNA content with a major peak in the tetraploid region. Atypical multinucleated cells have been described in cervical smears which have been interpreted as histiocytes [22] or endocervical cells [23]. However,

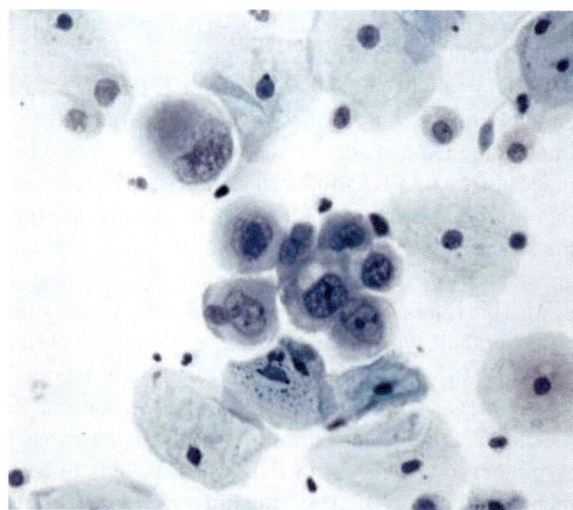

Fig. 8.35 Reactive-reparative cellular changes due to IUD usage: Abnormal morphology in a cervical smear from a patient with an IUD. Cellular nuclei are enlarged, showing irregularly distributed chromatin and recognizable nucleoli. These cells can mimic malignancy in a routine smear (Papanicolaou stain, original magnification; 40×). Reproduced from Kobayashi, Casslen and Stormby with permission [14]

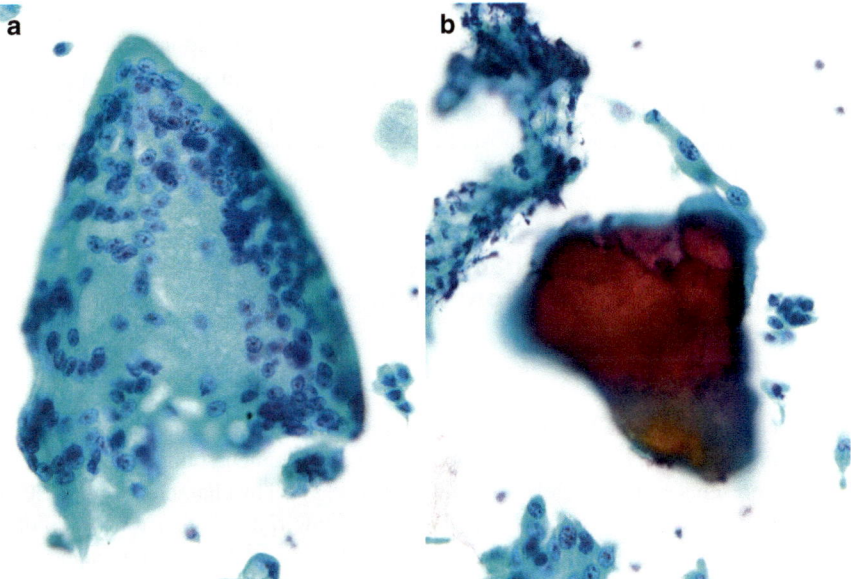

Fig. 8.36 Cellular changes due to IUD usage: Directly sampled endometrial smear obtained from IUD user. (**a**) A foreign-body-type giant cell can be seen. (**b**) Calcified body surrounded by reparative cells. (Papanicolaou stain, original magnification **a** and **b**; 40×)

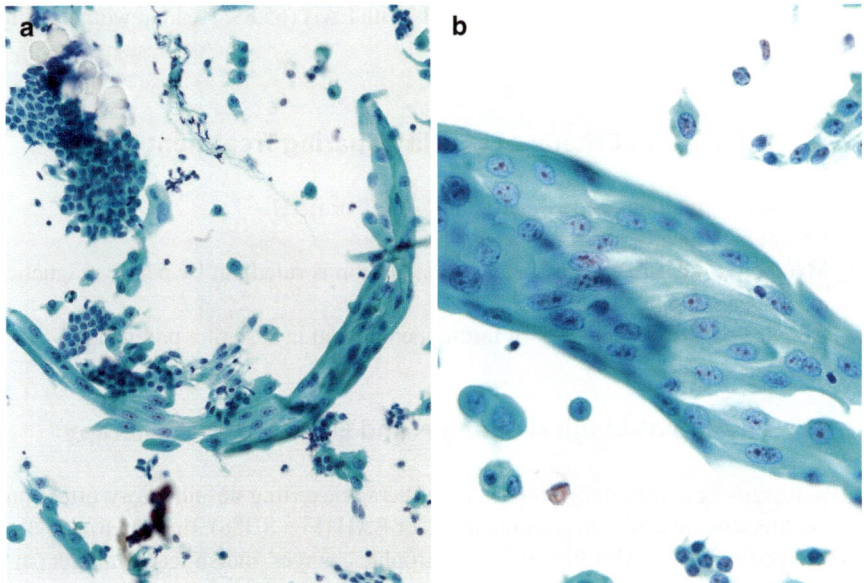

Fig. 8.37 Cellular changes due to IUD usage. (**a**) Reactive-reparative cells appeared in a sheet-like pattern. (**b**) Reactive-reparative cells show evident nucleoli in virtually every nucleus. (Papanicolaou stain, original magnification **a**; 20, **b**; 40×)

by consensus, they are now regarded as of endometrial origin [24–26]. It has been pointed out that there is some similarity between the atypical endometrial cells seen in smears from IUD users and the endometrial cells seen in smears from patients who have had miscarriage [16]. Clearly the atypical cells associated with IUD usage may result in an erroneous diagnosis of cancer and correct interpretation of these cells in both cervical and endometrial smears is of great importance.

8.7 Benign Reactive Change: MPA (Medroxyprogesterone Acetate)

8.7.1 Background

Endometrial cancer (EC) is the most common gynecological cancer in women of postmenopausal age. However, endometrial cancer in women younger than 40 years of age represents approximately 4–5% of all cases [27]. The characteristics of EC in young women are early-stage, low-grade, and endometrial endometrioid carcinoma (EEC). Endometrial atypical hyperplasia (EAH) is a well-known precancerous lesion, which may progress to EC within several years [28]. The standard treatment for EC is surgery with hysterectomy, however this procedure causes infertility in young women who may desire to preserve their ability to have children. Response rates and recurrence rates of fertility-sparing treatment with MPA (Medroxyprogesterone acetate) displayed considerable variation [29]. Complete response was significantly higher in patients with EAH (65.8%) than in women with EC (48.2%).

8.7.2 Adaptation Criteria of Fertility-Sparing Treatment [29]

1. Histologically diagnosed Grade1 (G1)—EEC or EAH.
2. Patient is clinically presumed as stage IA.
3. Myometrial invasion or any extra uterine lesion is ruled out by pelvic magnetic resonance imaging.
4. Absence of a significant risk of thromboembolism in case of a pregnancy.

8.7.3 Physiopathological Changes and Effects of MPA Therapy

MPA therapy is a high-dose progesterone therapy exerting an inhibitory effect on the proliferation of estrogen-dependent EC or EAH (Fig. 8.38a). Endometrial histological findings after MPA therapy show atrophic changes, indistinct chromatin patterns in the endometrial cells, stromal edema or decidual-like changes in endometrial stromal cells (Fig. 8.38b). During cytological follow-up of MPA therapy, when normal or benign appearing endometrial sheets or clusters are present in endometrial brushings, the patient is considered as having a complete response. However, when

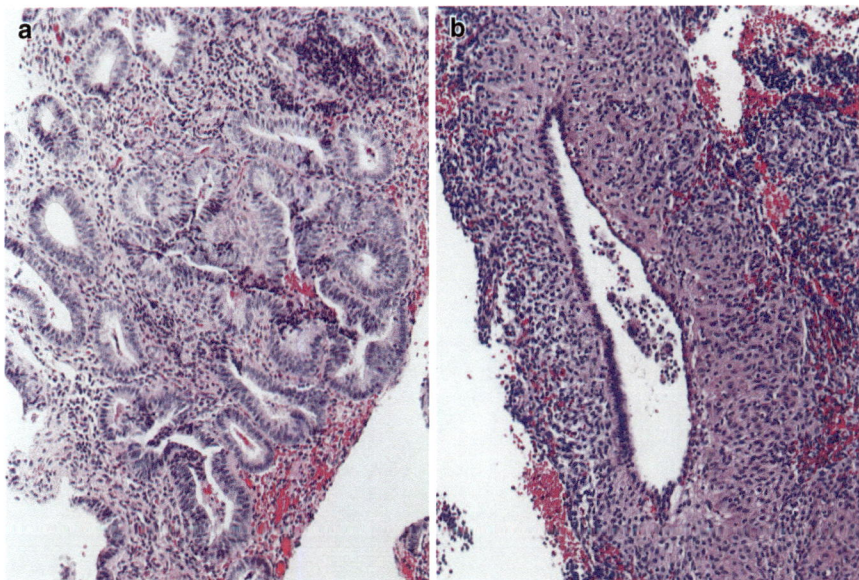

Fig. 8.38 Histological findings before and after MPA therapy: Before MPA therapy (**a**), a case has a diagnosis of EAH/ EIN. After MPA therapy (**b**), the histological diagnosis was consistent with complete response: a decrease of endometrial glands with decidua-like changes of endometrial stromal cells may be observed. (H&E stain, original magnification **a**; 10×, **b**; 20×)

endometrial cell clusters with overlapping nuclei of three-layers or more are observed in cytological smears, the patient is assessed as having progressive disease. Histological examination is then necessary.

8.7.4   Assessing the Efficacy of MPA Therapy: Cytological Criteria

Cytological findings in patients treated with MPA therapy may be classified as follows: 1), 2)

   1) Complete therapeutic efficacy (Fig. 8.39)

- Endometrial cells appear in a sheet pattern with <3 layers of nuclear overlapping.
- Endometrial cells may show secretory activity (honeycomb pattern) or oxyphilic cytoplasmic change.
- Nuclei show oval and uniform chromatin patterns.
- This case was diagnosed as negative for malignancy.

   2) No therapeutic efficacy (Fig. 8.40)

- Endometrial cell clusters show a gland-like structure with small irregular projections on the margin of the cluster (irregular nuclear protrusion pattern).

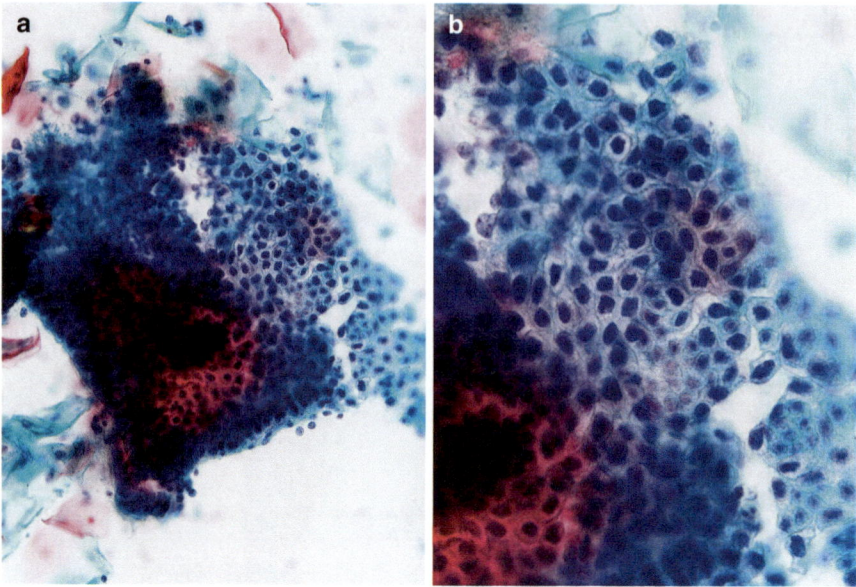

Fig. 8.39 Cytological findings after MPA therapy: a sheet-like pattern with honeycombed appearance may be observed. Before MPA therapy, the histological diagnosis of this case was EAH/EIN. After MPA therapy, endometrial cells showed sheets or clusters with <3 layers of nuclear overlapping. This case was diagnosed as negative for malignancy (TYS1). (Papanicolaou stain, original magnification **a**; 20, **b**; 40×)

- Endometrial cells show nuclear hyperchromasia with >3 layers of nuclear overlapping.
- If nuclear atypia is present but it is not of severe degree and malignancy cannot be ruled out, the case should be interpreted as ATEC-AE (TYS5).
- If nuclear atypia of severe degree is observed, the cytologic diagnosis of persisting EAH or EEC is performed and histological examination is necessary.

8.8   Benign Reactive Changes: TAM(Tamoxifen)

8.8.1   Background

Tamoxifen (TAM), an antiestrogen, is a competitive inhibitor of estradiol, and has been widely used for the adjuvant treatment of breast cancer. TAM produces objective tumor shrinkage in advanced breast cancer, and reduces the risk of relapse in women treated for breast cancer [30]. Although initially developed as an antiestrogen, TAM can also prevent postmenopausal osteoporosis as well as it may reduce cholesterol blood levels, due to its estrogen-agonist effects [31]. Its estrogen-agonist activity, however, can lead to significant side-effects such as endometrial cancer (EC) and thromboembolism. Therefore, long-term TAM therapy is included among

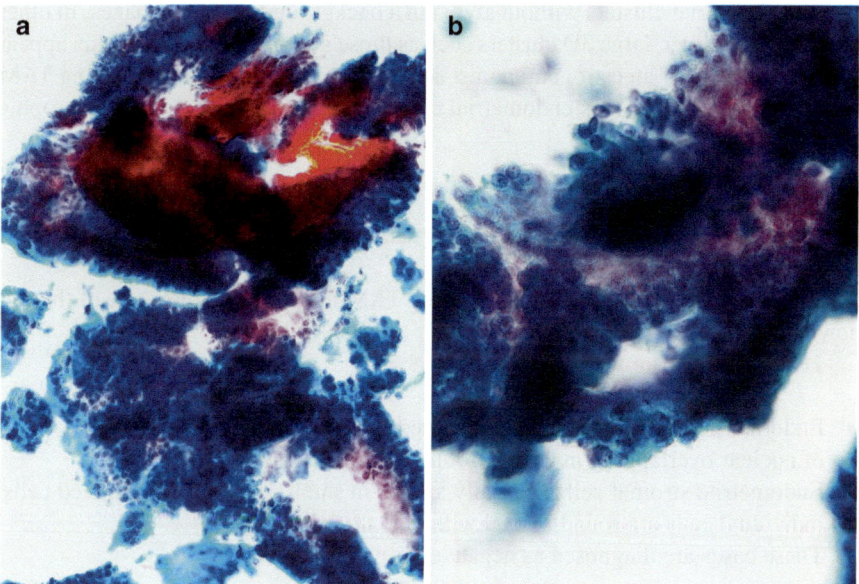

Fig. 8.40 Cytological findings before MPA therapy: The histological diagnosis of this case was EAH/EIN before MPA therapy. Atypical endometrial cells were observed after MPA therapy. Endometrial cell clusters show nuclear overlapping with >3 layers and small irregular projections on the margin of the cluster. However, nuclear atypia is not the severe grade. When malignancy cannot be ruled out, the case should be interpreted as ATEC-AE. (Papanicolaou stain, original magnification **a**; 20, **b**; 40×)

the risk factors for EC. Pathologic changes are often found in the endometrium of postmenopausal breast cancer patients receiving TAM therapy, the most frequent being endometrial polyps. In addition, some histologic changes may also occur, such as proliferative endometrium (PE), endometrial hyperplasia, and EC [30]. As it concerns TAM therapy for the prevention of breast cancer, women who received long-term TAM therapy had an increased risk of invasive endometrial cancer (Risk rate = 3.28, 95% CI = 1.87 to 6.03) [32]. However, in this study, the cancer risk is lower in women aged 49 year or younger, while it is higher in women aged 50 years or older (Risk rate = 5.33, 95% CI =2.47 to 13.17). Anyhow, another study has reported that women who received long-term TAM therapy had no increased risk of developing EC [33].

8.8.2   Cytological Findings in Endometrial Mucosa after TAM Therapy

In some postmenopausal breast cancer patients receiving TAM therapy, the endometrial mucosa does not show any pathologic change, and is referred to as atrophic endometrium (AE). Cytological findings in such cases include uniform sheets of small endometrial clusters without atypia in a background of macrophages. In other instances, however, large 3D clusters such as those seen in PE or hyperplasia appear due to the effect of an estrogen-agonist activity. Therefore, patients receiving TAM therapy may show various endometrial cytological findings ranging from atrophic to proliferative changes.

8.8.3 Cytological Endometrial Findings after TAM Therapy

Cytological findings in patients undergoing TAM therapy may show one of the following changes: 1)–3)

1) Atrophic endometrium (AE) (Fig. 8.41).

- Endometrial cells appear as monolayered sheets of various sizes with < 3 layers of nuclear overlapping in a background of macrophages.
- Endometrial stromal cells generally appear in small clusters or as isolated cells; individual cells are round to oval with bare nuclei.
- These cases are diagnosed as negative for malignancy.

2) Proliferative endometrium (PE) (Fig. 8.42)

- Endometrial cells appear in large clusters with <3 layers of nuclear overlapping.
- Nuclear findings are consisting with PE rather than AE.
- This case is diagnosed as negative for malignancy.

3) Reactive changes of endometrial cells with enlarged nuclei (Fig. 8.43)

- Endometrial cells show nuclear enlargement and anisonucleosis with an evident nucleolus.

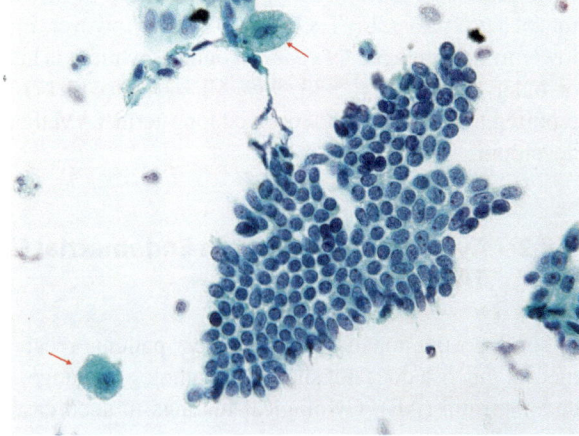

Fig. 8.41 Cytological findings of a postmenopausal woman after TAM therapy for breast cancer: Atrophic endometrium shows a sheet pattern with a background of macrophages (arrows). This case was diagnosed as negative for malignancy. (Papanicolaou stain, original magnification 40×)

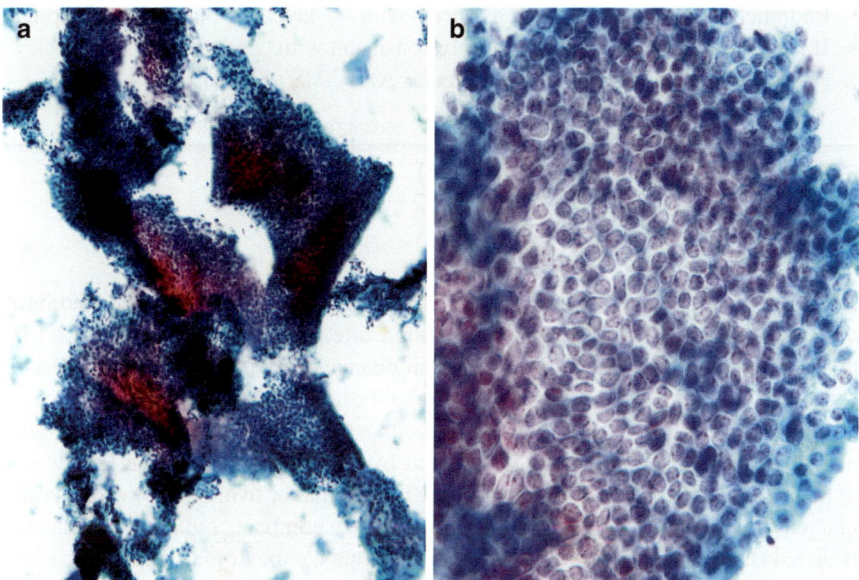

Fig. 8.42 Cytological findings of a postmenopausal woman after TAM therapy for breast cancer: Endometrial cells with proliferative changes appear as large clusters and nuclear atypia is not observed. This case was diagnosed as negative for malignancy. (Papanicolaou stain, original magnification **a**; 10×, **b**; 40×)

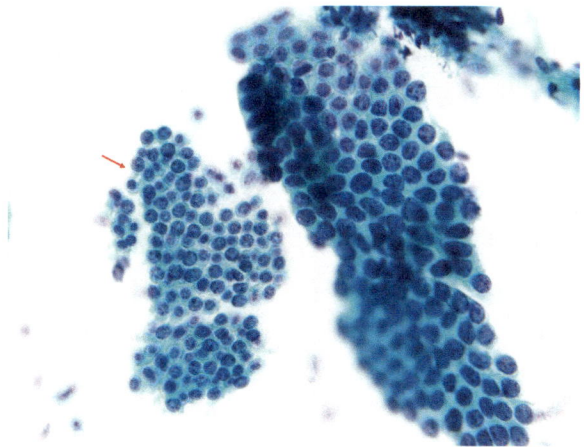

Fig. 8.43 Cytological findings of a postmenopausal woman after TAM therapy for breast cancer: A large sheet of atypical endometrial cells may be observed (right side of the image) showing nuclear enlargement and anisonucleosis with an evident nucleolus in a background of atrophic endometrial cells (arrow). However, due to the absence of high grade atypia and <3 layers of overlapping nuclei, this case was diagnosed as negative for malignancy. Whenever the presence of cytological atypias precludes the diagnosis of "negative for malignancy" (TYS1), the case should be interpreted as ATEC-US (TYS4). (Papanicolaou stain, original magnification 40×)

- Endometrial cells are observed as sheets with <3 layers of nuclear overlapping.
- If these findings should be not entirely consistent with "negative for malignancy" (TYS1), the case should be interpreted as ATEC-US (TYS4).

8.9   Arias-Stella Reaction (ASR)

8.9.1   Background

The Arias-Stella reaction (ASR) is a physiologic response to the presence of chorionic tissue either in the uterus or in an ectopic site [34]. In 1954, Arias-Stella [35] first described certain endometrial glandular changes as pathognomonic of intra- or extra-uterine pregnancy, or as result of trophoblastic disease. The characteristic histologic features of the ASR include large cells with abundant eosinophilic or vacuolated cytoplasm, nuclear enlargement, and hyperchromasia [36]. The appearance of the hypertrophic nuclei may vary widely from round or ovoid nuclei with a vesicular chromatin pattern to irregular nuclei with a compact, pyknotic appearance (Fig. 8.44). Some variants exhibit prominent intranuclear cytoplasmic invaginations or pseudoinclusions. The nuclei may protrude into the glandular lumen creating a hobnail appearance [37]. There may be loss of cell polarity with papillary projections and epithelial tufting. Usually, mitotic activity is absent [36]. When presenting in extra-uterine sites, the differentiation of ASR from other more ominous clear cell lesions may pose significant difficulties [38]. The incidence of the ASR in conjunction with an ectopic or other pregnancy has been reported as ranging from 5% to 74% [39, 40]. In 1981, Schneider [41] retrospectively studied hysterectomy specimens obtained during pregnancy; the ASR was found in the endocervix in 9% of cases. However, as only 3–4 blocks of the cervix had been obtained, and as the

Fig. 8.44 Endometrial histology showing ASR. ASR with hyperchromatic nuclei that bulge into the glandular lumen. The glands also show marked vacuolated hypersecretory cellular tufts. (H&E stain, original magnification; 20×)

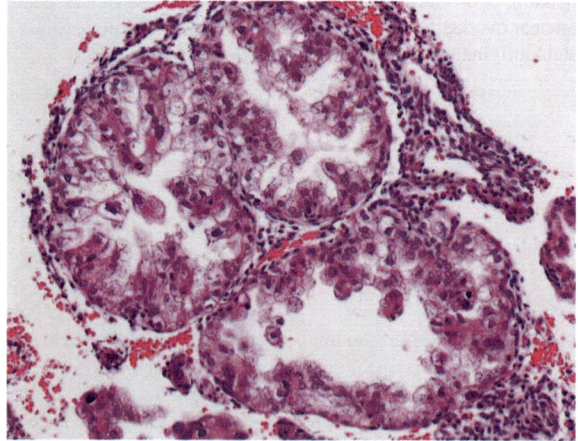

lesions were remarkably focal, often limited to 1 or 2 glands, he believed that the figure of 9% was likely to be an underestimate of the true prevalence of the condition in the endocervix. ARS is also well known to occur in non-pregnant patients taking various types of hormonal preparations. Felix et al. [42] reported that the finding of cervical ASR in a patient taking phytoestrogenic treatment highlights the fact that the source of hormonal exposure may not always be immediately evident; the unusual histology of the lesion led to an erroneous diagnosis of papillary serous carcinoma in the biopsy.

8.9.2   Cellular Changes of Arias-Stella Reaction (ASR)

Exfoliative cytology has proven to be an efficient technique for the detection of precancerous and cancerous squamous lesions of the cervix. However, this is not true for glandular lesions of endocervical and especially endometrial origin. Glandular changes resembling the ASR are very unusual in nonpregnant women. In this clinical settings, Arias-Stella cells have been rarely described and are difficult to differentiate from other atypical cell types in cervical and endometrial samples [43]. They are relatively easy to appreciate in endometrial samples containing tissue fragments, and in these cellular samples they lack the cytologic features of malignancy. Ferguson [44] described the cytological features of ASR, even before Arias-Stella's original article; he reported that atypical glandular cells were recognizable in vaginal smears in cases of abortion or endometritis, although their origin was unknown at that time. Others have described glandular changes in cytologic smears that might have been related to the ASR, but the corresponding tissue samples were lacking. This lack of histologic correlation has been a continuing concern until a case report by Mulvany et al. was published [45], in which cytologic and histologic samples confirming the features of the ASR were obtained. Bonoit and Kini [46] further studied the cytologic features of ASR in 1996. As a general rule, in case of a cytologic picture consistent with ASR the referring clinician must be informed that the above described changes may be related to pregnancy. It is of paramount importance that these cells are identified, as reactive and unequivocal diagnoses of malignancy are avoided.

A false diagnosis of malignancy, especially during pregnancy, will place the mother and the fetus at unnecessary risk because of further invasive procedures [45–48]. It has been pointed out that there is some similarity between the atypical endometrial cells seen in smears from IUD users and the endometrial cells in smears from patients who have had a miscarriage [48, 49]. The cytopathologists should be conversant with the degree of glandular atypia and with the possible occurrence of endocervical ASR [50]. The cytologic features of the ASR in cervicovaginal cytology smears obtained from women with tubal pregnancy were reported [49–51]. Based on the cytologic evidence, the ASR-associated cells were divided into 2 groups [49]. Striking cellular changes occurred in the cytoplasm, which contained many vacuoles; these changes seemed to reflect mucus production (Fig. 8.45). Some cells of this type also had intranuclear inclusions (Fig. 8.46). The second group of ARS-associated cells had naked nuclei and the nuclei were relatively hyperchromatic [51], they had homogeneous chromatin distribution, were oval or ellipsoid, varied markedly in size, and appeared singly or grouped (Fig. 8.47). These cellular changes mimic those found in epithelial atrophy accompanied by inflammatory changes. Nucleoli were present but were identified only with difficulty.

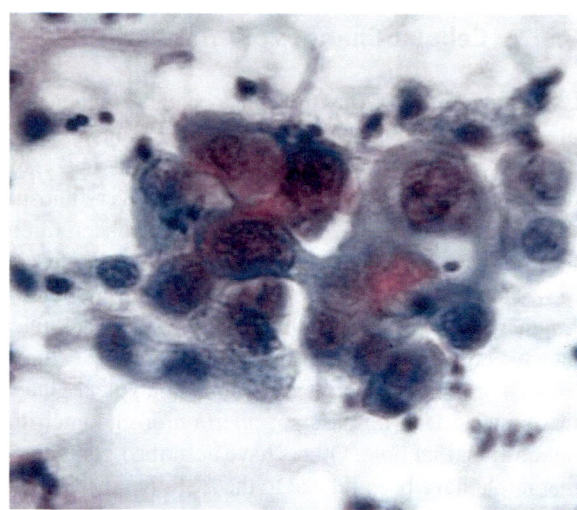

Fig. 8.45 Typical appearance of Arias-Stella cells arranged in a cluster. The nuclei are large and of variable size with irregularly clumped chromatin. (Papanicolaou stain, original magnification; 100×) Reprinted with permission from Kobayashi et al. [49]. Acta Cytol 1980)

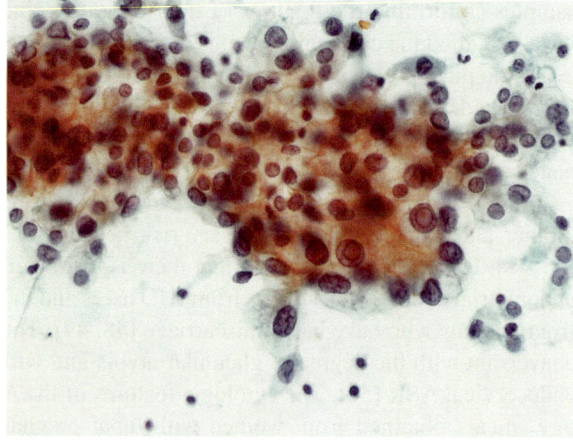

Fig. 8.46 Arias-Stella reaction: Endometrial smear showed atypical glandular cell clusters with enlarged nuclei and prominent intranuclear inclusions that could be mistaken for a glandular epithelial abnormality. (Papanicolaou stain, original magnification; 40×)

Fig. 8.47 Exfoliated naked-nuclei, ASR. Notice a relatively hyperchromatic and homogeneous chromatin distribution with an oval or ellipsoidal shape. Nuclei of this type vary greatly in size. (Papanicolaou stain, original magnification; 40×.) Reprinted with permission from Kobayashi and Okamoto [52]. (Am J Clin Pathol 2000)

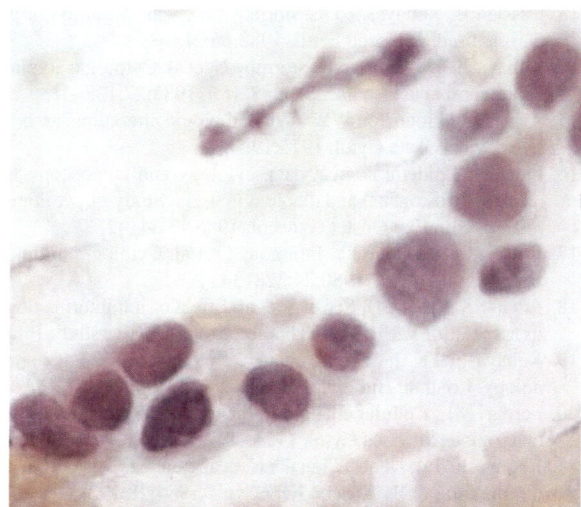

References

1. Mutter GL, Ferenczy A. Anatomy and histology of the uterine corpus. In: Kurman RJ, editor. Blaustein's pathology of the female genital tract. 5th ed. New York: Springer; 2001. p. 383–406.
2. Norimatsu Y, Shimizu K, Kobayashi TK, et al. Cellular features of endometrial hyperplasia and well-differentiated adenocarcinoma using the Endocyte sampler: diagnostic criteria based on the cyto-architecture of tissue fragments. Cancer. 2006;108:77–85.
3. Norimatsu Y, Kouda H, Kobayashi TK, et al. Utility of thin-layer preparations in the endometrial cytology: evaluation of benign endometrial lesions. Ann Diagn Pathol. 2008;12:103–11.
4. Kobayashi TK, Norimatsu Y, Buccoliero AM. Cytology of the body of the uterus. In: Gray W, Kocjan G, editors. Diagnostic cytopathology. 3rd ed. London: Churchill Livingstone; 2010. p. 689–719.
5. Anonymous. The IUD: An Important Method with Potential. Population Reports. 2006;33:3–5.
6. Buhling KJ, Zite NB, Lotke P, et al. Worldwide use of intrauterine contraception: a review. Contraception. 2014;89:162–73.
7. Serfaty D, Yaneva H. The endometrium and the IUD. In: Runnebaum B, Rabe T, Kiesel L, editors. Female contraception. Berlin, Heidelberg: Springer-Verlag; 1988. p. 325–40.
8. Casslen B, Kobayashi TK, Stormby N. The cellular composition of uterine fluid in IUD uses: a quantitative study. Contraception. 1981;24:685–93.
9. Yin M, Zhu P, Luo H, et al. The presence of mast cells in the human endometrium pre-and post-insertion of intrauterine devices. Contraception. 1993;48:245–54.
10. Kobayashi TK, Okamoto H, Harami K, et al. The presence of mast cells in IUD smears. Acta Cytol. 1980;24:268–9.
11. Myatt L, Bray MA, Gordon D, et al. Macrophages on intrauterine contraceptive devices produce prostaglandins. Nature. 1975;257:227–8.
12. Gupta PK, Malkani PK, Bhasin K. Cellular response in the uterine cavity after IUD insertion and structural changes on the IUD. Contraception. 1971;4:375–84.

13. Casslen B, Kobayashi TK, Stormby N. Cyclic variation of the cellular components in human uterine fluid. J Reprod Fertil. 1982;66:213–8.
14. Kobayashi TK, Casslen B, Stormby N, et al. Cytologic atypia in the uterine fluid of intrauterine contraceptive device users. Acta Cytol. 1983;27:138–41.
15. Ashton PR, Johnston WW. Cytopathologic alterations associated with intrauterine contraceptive devices. Acta Cytol. 1975;19:583.
16. Fiore N. Epidemiological data, cytology and colposcopy in IUD (intrauterine device), E-P (estro-progestogens) and diaphragm users. Study of cytological changes of endometrium IUD related. Clin Exp Obstet Gynecol. 1986;13:34–42.
17. Kobayashi TK, Yuasa Y, Fujimoto T, et al. Cytologic findings in post-partum and postabortal smears. Acta Cytol. 1980;24:328–34.
18. Erhan SS, Keser SH, Sensu S, et al. Effect of intrauterine device on cervicovaginal smears and its association with calcified bodies: a retrospective study. Int J Clin Exp Pathol. 2016;9:9372–9.
19. Kobayashi TK. Iatrogenic changes. In: Coleman DV, Chapman P, editors. Clinical cytotechnology. London: Butterworths; 1989. p. 425–40.
20. Fornari ML. Cellular changes in the glandular epithelium of patients using IUCD: a source of cytologic error. Atca Cytol. 1974;18:341–3.
21. Kobayashi TK, Ueno T, Tanaka N, et al. Nuclear DNA content of atypical glandular cells in the uterine fluid of IUD users. Acta Cytol. 1984;28:192–4.
22. Bibbo M. Look-alikes in cytology of the female genital tract. In: Wied GL, Koss LG, Reagan JW, editors. Compendium on diagnostic cytology. Tutorial of cytology Chicago. 4th ed. Illinois: Chicago. p. 194–7.
23. Koss LG. Histology and cytology of pregnancy and abortion. In: Koss LG, editor. Diagnostic cytology and its histologic bases. Philadelphia, PA: Lippincott; 1992. 281–294, 358–359.
24. Herting VW, Tauber PF. Endometrium-Zytologie bei Kupferhaltigen Intrauterinpessaren. Fortschr Med. 1978;96:311–4. (In German).
25. Gupta PK, Burroughs F, Luff RD, et al. Epithelial atypias associated with intrauterine contraceptive device (IUD). Acta Cytol. 1978;22:286–91.
26. Reagan JW, ABP N. Changes simulating adenocarcinoma. In: Wied GL, Koss LG, Reagan JW, editors. Compendium on diagnostic cytology, Tutorials of cytology. 4th ed. Illinois: Chicago; 1979. p. 173–5.
27. Biler A, Solmaz U, Erkilinc S, et al. Analysis of endometrial carcinoma in young women at a high-volume cancer center. Int J Surg. 2017;44:185–90.
28. Trimble CL, Kauderer J, Zaino R, et al. Concurrent endometrial carcinoma in women with a biopsy diagnosis of atypical endometrial hyperplasia: a gynecologic oncology group study. Cancer. 2006;106:812–9.
29. Ushijima K, Yahata H, Yoshikawa H, et al. Multicenter phaseII study of fertility-sparing treatment with medroxyprogesterone acetate for endometrial carcinoma and atypical hyperplasia in young women. J Clin Oncol. 2007;25:2798–803.
30. Deligdisch L, Kalir T, Cohen CJ, et al. Endometrial histopathology in 700 patients treated with tamoxifen for breast cancer. Gynecol Oncol. 2000;78:181–6.
31. Shiau AK, Barstad D, Loria PM, et al. The structural basis of estrogen receptor/coactivator recognition and the antagonism of this interaction by tamoxifen. Cell. 1998;95:927–37.
32. Fisher B, Costantino JP, Wickerham DL, et al. Tamoxifen for the prevention of breast cancer: current status of the National Surgical Adjuvant Breast and bowel project P-1 study. J Natl Cancer Inst. 2005;97:1652–62.
33. Cuzick J, Sestak I, Cawthorn S, et al. Tamoxifen for prevention of breast cancer: extended long-term follow-up of the IBIS-I breast cancer prevention trial. Lancet Oncol. 2015;16:67–75.
34. Mazur MT, Kurman RJ. Pregnancy, abortion, and ectopic pregnancy. In: Diagnosis of endometrial biopsies and Curettings. 2nd ed. New York: Springer; 2005. p. 34–66.
35. Arias-Stella J. Atypical endometrial changes associated with the presence of chorionic tissue. Arch Pathol. 1954;58:112–8.
36. Arias-Stella J. The Arias-Stella reaction: facts and fancies four decades after. Adv Anat Pathol. 2002;9:12–23.

37. Luks S, Simon RA, Lawrence WD. Arias-Stella reaction of the cervix: the enduring diagnostic challenge. Am J Case Rep. 2012;13:271–5.
38. Huettner PC, Gersell DJ. Arias-Stella reaction in nonpregnant women: a clinicopathologic study of nine cases. Int J Gynecol Pathol. 1994;13:241–7.
39. Novak ER. The endometrium. Clin Obstet Gynecol. 1974;17:31–49.
40. Birch HW, Collins CG. Atypical changes of genital epithelium associated with ectopic pregnancy. Am J Obstet Gynecol. 1961;81:198–208.
41. Schneider V. Arias-Stella reaction of the endocervix: frequency and location. Acta Cytol. 1981;21:224–8.
42. Felix A, Nogales FF, Arias-Stella J. Polypoid endometriosis of the uterine cervix with Arias-Stella reaction in a patient taking phytoestrogens. In J Gynecol Pathol. 2010;29:185–8.
43. Michael CW, Esfahani FM. Pregnancy-related changes: a retrospective review of 278 cervical smears. Diagn Cytopathol. 1997;17:99–100.
44. Ferguson JH. Some limitations of cytological diagnosis of malignant tumors. Cancer. 1949;2:845–52.
45. Mulvany NJ, Khan A, Ostor A. Arias-Stella reaction associated with cervical pregnancy: report of a case with a cytologic presentation. Acta Cytol. 1994;38:218–23.
46. Benoit JL, Kini SR. Arias-Stella reaction–like changes in endocervical glandular epithelium in cervical smears during pregnancy and postpartum sates: a potential diagnostic pitfall. Diagn Cytopathol. 1996;14:349–55.
47. Albukerk JN, Gnecco CA. Atypical cytology in tubal pregnancy. J Reprod Med. 1977;19:273–6.
48. Kobayashi TK, Fujimoto T, Okamoto H, et al. Cytologic evaluation of atypical cell in cervicovaginal smears from women with tubal pregnancies. Acta Cytol. 1983;27:28–32.
49. Kobayashi TK, Yuasa M, Fujimoto T, et al. Cytologic findings in postpartum and postabortal smears. Acta Cytol. 1980;24:328–34.
50. Shargo SS. The Arias-Stella reaction. A case report of a cytologic presentation. Acta Cytol. 1977;21:310–3.
51. Kobayashi TK, Okamoto H. Arias-Stella changes in cervicovaginal specimens. Cytopathology. 1997;8:289–90.
52. Kobayashi TK, Okamoto H. Cytopathology of pregnancy-induced cell patterns in cervicovaginal smears. Am J Clin Pathol. 2000;114:S6–S20.

Endometrial Hyperplasia without Atypia

9

Yoshinobu Maeda, Takeshi Nishikawa,
and Yoshiaki Norimatsu

9.1 Background

Endometrial hyperplasia without atypia is a condition in which the endometrial gland/stromal ratio increases far beyond that one observed in the proliferative phase of the endometrium. It is believed that prolonged unopposed estrogen exposure is necessary for the development of endometrial hyperplasia. In addition to hyperplastic changes, this pathological condition includes small numbers of potentially neoplastic lesions.

This condition has been referred to by multiple names such as cystic glandular hyperplasia and adenomatous hyperplasia, which have generated confusion in the understanding and terminology of this condition.

Kurman et al. reported that glandular structural and cellular atypia correlates with the risk of developing endometrioid carcinoma [1]. In 1994, the World Health Organization (WHO) classified endometrial hyperplasia into four subtypes in terms of complexity of glandular architecture and the presence or absence of nuclear atypia: simple hyperplasia without atypia, complex hyperplasia without atypia, simple hyperplasia with atypia, and complex hyperplasia with atypia [2]. This classification has been widely adopted. These four subtypes of endometrial hyperplasia were included in the third edition of the WHO classification in 2003 [3]. It has been reported that cell atypia is prevalent over glandular architecture in predicting the

Y. Maeda
Department of Diagnostic Pathology, Toyama Red Cross Hospital, Toyama, Japan

T. Nishikawa
Department of Diagnostic Pathology, Nara Medical University Hospital,
Kashihara, Nara, Japan

Y. Norimatsu (✉)
Department of Medical Technology, Faculty of Health Sciences, Ehime Prefectural University
of Health Sciences, Tobe-cho, Iyo-gun, Ehime, Japan
e-mail: ynorimatsu@epu.ac.jp

risk of malignancy of a given lesion, but some cases of endometrial hyperplasia without cell atypia also exhibit neoplastic potential.

In fact, 1–3% cases of endometrial hyperplasia without cell atypia progress to well-differentiated endometrioid carcinoma.

In the fourth edition of the WHO classification in 2014, conventional simple endometrial hyperplasia and complex endometrial hyperplasia without atypia were simplified to endometrial hyperplasia. The degree of endometrial proliferation (in response to prolonged estrogen stimulation) is one of the criteria to classify this lesion [4]. In the fifth edition of the WHO classification in 2020, this condition was simply classified as endometrial hyperplasia without atypia [5].

9.2   Definition

Endometrial hyperplasia without atypia (as in the 2020 WHO classification) is defined as the proliferation of endometrial glands of irregular size and shape without significant cytological atypia [5].

The endometrial mucosa of the uterine corpus consists of glandular lumens of varying shape (from tubular to elongated or dilated, according to the day of the menstrual cycle), each of them consisting of a "basalis" layer (containing weakly proliferative glands with dense spindled stroma abutting on the myometrium) and of a "functionalis," which is the most dynamic portion. This latter has a regular arrangement in the form of single gland units facing the uterine cavity in early proliferative endometrium. In endometrial hyperplasia without atypia, endometrial glands that vary in size and shape are admixed with branching and/or dilated glands with irregular proliferation and distribution. The ratio of glands to stroma increases compared to the normal proliferative phase endometrium, exceeding the ratio of 3:1 in hyperplasia. Furthermore, as the degree of complexity of the glandular architecture increases, the glands show more complex and irregular outlines. As the stroma decreases and glands become more crowded, a so-called back-to-back pattern may be observed. Glandular epithelial cells are columnar to tall shape, with oval to elongated nuclei and delicate cytoplasm. The nuclei are stratified but are regularly arranged along the basement membrane. Papillary infoldings into the glandular lumen are also occasionally observed. In addition to morule-like nests, ciliated cell metaplasia, and eosinophilic metaplasia, secretory changes may also be observed [6].

9.3   Diagnostic Criteria [7–12] (Figs. 9.1, 9.2, 9.3, 9.4, 9.5, 9.6, 9.7, 9.8 and 9.9)

- Number of clumps showing dilation and branching (as mentioned in Chap. 5, with maximum width of a gland is more than twice that of its minimum width and the cohesion of the endometrial stromal cells to the margins of the gland are characteristic).
- Increased branching and papillary infolding into the lumen;

Fig. 9.1 Endometrial hyperplasia without atypia. Dilated or branched gland pattern. (Papanicolaou stain, original magnification 4×)

Fig. 9.2 Endometrial hyperplasia without atypia. Dilated or branched gland pattern. (Papanicolaou stain, original magnification 4×)

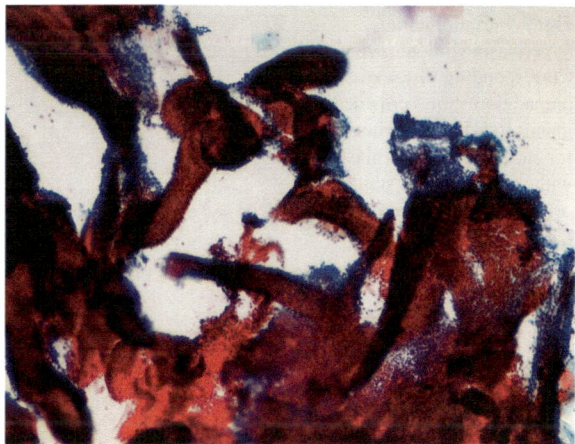

- Glandular cells appear in clusters of approximately 2–3 layers (not exceeding 3 layers), with cohesion of stromal cells around the clusters.
- Epithelial cells show a regular arrangement.
- Epithelial cells have generally uniform nuclei and no significant atypia.
- Occasionally, mitosis is observed.
- When 5 or more clusters of dilated glands are observed, complex endometrial hyperplasia should be suspected.

9.4   Explanatory Note

As mentioned in the diagnostic criteria, endometrial hyperplasia is characterized by the appearance of clusters of dilated glands without cellular atypia. However, owing to the varying degrees of appearance of dilated and branched clusters, some

Fig. 9.3 Endometrial hyperplasia without atypia. Epithelial cells show no atypia, no significant nuclear overlapping not more than three layers. (Papanicolaou stain, original magnification 40×)

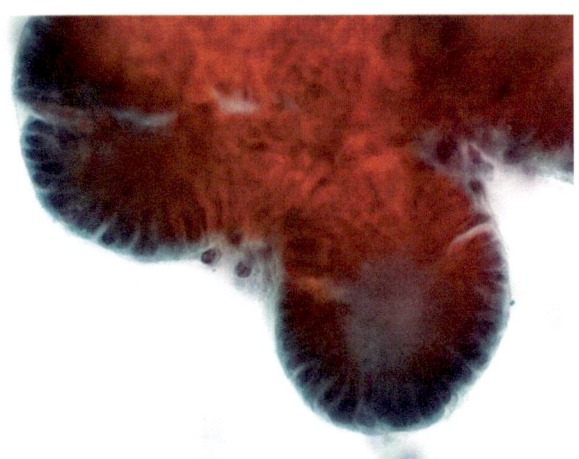

Fig. 9.4 Endometrial hyperplasia without atypia. Corresponding histologic preparation shows crowded glands to fulfill the criteria for endometrial hyperplasia without atypia. (HE stain, original magnification 4×)

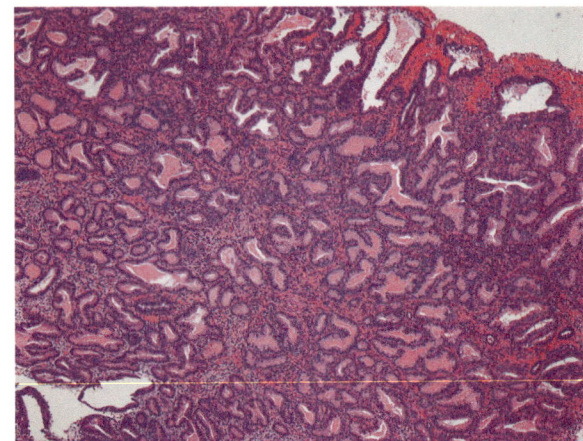

Fig. 9.5 Endometrial hyperplasia without atypia. Same sample as Fig. 9.4. Glands show dilation, branching, and simple tubular pattern. (HE stain, original magnification 20×)

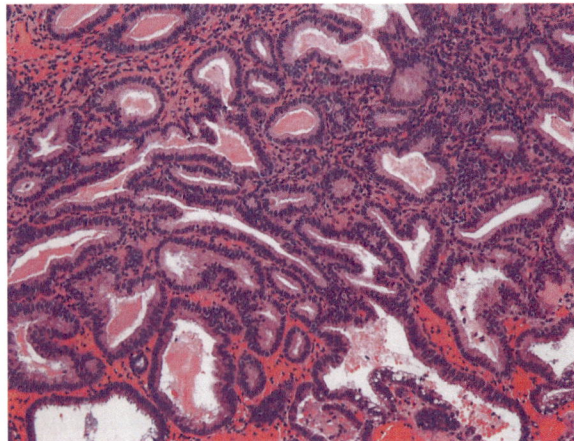

Fig. 9.6 Endometrial hyperplasia without atypia. Dilated or branched gland pattern. (Papanicolaou stain, original magnification 10×)

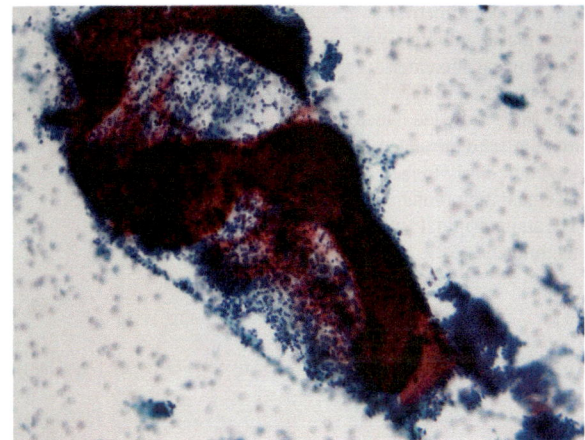

Fig. 9.7 Endometrial hyperplasia without atypia. Histological specimen corresponding to Fig. 9.6, glands are mildly crowded, and are dilated. (HE stain, original magnification 10×)

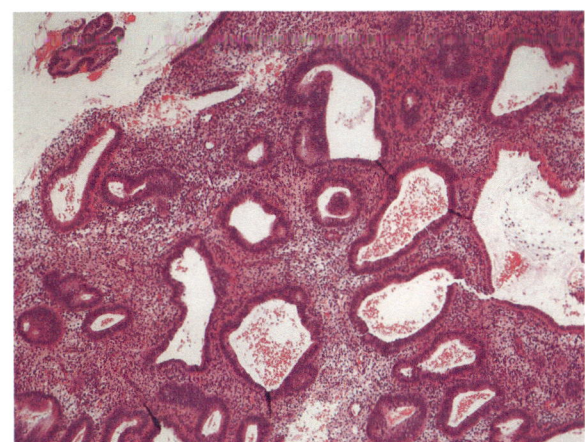

Fig. 9.8 Endometrial hyperplasia without atypia. Dilated or branched gland pattern. (Papanicolaou stain, original magnification 10×)

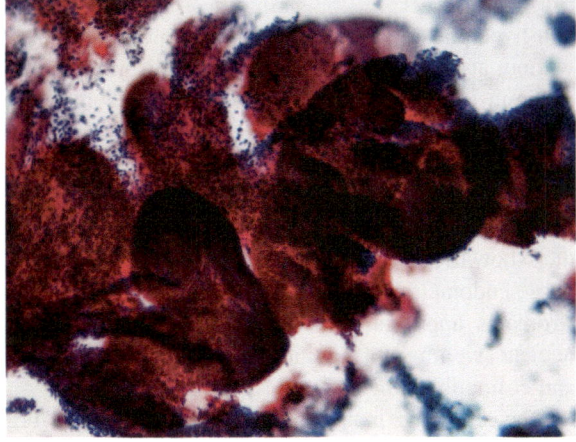

Fig. 9.9 Endometrial hyperplasia without atypia. Histological specimen corresponding to Fig. 9.8, glands are crowded. Nuclei are elongated, arranged perpendicularly to the basement membrane. (HE stain, original magnification 20×)

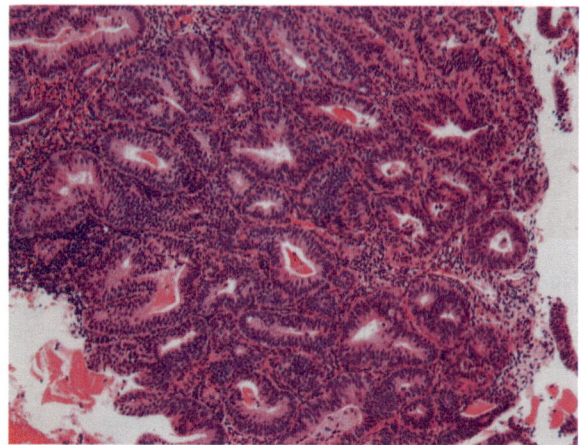

Fig. 9.10 Disordered proliferative phase endometrium. Dilated glands and irregular shaped glands are present, with typical proliferative phase endometrial glands. (HE stain, original magnification 4×)

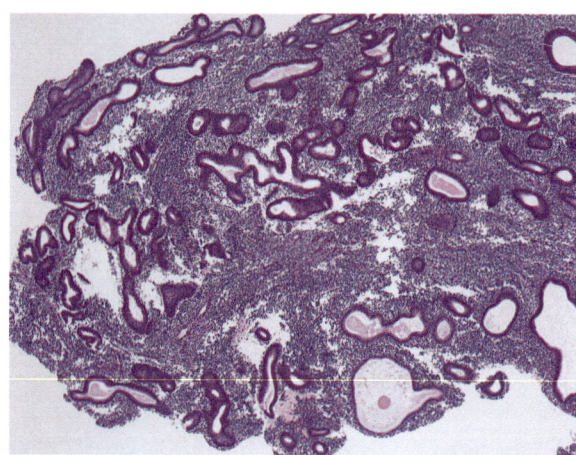

non-hyperplastic lesions are difficult to distinguish from endometrial hyperplasia without atypia. These lesions include "disordered proliferative phase (DPP) endometrium as a benign reactive change." DPP is a lesion caused by the repeated, sequentially occurring anovulatory cycle and appears as one of the functional bleeding disorders.

Histologically, the proliferative endometrial glands show focal and irregular cystic dilation and irregular branching due to prolonged estrogen exposure. Although they show irregular shape and little architectural disorder, some normal proliferative phase endometrial glands are often intermingled to them and the ratio of glands to stroma is approximately 1:1. Endometrial glandular epithelial cells show mild nuclear swelling and stratification and may sometimes show ciliated cell metaplasia and eosinophilic metaplasia [13] (Fig. 9.10).

The cell image includes the following findings: (Figs. 9.11 and 9.12)

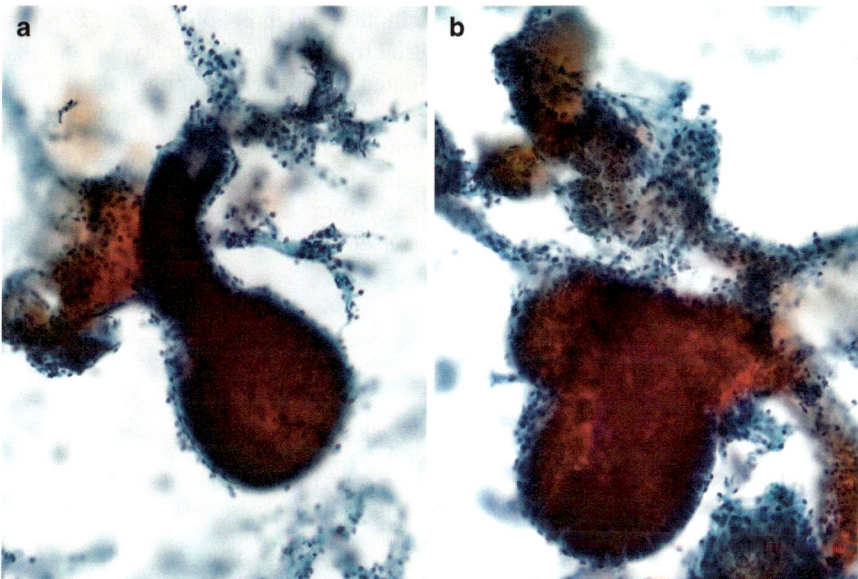

Fig. 9.11 Disordered proliferative phase endometrium. Corresponding cytologic preparation Fig. 9.10, (**a**): dilated gland pattern should been seen. (**b**): dilated and branched gland pattern with cohesion of endometrial stromal cells should been seen. (Papanicolaou stain, original magnification **a, b**: 20×)

Fig. 9.12 Disordered proliferative phase endometrium. Cytological specimen corresponding to Fig. 9.10, (**a**): dilated gland pattern should been seen. Nuclear atypia is absent, and nuclear overlapping does not excess three layers. (Papanicolaou stain, original magnification 40×)

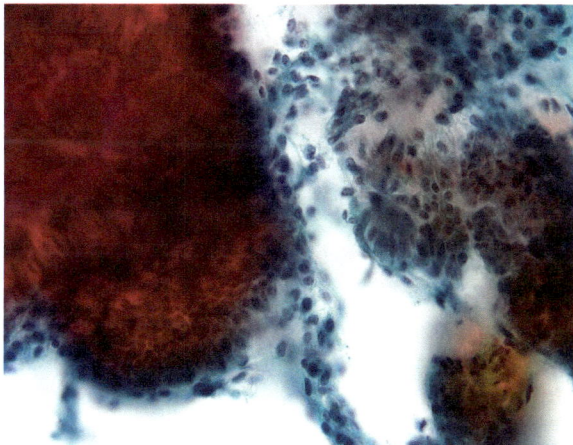

- Often lacking characteristic findings;
- Epithelial cell clusters show an increase in diameter indicating dilatation;
- Nuclear overlapping in clusters is inconspicuous, and stromal cell cohesion is observed in the periphery of the glands;
- Epithelial cell clusters of endometrial glandular cells in the proliferative phase are Admixed to the former;

In this chapter we have illustrated the definition and morphology of DPP, which can usually be differentiated from endometrial hyperplasia without atypia.

However, in cytological preparations, dilated and branching epithelial cell clusters may be commonly observed in both lesions. Even though the two lesions have different histopathological features, in cytologic preparations of directly sampled endometrial mucosa, it may be difficult to discriminate clearly between these two lesions. We believe that a comprehensive diagnostic judgment should be reached only when clear-cut diagnostic criteria for each of the illustrated conditions are found and verified.

References

1. Kurman RJ, Kaminski PF, Norris HJ. The behavior of endometrial hyperplasia:a long-term study of untreated hyperplasia in 170 patients. Cancer. 1985;56:403–12.
2. Scully RE, Bonfiglio TA, Kurman RJ, et al., editors. WHO international histological classification of tumours. Histological typing of female genital tract tumours. 2nd ed. Berlin: Springer-Verlag; 1994. p. 13–8.
3. Tavassoli FA, Devilee P. WHO classification of tumours. pathology & genetics. In: Tumours of the breast and female genital organs. 3rd ed. Lyon: IARC Press; 2003. p. 218–32.
4. Zaino R, Carinelli SG, Ellenson LH, et al. Epithelial tumours and precursors. In: Kurman RJ, Carangiu ML, Herrington CS, et al., editors. WHO classification of tumor of female reproductive organs. 4th ed. Lyon: IARC Press; 2014. p. 125–35.
5. Ellenson LH, Matias-Guiu X, Mutter GL. Endometrial hyperplasia without atypia. In: WHO Classification of Tumor Editorial Board, editor. WHO classification of tumor of female reproductive organs. 5th ed. Lyon: IARC Press; 2020. p. 248–9.
6. Ellenson LH, Ronnett BM, Kurman RJ. Precursors of endometrial carcinoma. In: Kurman RJ, Ellenson LH, Ronnett BM, editors. Blaustein's pathology of the female genital tract. 7th ed. New York: Springer-Verlag; 2019. p. 439–72.
7. Norimatsu Y, Shimizu K, Kobayashi TK, et al. Cellular features of endometrial hyperplasia and well differentiated adenocarcinoma using the Endocyte sampler: diagnostic criteria based on cytoarchitecture of tissue fragments. Cancer. 2006;108:77–85.
8. Papaefthimiou M, Symiakaki H, Mentzelopoulou P, et al. Study on the morphology and reproducibility of the diagnosis of endometrial lesions utilizing liquid-based cytology. Cancer. 2005;105:56–64.
9. Norimatsu Y, Yanoh K, Kobayashi TK. The role of liquid-based preparation in the evaluation of endometrial cytology. Acta Cytol. 2013;57:423–35.
10. Norimatsu Y, Moriya T. A study of endometrial cytology in benign cases: differentiation from endometrial hyperplasia. J Jpn Soc Clin Cytol. 2002;41:313–20. (in Japanese with English abstract).
11. Shimizu K, Norimatsu Y, Ogura S, et al. Analysis of suspicious endometrial cytology: by a new criteria based on architectural atypia. J Jpn Soc Clin Cytol. 2002;41:313–20. (in Japanese with English abstract).
12. Yoshida S, Kusunoki N, Ishiyama K, et al. A study of diagnostic criteria for endometrial cytology. J Jpn Soc Clin Cytol. 2008;47:227–35. (in Japanese with English abstract).
13. Lastra RR, McCluggage WG, Ellenson LH. Benign disease of the endometrium. In: Kurman RJ, Ellenson LH, Ronnett BM, editors. Blaustein's pathology of the female genital tract. 7th ed. New York: Springer-Verlag; 2019. p. 375–437.

Endometrial Atypical Hyperplasia/ Endometrioid Intraepithelial Neoplasia

10

Yoshinobu Maeda, Akihiko Kawahara, and Yoshiaki Norimatsu

10.1 Background

In 1985, Kurman et al. reported that the frequencies of endometrial hyperplasia progressing to endometrial carcinoma were 1% for endometrial hyperplasia without atypia, simple, 3% for endometrial hyperplasia without atypia, complex, 8% for endometrial hyperplasia with atypia, simple, and 29% for endometrial hyperplasia with atypia, complex, respectively [1]. Therefore, it was assessed that the risk of developing endometrial carcinoma correlates with the degree of glandular architectural abnormalities and cellular atypia, and the latter two aforementioned subtypes were categorized as atypical endometrial hyperplasia.

The WHO classification in 1994 introduced the four subtypes of endometrial hyperplasia, namely simple hyperplasia without atypia, complex hyperplasia without atypia, simple hyperplasia with atypia, and complex hyperplasia with atypia; these subtypes were also included in the third edition WHO classification in 2003 [2, 3].

In addition to the risk of developing endometrial carcinoma, endometrial hyperplasia with atypia is associated with the risk of coexisting endometrial carcinoma. According to some reports, the frequency of the coexistence of endometrioid carcinoma in patients who underwent a hysterectomy after being diagnosed with endometrial hyperplasia with atypia in biopsy or curettage specimens was approximately 40%. This is the reason why most endometrial hyperplasias with atypia, are

Y. Maeda
Department of Diagnostic Pathology, Toyama Red Cross Hospital, Toyama, Japan

A. Kawahara
Department of Diagnostic Cytopathology, Kurume University Hospital, Fukuoka, Japan

Y. Norimatsu (✉)
Department of Medical Technology, Faculty of Health Sciences, Ehime Prefectural University of Health Sciences, Tobe-cho, Iyo-gun, Ehime, Japan
e-mail: ynorimatsu@epu.ac.jp

presumed to have a neoplastic potential, and it is considered clinically important to separate endometrial hyperplasias without atypia from them [4, 5].

In 2000, Mutter et al. originally defined endometrial intraepithelial neoplasia (EIN) as a new pathologic entity having clonal and neoplastic features. EIN is diagnosed based on the following three criteria (1).: in terms of architecture, the ratio of glandular to stroma exceeds 55%, and individual glands exhibit a severe degree of cytological atypia; (2). The cytology of EIN lesions is different from that of the background endometrial mucosa (polymorphism and loss of polarity); and (3) the lesion size is 1 mm or more [6].

According to the conventional criteria, 5%, of simple endometrial hyperplasias, 44% of complex endometrial hyperplasias without atypia, and 79% of endometrial atypical hyperplasias, respectively, represent EIN. In fact, 38% of endometrial atypical hyperplasias fulfilling the EIN criteria progressed to endometrioid carcinoma, but no case of endometrial atypical hyperplasia not fulfilling the criteria progressed to endometrioid carcinoma. Therefore, EIN and endometrial atypical hyperplasia are not essentially the same lesion [6–8].

In the fourth edition of the WHO classification in 2014, endometrial hyperplasia was positioned as a precursor lesion, and only two subcategories in the classification were considered: endometrial hyperplasia without atypia and atypical hyperplasia/EIN. It is in this way rather surprising how conventional endometrial atypical hyperplasia and EIN, which have different pathobiology as mentioned above, were clustered into the same category. The presence or absence of cellular atypia is the most important finding for the evaluation and classification of endometrial hyperplasia in this classification [9].

10.2 Definition

Endometrial hyperplasia with atypia may currently be defined as an "overgrowth of endometrial glands with cell atypia." In the fifth edition of the WHO classification in 2020, endometrial atypical hyperplasia (EAH)/EIN is defined as the simultaneous appearance of cytological atypia in endometrial glands and an increased ratio of endometrial glands to stroma (crowded gland architecture) within a morphologically defined region, distinct from the surrounding endometrium or from entrapped normal glands [10] (Figs. 10.1 and 10.2).

The glandular architecture is more irregular than that of complex endometrial hyperplasia without atypia and often shows a back-to-back pattern. Papillary infoldings into the glandular lumen are also common. Another feature associated with endometrial hyperplasia with atypia is the presence of various metaplastic changes such as eosinophilic metaplasia, ciliated cell metaplasia, squamous metaplasia, and secretory change.

In addition, it is necessary to exclude endometrioid carcinoma grade 1 (G1). It is known that endometrial atypical hyperplasia sometimes shows more severe nuclear atypia than that of endometrial carcinoma G1. If the degree of nuclear atypia in

Fig. 10.1 EAH/
EIN. Lesion consists of
crowded tubular glands
with little intervening
stroma and exceeds 1 mm
in maximum size. (HE
stain, original
magnification 10×)

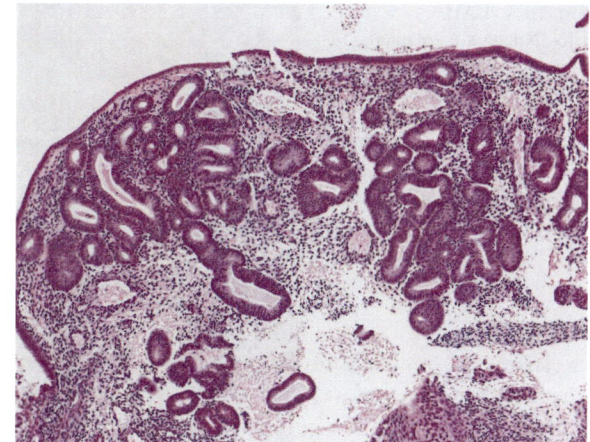

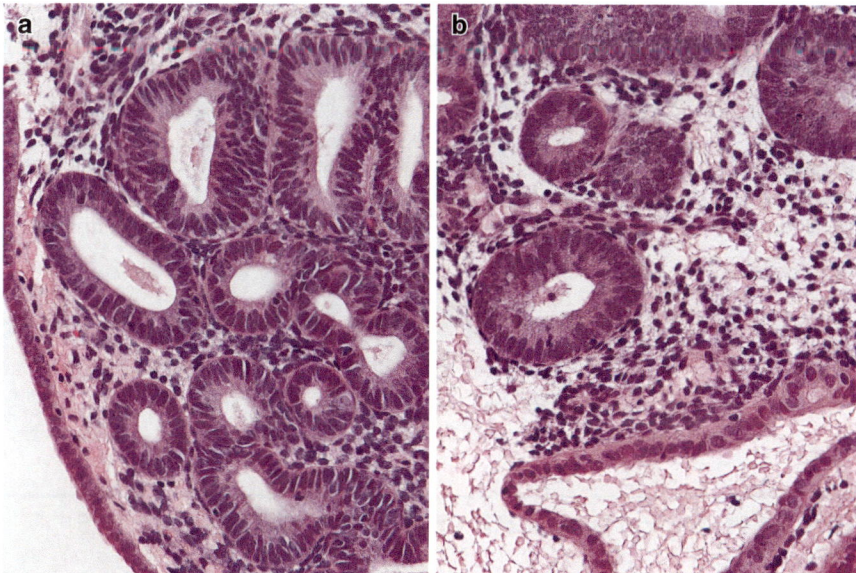

Fig. 10.2 EAH/EIN. (**a**): The nuclei are enlarged, of round-oval or square shape hyperchromatic, with granular chromatin. (**b**): The cytologic features of crowded glands differ from neighboring normal endometrial gland (Bottom). (HE stain, original magnification, **a** and **b**: 20×)

lesions without stromal invasion is similar to or more severe than that of G1 endometrial carcinoma, it should be classified as endometrial atypical hyperplasia. This means that endometrial hyperplasia with atypia even includes the concept of endometrial carcinoma in situ that is not comprised in the classification [11–13] (Fig. 10.3).

Fig. 10.3 EAH/
EIN. Nuclei are enlarged,
of rounded/irregular shape,
with granular chromatin
and conspicuous nucleoli,
and show stratification and
loss of polarity. (HE stain,
original magnification 40×)

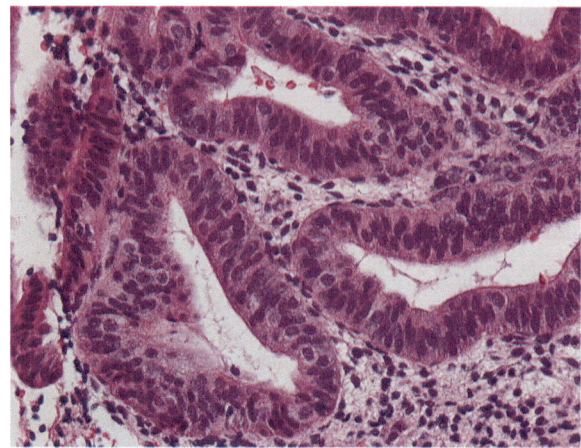

Fig. 10.4 EAH/EIN. This
cytological preparation
was obtained from a
40-y-o woman with
irregular genital bleeding.
Many clusters show
dilations and branching.
(Papnicolaou stain, original
magnification 4×)

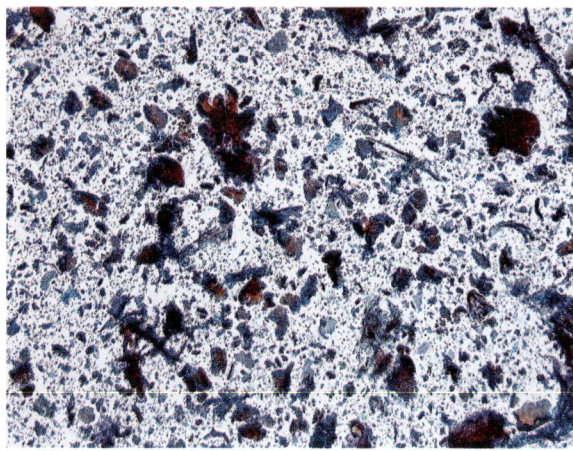

10.3 Cytological Diagnostic Criteria [14–18] (Figs. 10.4, 10.5, 10.6, 10.7, 10.8, 10.9, 10.10, 10.11, 10.12 and 10.13)

- Many clusters showing dilation and branching (maximum width of the gland lumen being more than twice the minimum width).
- An irregular protrusion pattern is sometimes present.
- The nuclear overlapping in epithelial cell clusters exceeds three layers, and the cohesion of stroma cells around the clusters may be indistinct.
- Occasionally, epithelial cell clusters show an increase of glandular complexity.
- The arrangement in the epithelial cell clusters may be irregular, and the nucleus may protrude toward the periphery of clusters.
- Glandular epithelial cells with nuclear swelling, anisonucleosis, increased chromatin granularity, and conspicuous nucleoli are observed.
- Sometimes, mitoses are also observed.

Fig. 10.5 EAH/
EIN. Dilated or branched
gland pattern. (Papnicolaou
stain, original
magnification 20×)

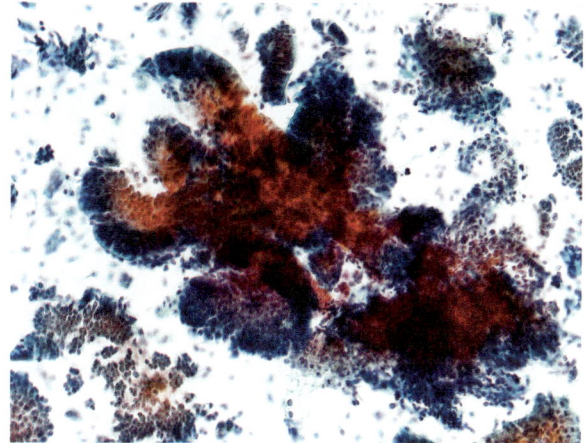

Fig. 10.6 EAH/
EIN. Similar to Fig. 10.5,
it shows a dilated or
branched gland pattern.
(Papnicolaou stain, original
magnification 20×)

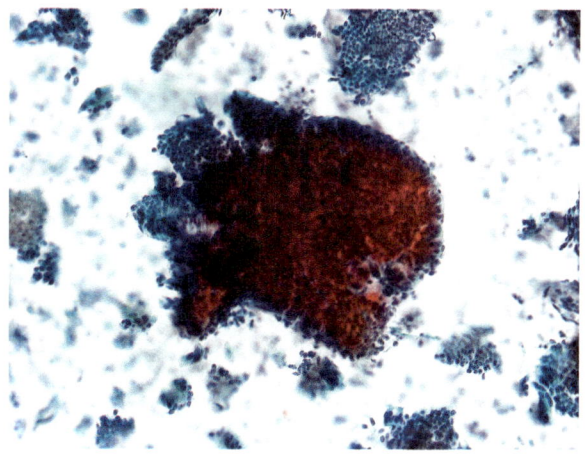

Fig. 10.7 EAH/
EIN. Epithelial cluster
shows significant nuclear
overlapping and cohesion
of stromal cells in the
periphery of cluster
(arrows). (Papnicolaou
stain, original
magnification 40×)

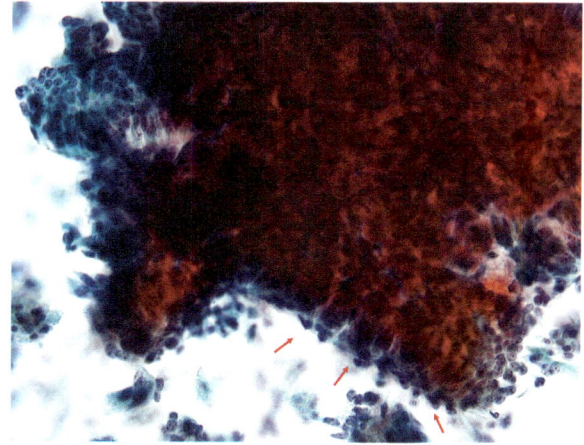

Fig. 10.8 EAH/
EIN. Nuclei show three
degrees of overlapping,
vary from rounded to ovoid
in shape and display fine
granular chromatin and
conspicuous nucleoli.
(Papanicolaou stain,
original magnification 60×)

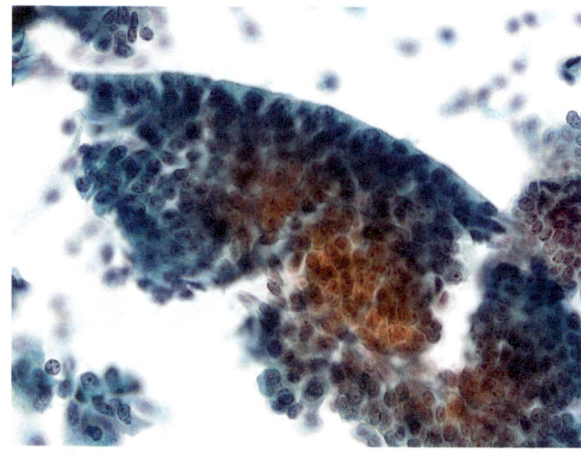

Fig. 10.9 EAH/
EIN. Atypical epithelial
cells with vacuolated
cytoplasm are seen in
cluster (arrows). These
cells suggest secretory
change. (Papanicolaou
stain, original
magnification 60×)

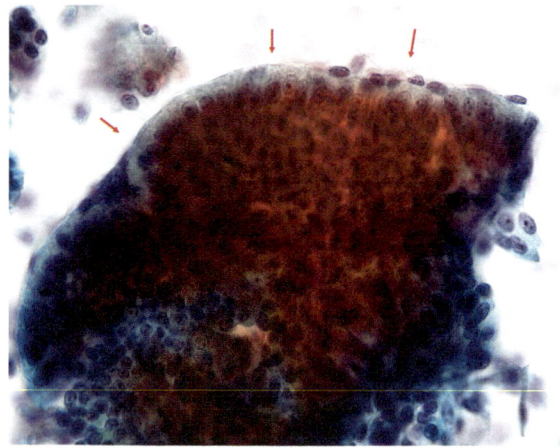

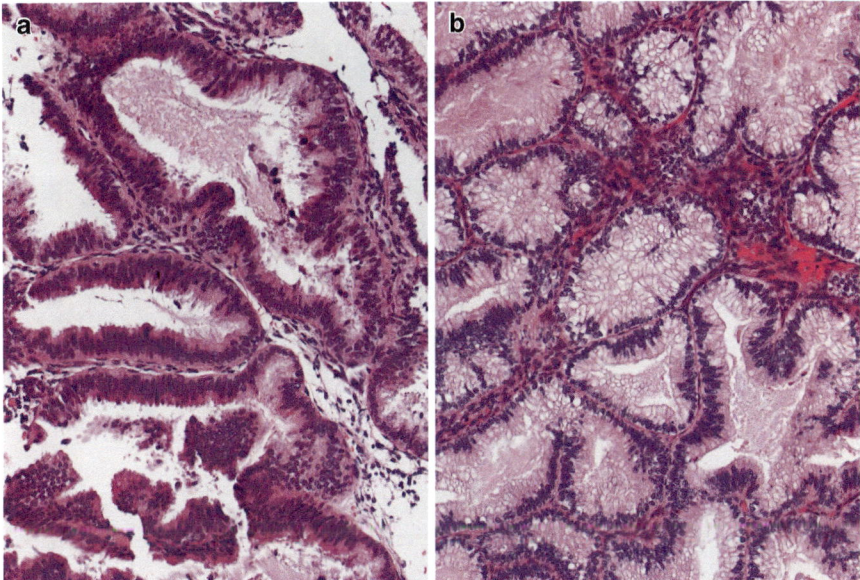

Fig. 10.10   EAH/EIN. Histological images corresponding to Fig. 10.8(a) and Fig. 10.9(b). Dilated and irregular shaped glands are present (**a**). Secretory change, crowded glands have supranuclear vacuoles (**b**). (HE stain, original magnification **a** and **b**: 20×)

Fig. 10.11   EAH/EIN. Atypical epithelial cells with cilia are intermingled in cluster (arrows). (Papanicolaou stain, original magnification 60×)

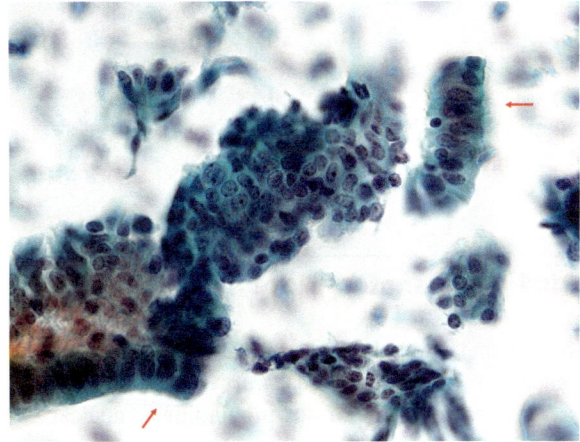

Fig. 10.12 EAH/EIN.
Nuclei are enlarged, have
large evident nucleoli and a
few mitotic figures
(arrows). (Papanicolaou
stain, original
magnification 60×)

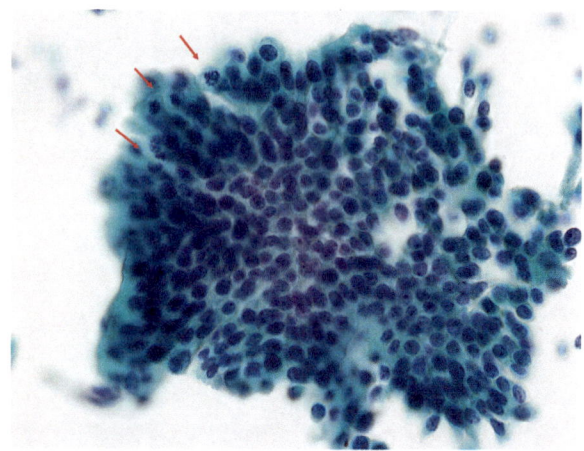

Fig. 10.13 EAH/
EIN. Histlogical images
corresponding to
Fig. 10.12. (HE stain,
original magnification 40×)

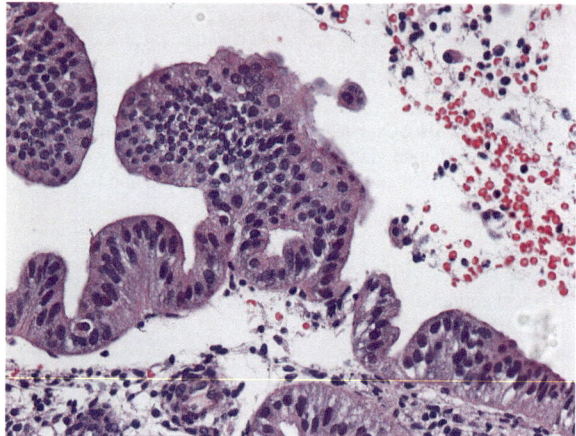

10.4 Explanatory Note

In endometrial hyperplasia with atypia, the definition of cellular atypia has not yet
reached a definitive consensus. It remains somewhat subjective and problematic. It
is well known that the following findings; "nuclear swelling showing round to oval
shape," "anisonucleosis," "increased chromatin granularity," "conspicuous and
enlarged nucleoli," and "pseudostratification and loss of polarity of the nucleus"
were mentioned as parameters for cellular atypia, but no complete agreement even
among gynecological pathologists has ever been achieved. Kendall et al. reported
that "clear and large nucleoli, and increased number of nucleoli" are helpful in distinguishing endometrial hyperplasia with atypia from endometrial hyperplasia without atypia [19].

Since the evaluation of cellular atypia is somewhat subjective, the current diagnostic criteria for EAH/EIN emphasize the distinction from the surrounding endometrium. Any case with a mild degree of nuclear atypia could theoretically be included in the classification of EAH/EIN (Fig. 10.14). Therefore, the differential diagnosis between endometrial hyperplasia and EAH/EIN is more difficult in the corresponding cytological preparations. The ATEC category will hence be useful in the cytological diagnosis when the differential is EAH/EIN [20, 21].

EAH/EIN has a number of molecular genetic alterations including microsatellite instability, *PAX2* inactivation, *PTEN* and β-catenin mutation. Immunocytochemically, loss of expression for *PTEN*, *PAX2* and mismatch repair protein may be helpful ancillary findings in the cytological diagnosis of EAH/EIN [10, 22–24] (Figs. 10.15, 10.16, 10.17 and 10.18).

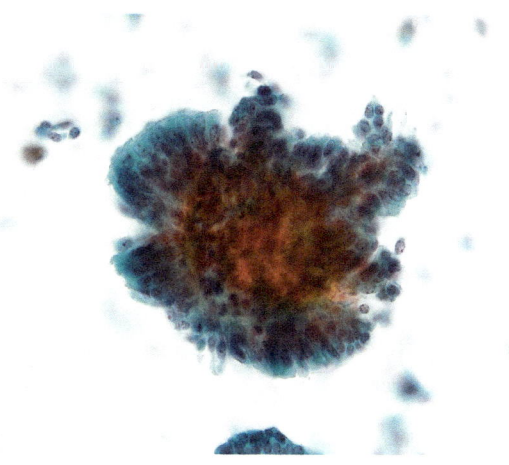

Fig. 10.14 EAH/EIN. This cytological preparation is obtained from a 50-y-o woman with irregular genital bleeding. An irregularly shaped cluster with nuclear protrusion at the periphery is seen. (Papanicolaou stain, original magnification 40×)

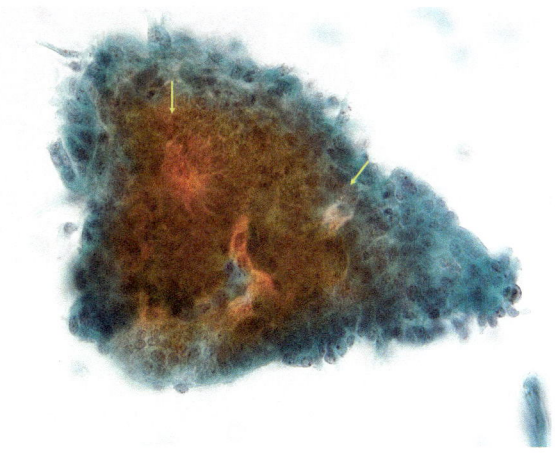

Fig. 10.15 EAH/EIN. Same sample as Fig. 10.14. Irregularly shaped lumens are seen within a cellular cluster (arrows). (Papanicolaou stain, original magnification 40×)

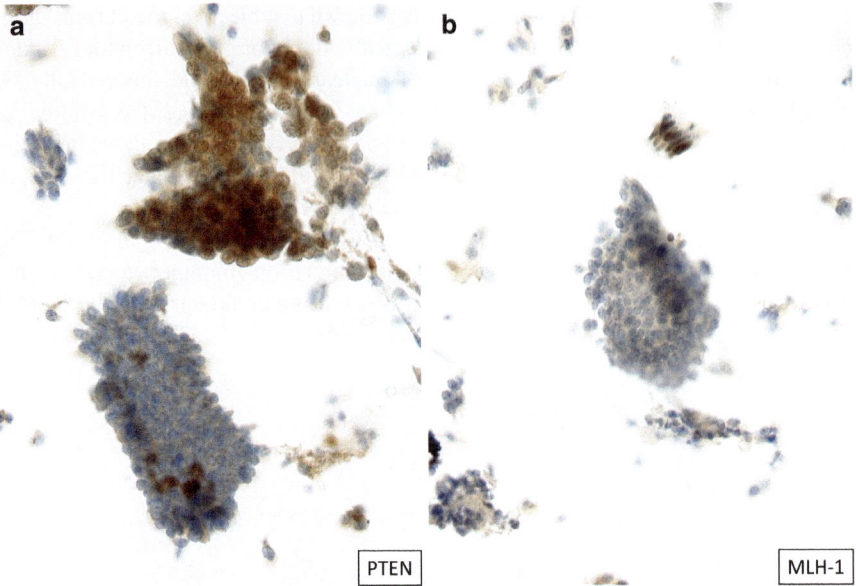

Fig. 10.16 EAH/EIN. Same sample as Figs. 10.14 and 10.15. (**a**): On immunocytochemistry, both PTEN expressing clusters (upper left) and clusters with loss of PTEN expression (lower left), with a small number of PTEN expressing stromal cells can be seen. (**b**): tumor cells show complete loss of MLH-1 expression. (ICC, original magnification **a** and **b**: 40×)

Fig. 10.17 EAH/EIN. Histological specimen corresponding to Figs. 10.14 and 10.15. (HE stain, original magnification 40×)

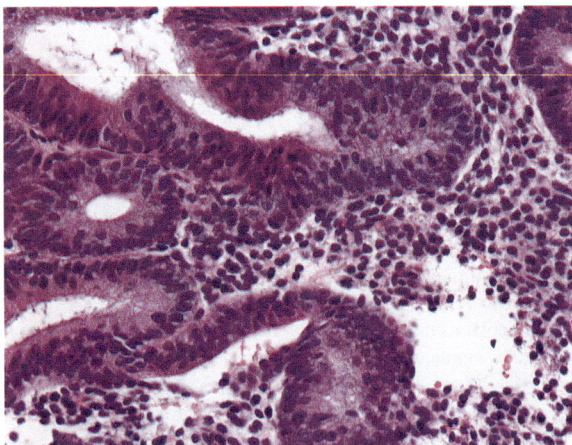

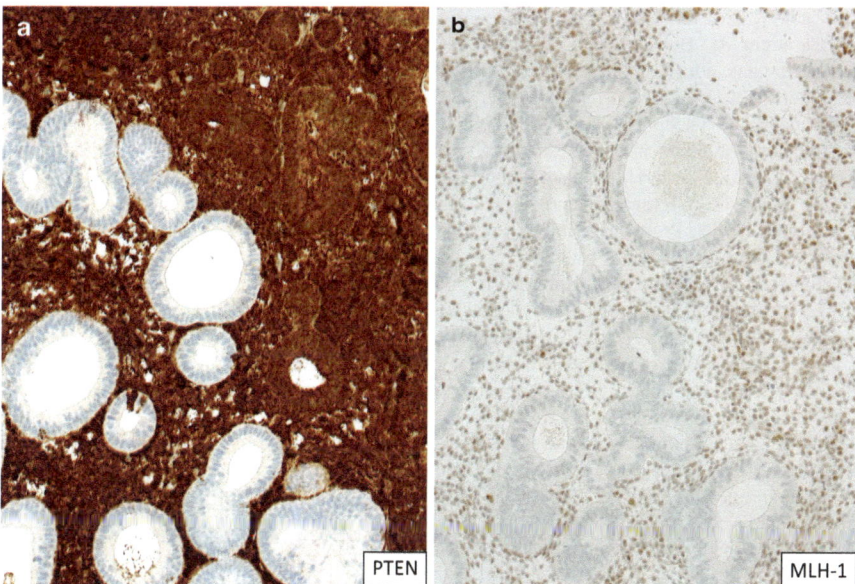

Fig. 10.18   EAH/EIN. Same sample as Fig. 10.17. (**a**): EAH/EIN with a discrete area with PTEN loss of expression (left). (**b**): Atypical glands with complete loss of MLH-1 expression. (IHC, original magnification **a** and **b**: 20×)

References

1. Kurman RJ, Kaminski PF, Norris HJ. The behavior of endometrial hyperplasia: a long-term study of untreated hyperplasia in 170 patients. Cancer. 1985;56:403–12.
2. Scully RE, Bonfiglio TA, Kurman RJ, et al., editors. WHO international histological classification of Tumours. Histological typing of female genital tract Tumours. 2nd ed. Berlin: Springer-Verlag; 1994. p. 13–8.
3. Tavassoli FA, Devilee P. WHO classification of Tumours. Pathology & Genetics. In: Tumours of the breast and female genital organs. 3rd ed. Lyon: IARC Press; 2003. p. 218–32.
4. Trimble CL, Kauderer J, Zaino R, et al. Concurrent endometrial carcinoma in women with a biopsy diagnosis of atypical endometrial hyperplasia: a gynecologic oncology group study. Cancer. 2006;106:812–9.
5. Rakha E, Wong SC, Soomro I, et al. Clinical outcome of atypical endometrial hyperplasia diagnosed on an endometrial biopsy: nstitutional experience and review of literature. Am J Surg Pathol. 2012;36:1683–90.
6. Mutter GL. Diagnosis of premalignant endometrial disease. J Clin Pathol. 2002;55:326–31.
7. Baak JP, Mutter GL, Robboy S, et al. The molecular genetics and morphometry-based endometrial intraepithelial neoplasia classification system predicts disease progression in endometrial hyperplasia more accurately than 1994 World Health Organization classification system. Cancer. 2005;103:2304–12.
8. Owings RA, Quick CM. Endometrial intraepithelial neoplasia. Arch Pathol Lab Med. 2014;138:484–91.

9. Zaino R, Carinelli SG, Ellenson LH, et al. Epithelial tumours and precursors. In: Kurman RJ, Carangiu ML, Herrington CS, et al., editors. WHO classification of tumor of female reproductive organs. 4th ed. Lyon: IARC Press; 2014. p. 125–35.

10. Ellenson LH, Matias-Guiu X, Mutter GL. Endometrial atypical hyperplasia/endometrioid intraepithelial neoplasia. In: WHO Classification of Tumor Editorial Board, editor. WHO classification of tumor of female reproductive organs. 5th ed. Lyon: IARC Press; 2020. p. 250–1.

11. Kurman RJ, Norris HJ. Evaluation of criteria for distinguishing atypical endometrial hyperplasia from well-differentiated carcinoma. Cancer. 1982;49:2547–59.

12. Silverberg SG. Problems in the differential diagnosis of endometrial hyperplasia and carcinoma. Mod Pathol. 2000;13:309–27.

13. Ellenson LH, Ronnett BM, Kurman RJ. Precursors of endometrial carcinoma. In: Kurman RJ, Ellenson LH, Ronnett BM, editors. Blaustein's pathology of the female genital tract. 7th ed. New York: Springer-Verlag; 2019. p. 439–72.

14. Papaefthimiou M, Symiakaki H, Mentzelopoulou P, et al. Study on the morphology and reproducibility of the diagnosis of endometrial lesion utilizing liquid-based cytology. Cancer. 2005;105:56–64.

15. Norimatsu Y, Shimizu K, Kobayashi TK, et al. Cellular feature of endometrial hyperplasia and well-differentiated adenocarcinoma using the endocyte sampler: diagnostic criteria based on cyto-architecture of tissue fragments. Cancer. 2006;108:77–85.

16. Shimizu K, Norimatsu Y, Ogura S, et al. Analysis of suspicious endometrial cytology: by a new criteria based on the architectural atypia. J Jpn Soc Clin Cytol. 2002;41:313–20. (in Japanese with English abstract).

17. Yoshida S, Kusunoki N, Ishiyama K, et al. A study of diagnostic criteria for endometrial cytology. J Jpn Soc Clin Cytol. 2008;47:227–35.(in Japanese with English abstract).

18. Norimatsu Y, Yanoh K, Kobayashi TK. The role of liquid-based preparation in the evaluation of endometrial cytology. Acta Cytol. 2013;57:423–35.

19. Kendall BS, Ronnett BM, Isacson C, et al. Reproducibility of the diagnosis of endometrial hyperplasia, atypical hyperplasia, and well differentiated carcinoma. Am J Surg Pathol. 1998;22:1012–9.

20. Yanoh K, Norimatsu Y, Munakata S, et al. Evaluation of endometrial cytology prepared with the Becton Dickinson SurePath™ method: a pilot study by the Osaki study group. Acta Cytol. 2014;58:153–61.

21. Fulciniti F, Yanoh K, Karakitsos P, et al. The Yokohama system for reporting directly sampled endometrial cytology: the quest to develop a standardized terminology. Diagn Cytopathol. 2018;46:400–12.

22. Norimatsu Y, Moriya T, Kobayashi TK, et al. Immunohistochemical expression of PTEN and β-catenin for endometrial intraepithelial neoplasia in Japanese women. Ann Diagn Pathol. 2007;11:103–8.

23. Monte NM, Webster KA, Neuberg D, et al. Joint loss of PAX2 and PTEN expression in endometrial precancers and cancer. Cancer Res. 2010;70:6225–32.

24. Chapel DB, Patil SA, Plagov A, et al. Quantitative next-generation sequencing-based analysis indicates progressive accumulation of microsatellite instability between atypical hyperplasia/endometrial intraepithelial neoplasia and paired endometrioid endometrial carcinoma. Mod Pathol. 2019;32:1508–20.

Malignant Neoplasm

11

Yoshinobu Maeda, Akihiko Kawahara, Takeshi Nishikawa, and Yoshiaki Norimatsu

11.1 Endometrial Endometrioid Carcinoma

11.1.1 Background

Endometrial endometrioid carcinoma (EEC) is the most common histological type of endomerial carcinoma (EC), accounting for more than 75% of all endometrial carcinomas. The incidence of endometrial carcinomas varies globally, with age-standardized incidence rates varying from 1 to 25 cases per 100,000 person-years in 2018. In Japan, the incidence of endometrial carcinomas has steadily increased in recent years.

The median patient age at the onset of EEC was 63 years [1]. It is well known that irregular genital bleeding is observed in postmenopausal women. The highest incidence rates occur in North America and Europe. The lowest incidence rate (4–5 times lower) is found in countries with low human development index [1, 2].

A major cause of the development of EEC is prolonged exposure to unopposed estrogen stimulation associated with an ovulation disorder such as polycystic ovarian syndrome, estrogen replacement therapy, tamoxifen treatment for breast cancer, and estrogen-producing neoplasms (e.g., ovarian thecoma and granulosa cell tumor.)

Y. Maeda
Department of Diagnostic Pathology, Toyama Red Cross Hospital, Toyama, Japan

A. Kawahara
Department of Diagnostic Cytopathology, Kurume University Hospital, Fukuoka, Japan

T. Nishikawa
Department of Diagnostic Pathology, Nara Medical University Hospital, Kashihara, Nara, Japan

Y. Norimatsu (✉)
Department of Medical Technology, Faculty of Health Sciences, Ehime Prefectural University of Health Sciences, Tobe-cho, Iyo-gun, Ehime, Japan
e-mail: ynorimatsu@epu.ac.jp

© The Author(s), under exclusive license to Springer Nature Singapore Pte Ltd. 2022
Y. Hirai, F. Fulciniti (eds.), *The Yokohama System for Reporting Endometrial Cytology*, https://doi.org/10.1007/978-981-16-5011-6_11

Early menarche, late menopause, nulliparity, obesity, and diabetes are well-known risk factors for EEC. In addition to these factors, Lynch syndrome with a mutation in DNA mismatch repair genes and Cowden syndrome caused by *PTEN* mutation are associated with familial endometrioid carcinoma [1, 2].

In the 1980s, ECs were classified as estrogen-dependent type I or estrogen-independent type II tumors by Bokhman [3]. Representative subtypes of type I are approximately corresponding to EECs, grade1 (G1) and grade2 (G2), which develop from endometrial atypical hyperplasia/endometrioid intraepithelial neoplasia (EIN). On the other hand, EEC, grade3 (G3) is classified as type II, which arises de novo from atrophic endometrium [2, 4].

In 2013, The Cancer Genome Atlas (TCGA) study divided endometrioid carcinoma into four subgroups, integrating genomic profiles, such as "ultramutated," "hypermutated," "copy number low," and "copy number high." [5]

11.1.2 Definition

EEC is a malignant epithelial neoplasm displaying varying proportions of glandular, papillary, and solid architecture, with neoplastic cells showing endometrioid differentiation. These tumors are referred to as "endometrioid" due to their similarity to proliferative phase endometrium [1, 2].

EEC is typically composed of columnar cells with eosinophilic and granular cytoplasm and has a low account of mucin. Histologically the tumor displays glandular, papillary with fine fibrovascular stroma, and solid pattern (Fig. 11.1). Nuclear

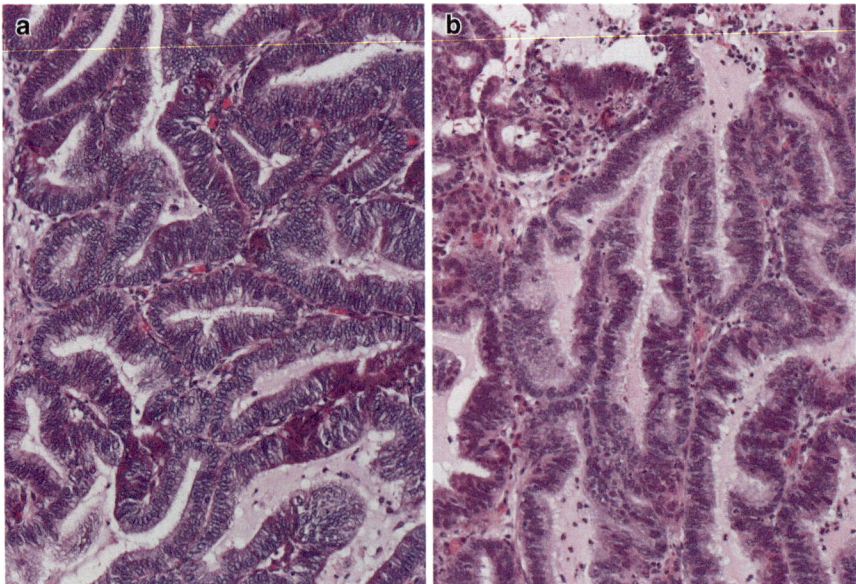

Fig. 11.1 EEC, G1. Histologic preparation shows irregularly shaped glands (**a**) and papillary architecture (**b**). (HE stain, original magnification **a** and **b**: 20x)

pseudostratification is usually observed and nuclear atypia is mild to moderate. Nucleoli are mostly inconspicuous [1, 2].

EECs were divided into three grades (G1, G2 or G3) according to the FIGO grading criteria. They are based on histological architecture and cellular atypia.

The architectural grade was determined according to the presence of a solid component without squamous differentiation. EEC, G1; the proportion of solid components is no more than 5%. EEC, G2; the proportion of solid growth is 6–50%, and EEC, G3; the glandular structure remains irregularly in some areas but is extremely obscured, and more than half is composed of the solid component.

Alternatively, if the rate of solid component is less than 5% and 6–50%, but the cell atypia is remarkable, raise G2, G3 instead of G1, G2, respectively [1, 2].

EECs have some histological and cytological variants. Squamous differentiation is composed of keratinizing cells and/or eosinophilic cells, including as morules occurring in 10–25% of endometrioid carcinomas. Other histological patterns include a secretory pattern in which the majority of tumor cells resemble early secretory phase endometrial glands, ciliated pattern, microglandular pattern, spindle cell pattern, sertoliform pattern, and mucinous pattern in various proportions in tumors [1, 2].

11.1.3 Cytologic Diagnostic Criteria [6–10] (Figs. 11.2, 11.3, 11.4, 11.5, 11.6, 11.7, 11.8 and 11.9)

- Almost all clusters show an irregular protrusion pattern.
- The nuclear overlap in epithelial cell clusters exceeds three layers, and the cohesion of stroma cells around the clusters is absent.
- Usually, epithelial cell clusters show glandular complexity with an increasing number of lumens, observed as a cribriform pattern in histologic preparation.

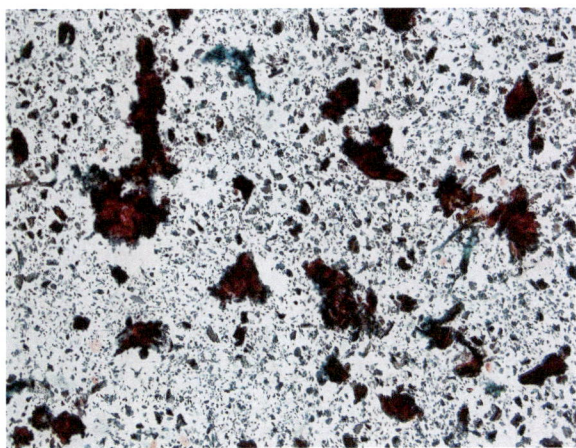

Fig. 11.2 EEC, G1. Same sample as Fig. 11.1. This cytological preparation was obtained from a 40 y-o woman with irregular genital bleeding. Many clusters show various sizes and irregular shapes. (Papanicolaou stain, original magnification 4×)

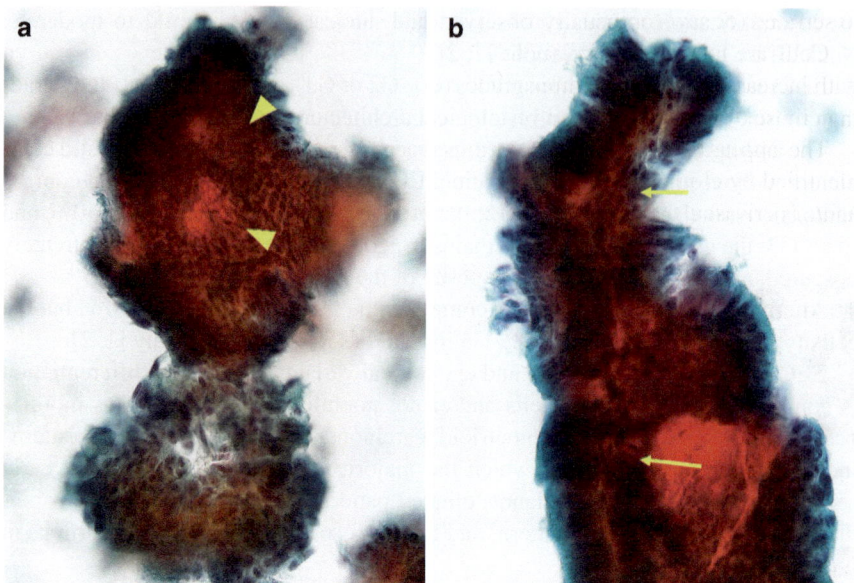

Fig. 11.3 EEC, G1. Same sample as Fig. 11.1. (**a**): An irregularly shaped cluster with lumens (arrowheads) is seen. (**b**): Fine strands consisted spindle cells (arrows) are seen in cellular clusters and nuclei of tumor cells are arranged perpendicular to strands. (Papanicolaou stain, original magnification **a** and **b**: 40×)

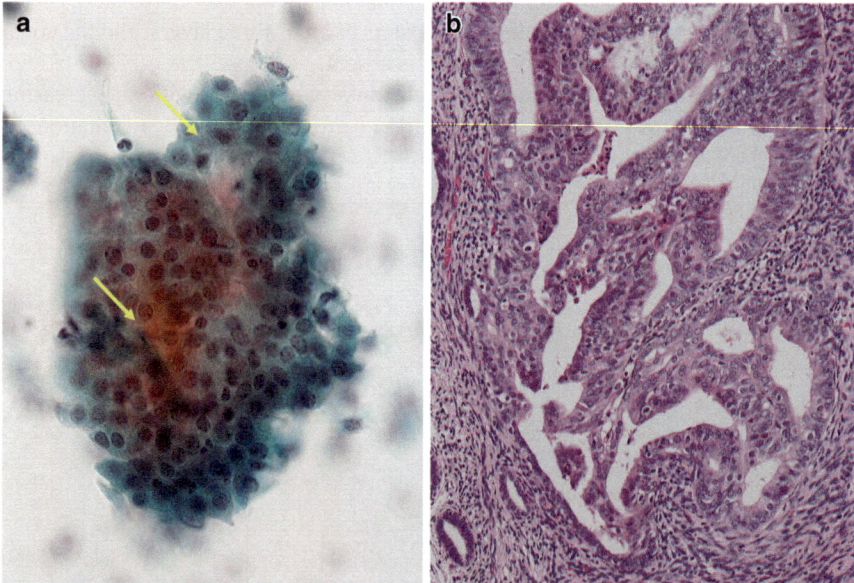

Fig. 11.4 EEC, G1. An irregularly shaped cluster with lumens (arrows) and nuclear overlapping is seen. Nuclei are enlarged, have conspicuous nucleoli. (**a**) Corresponding to histologic preparation (**b**), dilated and irregular shaped glands are present. (**a**: Papanicolaou stain, original magnification 60×, **b**: HE stain, original magnification 20×)

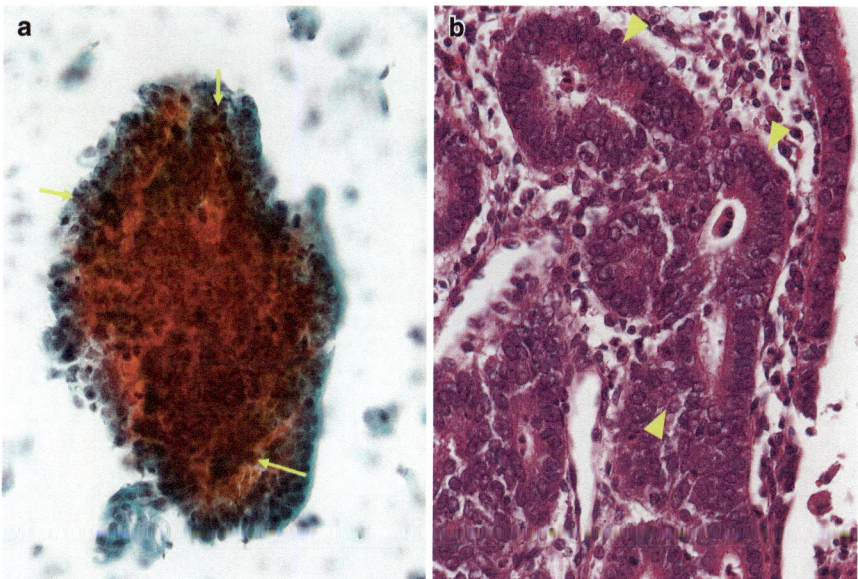

Fig. 11.5 EEC, G1. Irregularly shaped lumens (arrows) increase numbers in cellular clusters. (**a**) Corresponding to histologic preparation (**b**), irregular shaped fused glands are present (arrowheads). (**a**: Papanicolaou stain, original magnification 40×, **b**: HE stain, original magnification 40×)

Fig. 11.6 EEC, G1, with squamous differentiation. Large and irregular clusters can be seen. (Papanicolaou stain, original magnification 10×)

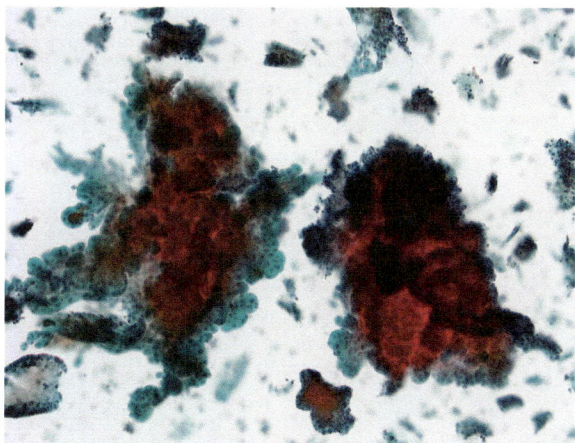

- The arrangement in the epithelial cell clusters becomes irregular, and the nucleus frequently protrudes toward the periphery of the clusters.
- Glandular epithelial cells with nuclear swelling, anisonucleosis, increased chromatin granularity, and conspicuous nucleoli are observed.
- Mitosis can be occasionally observed.
- Hemorrhagic and necrotic exudate can be seen in the background.

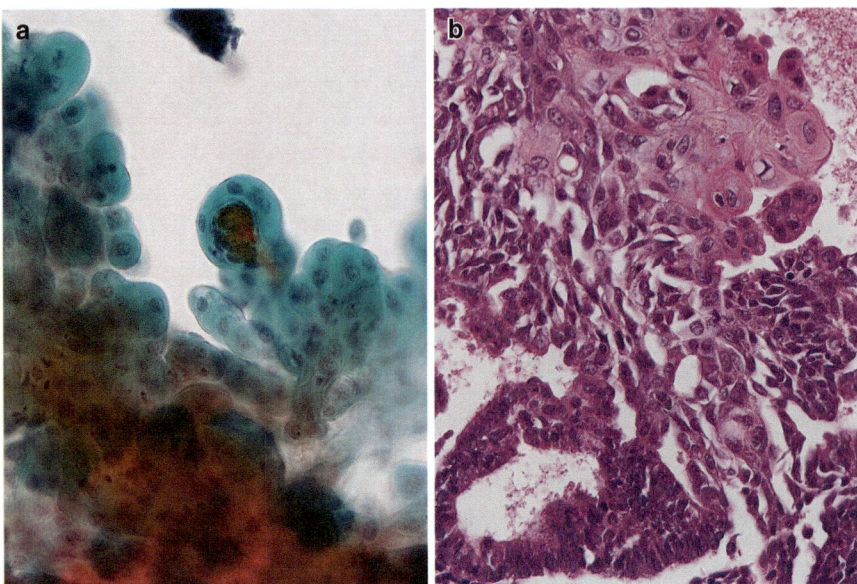

Fig. 11.7 EEC, G1, with squamous differentiation. (**a**): Tumor cells with light-green cytoplasm show a low N/C ratio, and nuclei are located centrally. (**b**): Corresponding to histologic preparation shows squamous differentiation with single-cell keratinization (upper right). (**a**: Papanicolaou stain, original magnification 40×, **b**: HE stain, original magnification 40×)

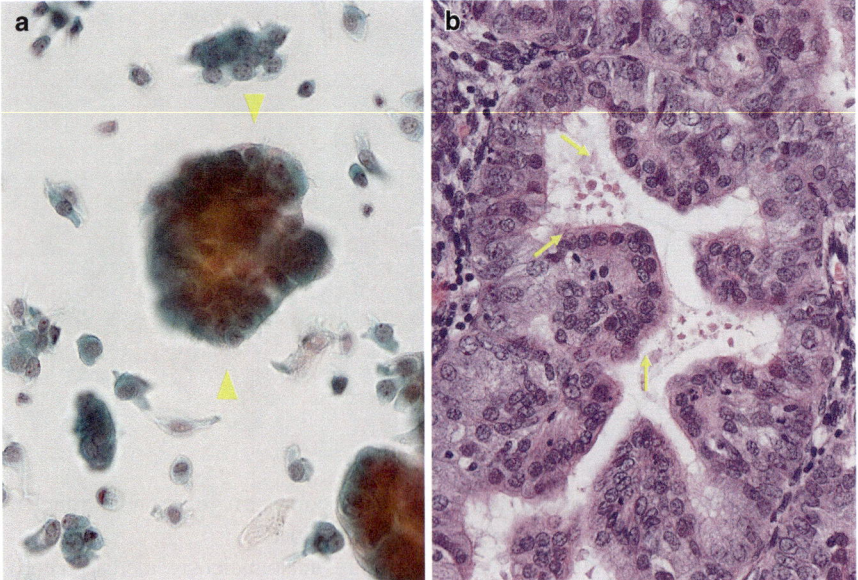

Fig. 11.8 EEC, G1, with ciliated change. Neoplastic epithelial cells (**a**) with cilia are intermingled in cellular clusters (arrowheads). Corresponding to histologic preparation (**b**), ciliated neoplastic cells are seen along luminal aspect (arrows). (**a**: Papanicolaou stain, original magnification 60×, **b**: HE stain, original magnification 40×)

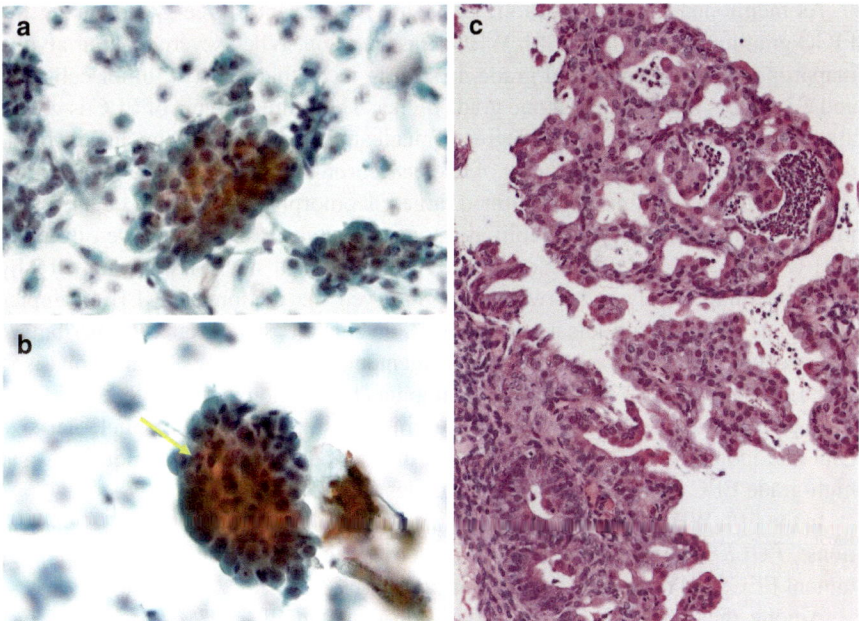

Fig. 11.9 EEC, G1, with microglandular pattern. Medium-sized epithelial cluster shows mild nuclear overlapping (**a**) and tiny irregular lumen (arrow) is seen in cluster (**b**). Corresponding to histologic preparation (**c**) show microglandular pattern and contain a number of neutrophils in lumens. (**a** and **b**: Papanicolaou stain, original magnification 40×, **c**: HE stain, original magnification 20×)

The method of evaluation of neoplastic epithelial clusters is mentioned in Chap. 5, as an algorithmic interpretational approach of endometrial cytology for the Yokohama System.

11.1.4 Explanatory Note

Several previous studies have identified genetic alterations of ECs, such as microsatellite instability and mutation in the *PTEN, PIK3CA, CTNNB1, ARID1A, KRAS, TP53* genes. In the Bokhman classification, each subtype shows characteristic frequencies of molecular alterations, with type I tumors having more mutation in genes for *PTEN, PIK3CA, CTNNB1, ARID1A, KRAS*, whereas type II having more *TP53* mutations [4].

Profiling the notable pattern of somatic genomic alterations, based on TCGA study revealed that EECs were divided into four molecular subtypes: ultramutated (POLE hotspot mutation), hypermutated (microsatellite instability), copy number low, and high copy number [7]. These four subtypes show characteristic gene mutations, histological features, clinical features, and prognosis [4, 5, 11, 12].

As mentioned in the definition, EECs are divided into three grades using the FIGO grading criteria in the fifth WHO Classification. When severe cellular atypia, inappropriate for architectural grade, is seen in more than 50% of tumor cells, G1 and G2 tumors are considered one grade higher. The cellular atypia of EEC is generally evaluated according to the degree of nuclear size, shape, anisonucleosis, pseudostratification and loss of polarity of nucleus, chromatin distribution, and nucleolus size and numbers. Zaino et al. defined large, pleomorphic nuclei with coarse chromatin, and large irregular nucleoli, as the notable atypia to raise a grade of tumors [13] (Fig. 11.10). Recently Norimatsu et al. evaluated nuclear morphometry by using an image analysis software, and observed that endometrial LBC samples exhibit an increase in nuclear enlargement, anisonucleosis, chromatin distribution and structure, nuclear shape, nuclear arrangement, and nucleolar size in comparison with EEC, G1, EEC, G3 and serous carcinoma [14]. Although the evaluation of cellular atypia is somewhat subjective, the objective measurement of nucleolar size could be indicative of cellular atypia and distinction between low-grade EEC and high-grade EEC in endometrial LBC samples [14].

In the fifth WHO Classification, EECs are divided into four molecular classifications: *POLE*-ultramutated EEC, mismatch repair (*MMR*)-deficient EEC, *p53*-mutant EEC, and no specific molecular profile (NSMP) EEC.

Among these four subgroups, *POLE*-ultramutated EEC, *MMR*-deficient EEC, and *p53*-mutant EEC exhibit high-grade histological appearance, and NSMP EC are

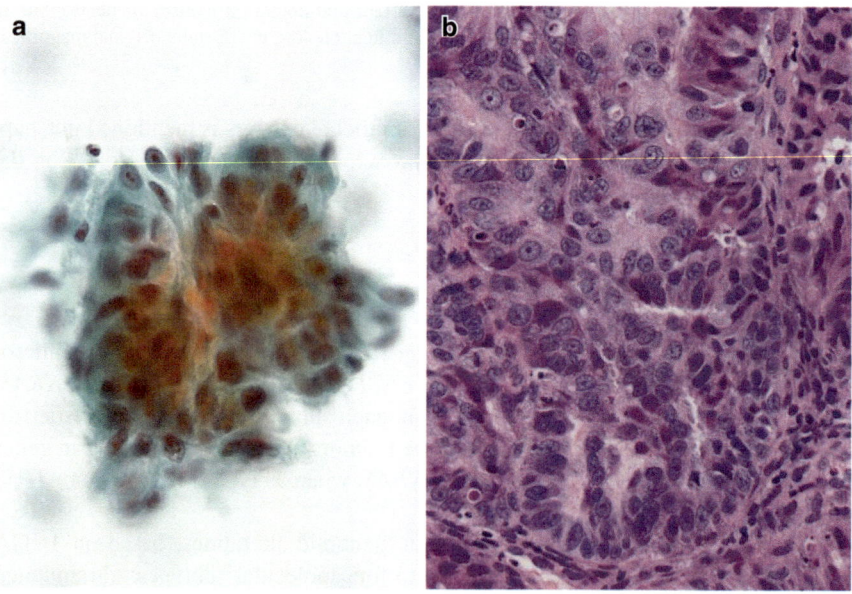

Fig. 11.10 EEC, G2, with severe nuclear atypia. Many tumor cells have enlarged nuclei with large and eosinophilic Nucleoli (**a**). Corresponding to histologic preparation shows an inconspicuous glandular pattern. (**b**) (**a**: Papanicolaou stain, original magnification 60×, **b**: HE stain, original magnification 40×)

mostly as low-grade feature with squamous differentiation or morules. However, the frequency of NSMP EC is approximately 30–40%, and other low-grade EECs belong to three different subgroups (Figs. 11.11, 11.12, 11.13 and 11.14). In contrast, high-grade EECs were found in all four subgroups. Although the morphological features of high-grade EECs are overlapped between these subgroups, clinical outcomes show distinctive differences [15]. However, *POLE*-ultramutated EEC has an excellent prognosis. This subtype shows frequently increasing nuclear size, irregular nuclear contours, striking hyperchromasia, prominent nucleoli [16, 17]. As mentioned above, accurate evaluation of the degree of nuclear atypia is considered an indicative finding in estimating the biological features of tumors [13, 18], but in the diagnosis of EEC, an approach from the aspect of tumor morphology alone may be insufficient [19, 20]. The algorithm for diagnosis of EEC, using molecular and immunohistochemical surrogate markers for each subgroup such as *POLE* hotspot mutation, *MSI* assay, *MMR*-deficient, *TP53* mutation, and *p53* immunohistochemistry, has also been proposed [21] (Figs. 11.15, 11.16, 11.17, 11.18, 11.19 and 11.20).

Recently in LBC endometrial sample, *PTEN* mutation and loss of expression, *p53* overexpression and β catenin nuclear expression could be evaluated by immunocytochemistry or molecular techniques [22–24]. Application of DNA analysis using LBC endometrial samples has been reported [25], and it will be possible to consider cytological approaches including immunocytochemical and molecular analysis in near future.

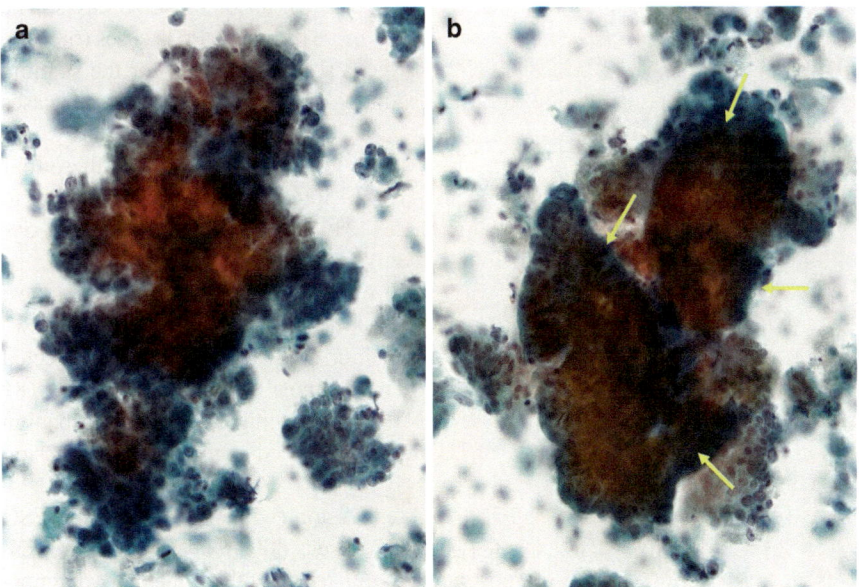

Fig. 11.11 EEC, G2. This cytological preparation was obtained from a 50 y-o woman. Clusters show irregular shapes and marked overlapping (**a** and **b**). Irregularly shaped lumens are seen within a cellular cluster (arrows). (**b**) (Papanicolaou stain, original magnification **a** and **b**: 40×)

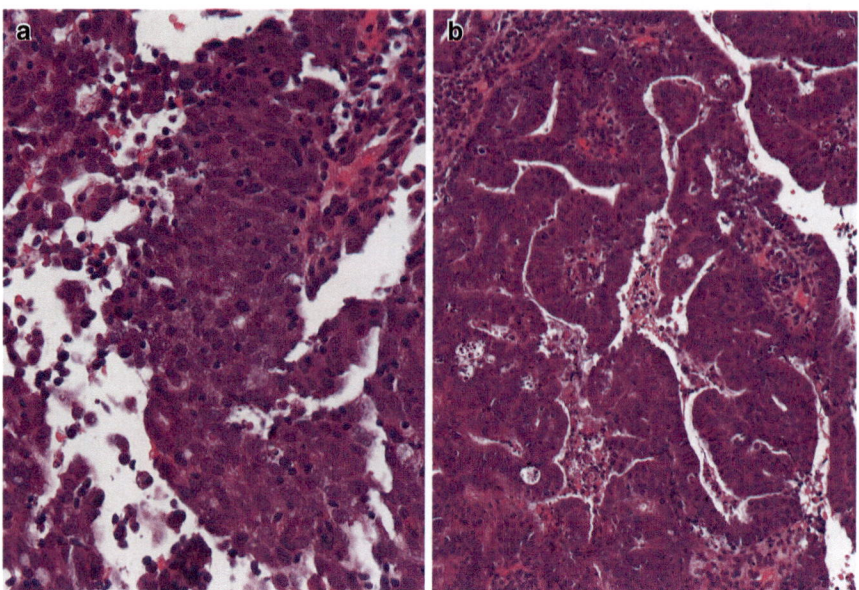

Fig. 11.12 EEC, G2. Histological specimen corresponding to Fig. 11.11. Solid nest with lympho-cytes infiltration is present (**a**). Complex papillary and glandular architecture can be seen (**b**). (HE stain, original magnification **a**: 40×, **b**: 20×)

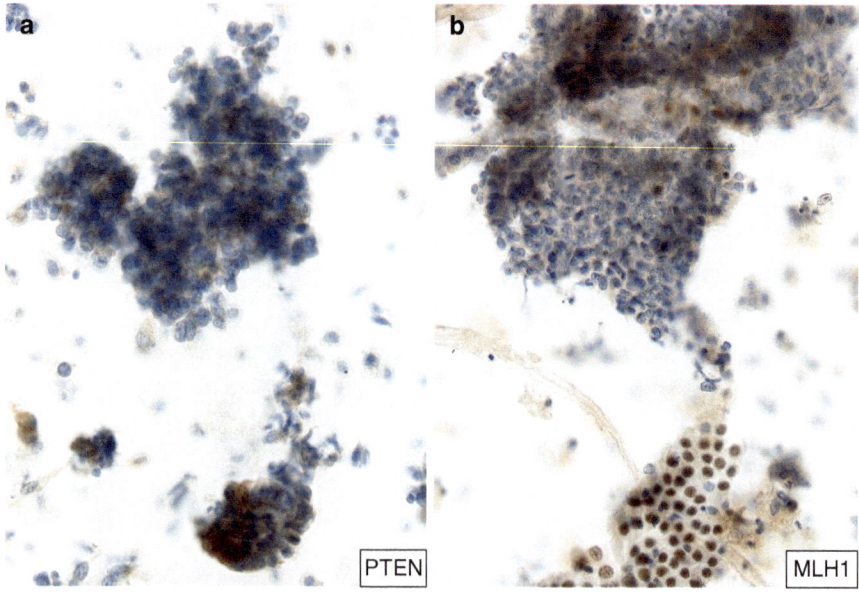

Fig. 11.13 EEC, G2. Same sample as Fig. 11.12. (**a**): On immunocytochemistry (ICC), clusters with loss of PTEN expression (upper), with a small number of PTEN expressing stromal cells can be seen (lower). (**b**): tumor cells show complete loss of MLH-1 expression (upper). MLH-1 expressing atrophic endometrial epithelial cells can be seen (lower). (ICC, original magnification **a** and **b**: 40×)

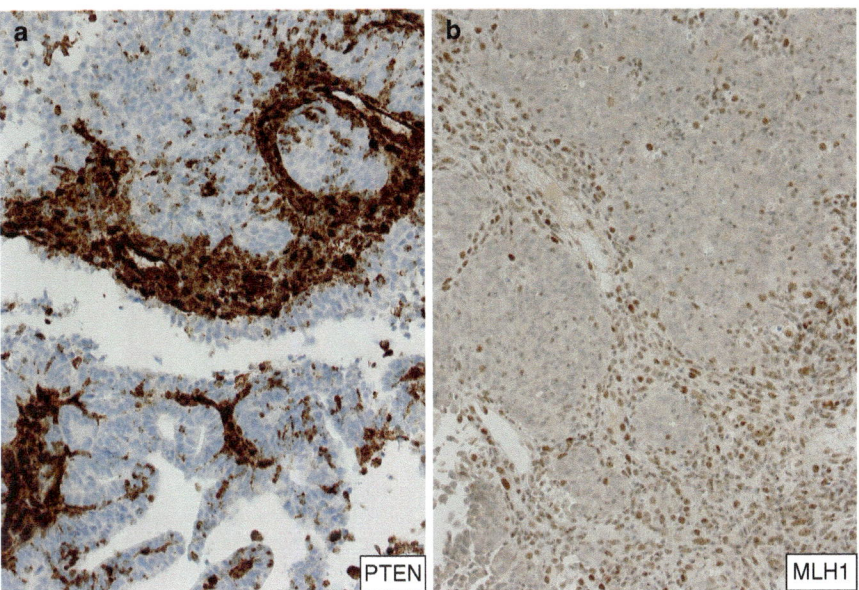

Fig. 11.14 EEC, G2. Same sample as Fig. 11.11. On immunohistochemistry (IHC), neoplastic glands show complete PTEN loss of expression (**a**) and loss of MLH-1 expression (**b**) (IHC, original magnification **a** and **b**: 20×)

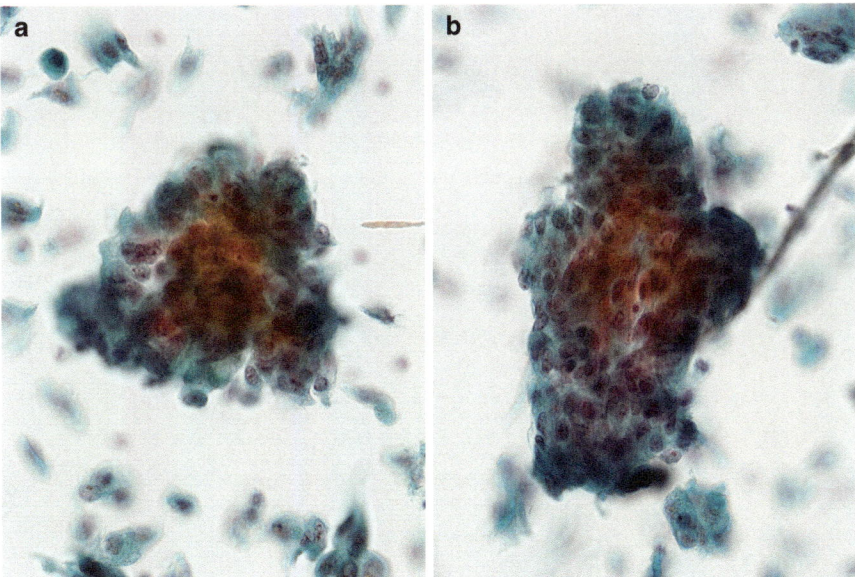

Fig. 11.15 EEC, G3. Nuclear overlapping and loose connection in cluster can be seen (**a**). Nuclei show enlarged, various shapes, and display fine granular chromatin and conspicuous nucleoli (**b**). (Papanicolaou stain, original magnification **a** and **b**: 60×)

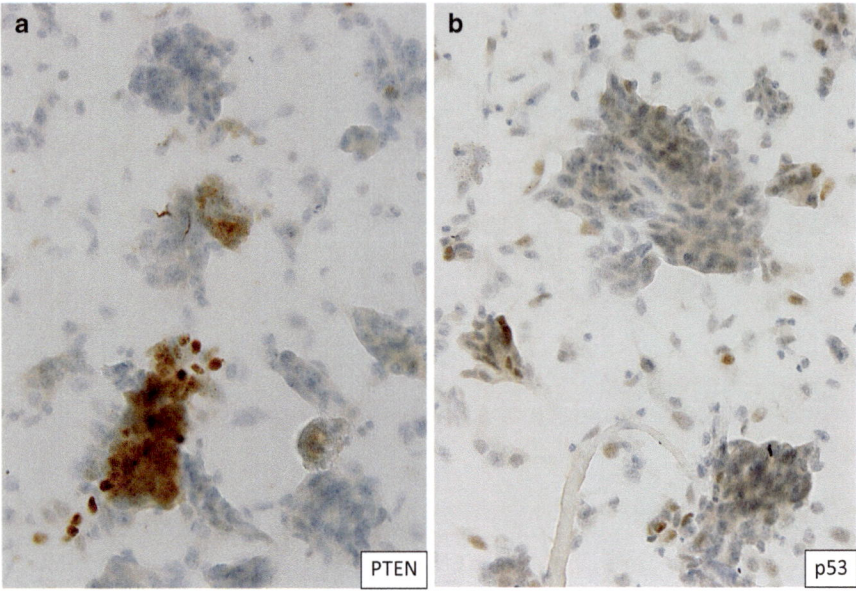

Fig. 11.16 EEC, G3 (same sample as Fig. 11.15) (**a**): On ICC, neoplastic clusters with loss of PTEN expression, with numbers of PTEN expressing stromal cells, can be seen (bottom) (**b**): almost neoplastic cells exhibit weak expression of *p*53. (ICC, original magnification **a** and **b**: 40×)

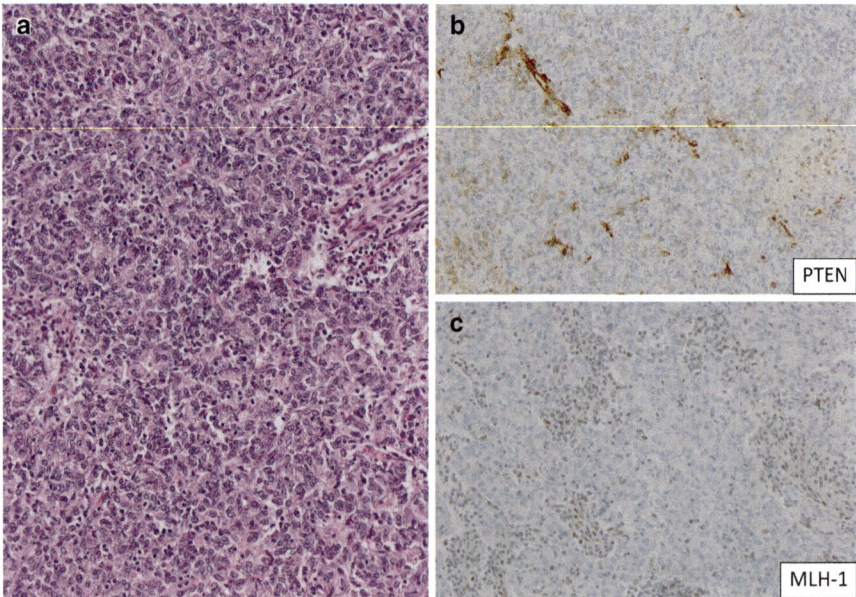

Fig. 11.17 EEC, G3 (same sample as Fig. 11.15). Corresponding to histologic preparation (**a**) shows solid nests with lymphocytes infiltration. (**b**): Tumor nests show PTEN loss of expression. (**c**): Tumor nests show with complete loss of MLH-1 expression. (**a**: HE stain, original magnification 20×, **b** and **c**: IHC, original magnification 20×)

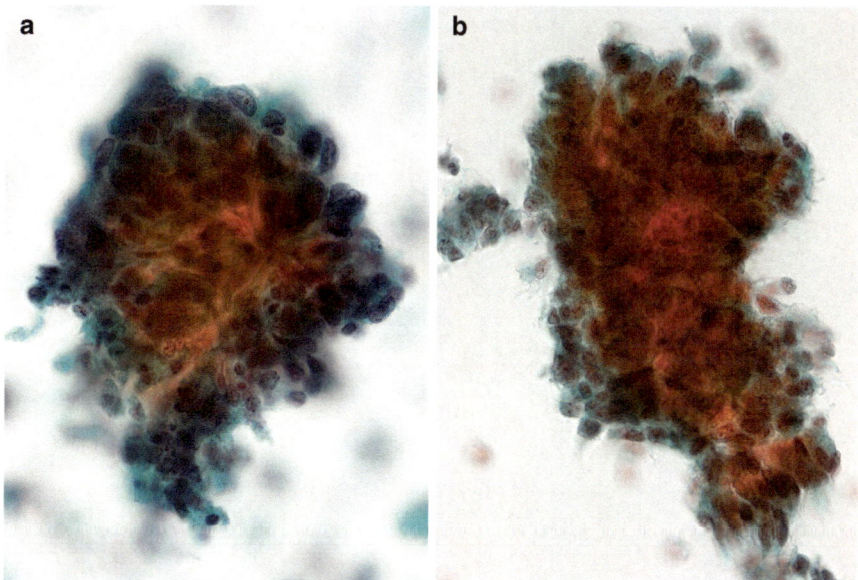

Fig. 11.18 EEC, G3. Cluster shows an irregular shape. Significant nuclear overlapping in cluster can be seen. Nuclei show enlarged, various shapes, and display granular chromatin and conspicuous nucleoli. (Papanicolaou stain, original magnification **a** and **b**: 60×)

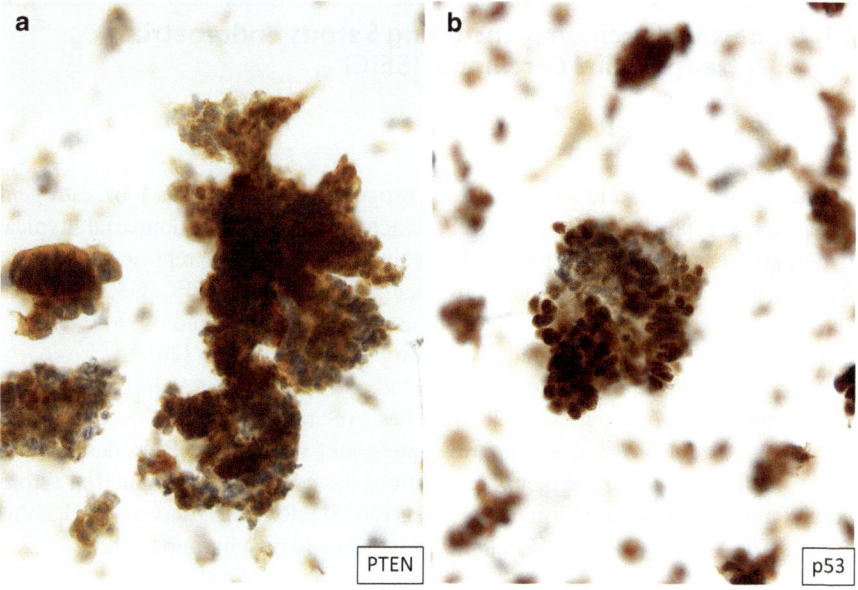

Fig. 11.19 EEC, G3 (same sample as Fig. 11.18). (**a**): On ICC, PTEN expressing clusters. (**b**): almost all tumor cells exhibit strong and diffuse nuclear expression of *p*53. (ICC, original magnification **a** and **b**: 40×)

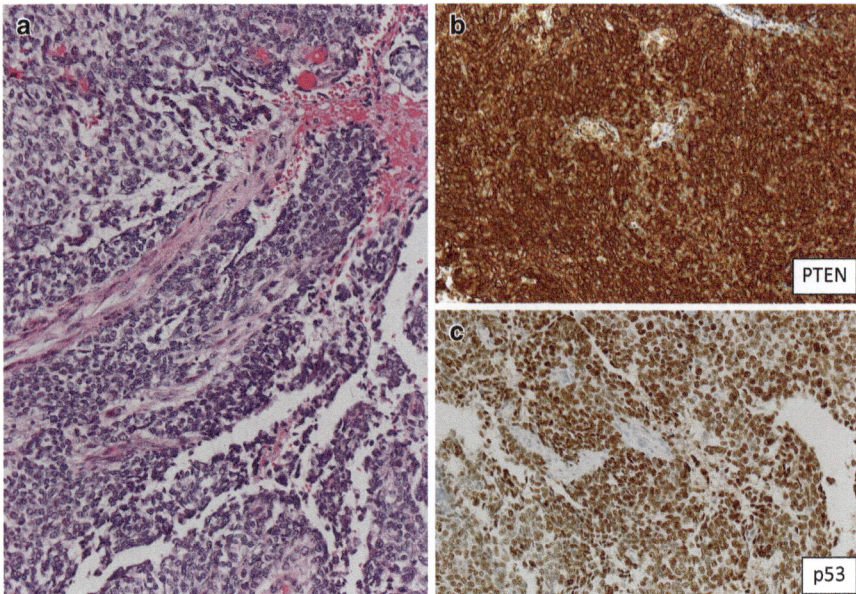

Fig. 11.20 EEC, G3 (same sample as Fig. 11.18). Corresponding histological specimen (**a**) shows sheet-like solid nests. (**b**): Tumor nests show PTEN expression. (**c**): Tumor nests show exhibit strong and diffuse nuclear expression of *p*53. (**a**: HE stain, original magnification 20×, **b** and **c**: IHC, original magnification 20×)

11.2 Serous Carcinoma, Including Serous Endometrial Intraepithelial Carcinoma (SEIC)

11.2.1 Background

In the 1980s, ECs were classified as estrogen-dependent Type I or estrogen-independent Type II. G1 and G2 EECs, which develop from endometrial atypical hyperplasia/endometrioid intraepithelial neoplasia (EIN), are representative subtypes of Type I. On the other hand, serous carcinoma (SC) and EEC, G3, are typical subtypes of Type II. However, Type II tumors are infrequent and often develop in postmenopausal women with underlying atrophic endometrium [26].

SC was first described by Hendrickson et al. in 1982, and has aggressive biological features and poor prognosis [27, 28]. It has a relatively low prevalence, accounting for 2–10% of all ECs, and approximately half of all EC-related deaths [29]. Some studies have reported that *p*53 mutations are common in endometrial serous carcinoma, and occur early in carcinogenesis [30, 31]. Recently, the Cancer Genome Atlas (TCGA) study placed SC in the copy-number-high subgroup [32].

11.2.2 Definition

In the fifth edition of the WHO classification in 2020, SC is defined as a carcinoma with diffuse, marked nuclear pleomorphism, and a typical papillary and/or glandular growth pattern. In addition to arising in the atrophic endometrium, development within endometrial polyps is also possible [33].

SC shows papillary structures with delicate fibrovascular stroma or thick fibrous strands and, sometimes, tubular structures or slit-like spaces. Tubular structures composed of columnar tumor cells needing to be differentiated from ECC are sometimes recognized. A solid pattern can also be present. Tumor cells are polygonal to columnar and show high-grade nuclear atypia, with a high N/C ratio. Psammoma bodies are occasionally encountered [34].

11.2.3 Cytologic Diagnostic Criteria (Figs. 11.21, 11.22, 11.23, 11.24, 11.25, 11.26 and 11.27)

- Frequent hemorrhagic background.
- Frequent occurrence of small to medium-sized 3D clusters showing irregular structure.
- Nuclear overlapping of three or more layers and irregular cellular arrangement in the clusters.
- Light-green cytoplasm in almost all tumor cells.
- Nuclei show the increased size and marked pleomorphism with coarse nuclear chromatin and large and eosinophilic nucleoli; cells with bizarre nuclei and/or multinucleated syncytial tumor cells are frequently found.
- Mitotic activity is usually high and atypical mitoses are easily recognized.
- Psammoma bodies are present in approximately 30% of cases.

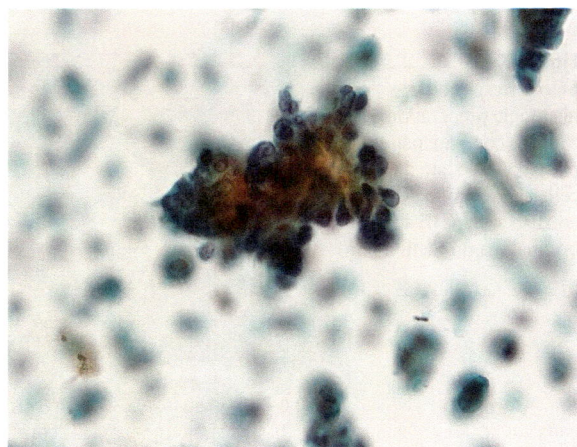

Fig. 11.21   SC. Irregularly shaped 3D cluster of tumor cells showing disordered cellular arrangement. (Papanicolaou stain, original magnification 40×)

Fig. 11.22 Serous carcinoma. Tumor cell clusters are small to medium-sized and show nuclear overlapping of three or more layers. (Papanicolaou stain, original magnification 40×)

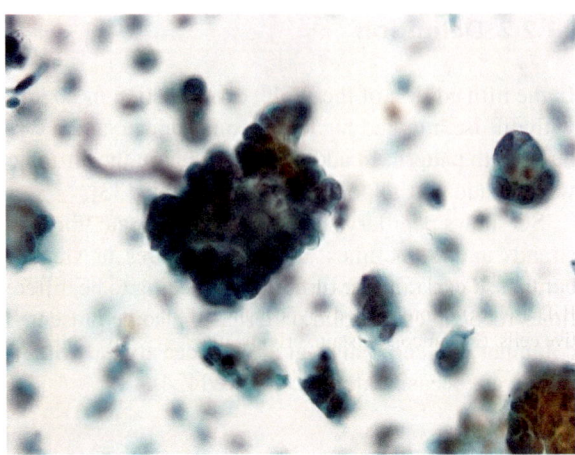

Fig. 11.23 SC. Tumor cells with light-green cytoplasm show increased N/C ratio, Hyperchromasia, conspicuous nucleoli, and pleomorphism. (Papanicolaou stain, original magnification 40×)

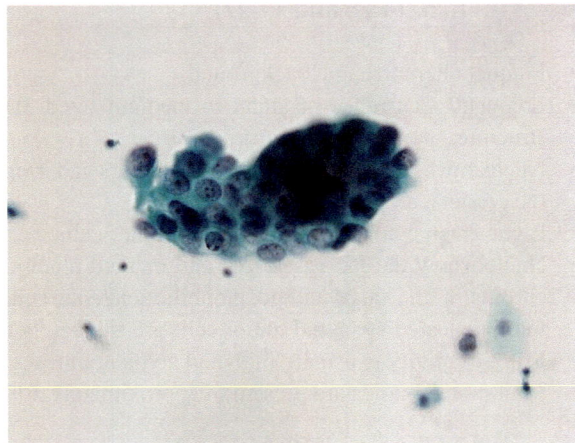

Fig. 11.24 SC. Corresponding histologic preparation shows a complex papillary pattern. Dissociated tumor cells and necrotic debris are also seen. (HE stain, original magnification 20×)

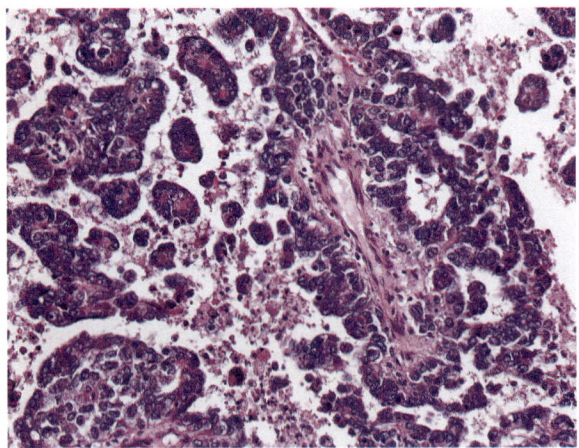

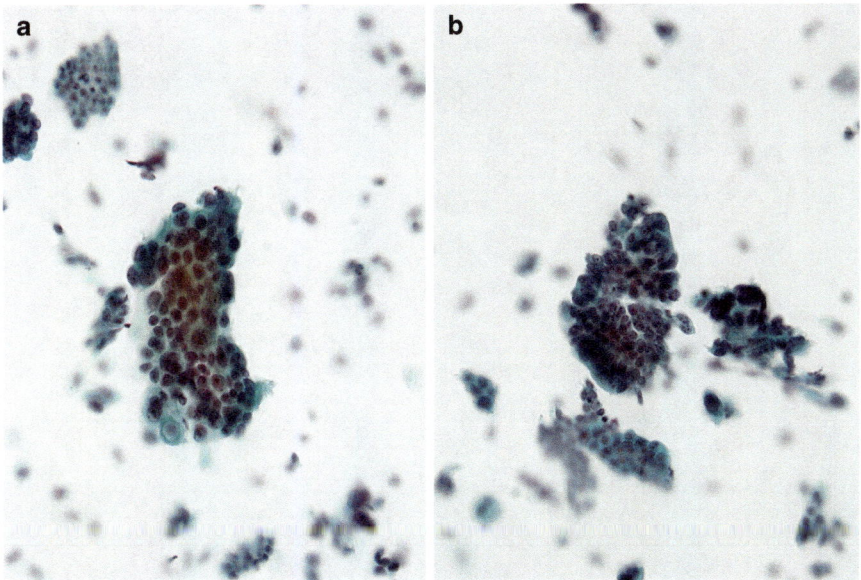

Fig. 11.25 SC (tiny lesion). 60 y-o patient (**a**): a medium-sized irregular cluster of tumor cells from small lesion of serous carcinoma shows pleomorphic, enlarged nuclei. (**b**): small-sized clusters show nuclear overlapping of more than three layers. (Papanicolaou stain, original magnification **a** and **b**: 40×)

Fig. 11.26 SC (tiny lesion). Corresponding histologic preparation shows complex papillary and tubular structures. Tumor is confined to an endometrial polyp and 4 mm in maximum size. (HE stain, original magnification 4×)

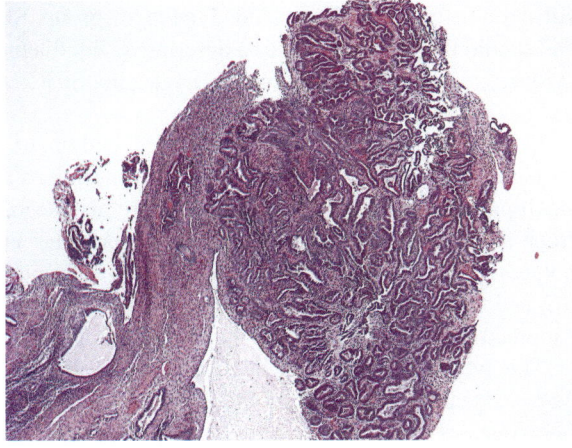

In the fourth edition of the WHO classification, serous endometrial intraepithelial carcinoma (SEIC) is described as an immediate precursor lesion of SC that has no stromal invasion [35]. Similar to SC, the background consists of atrophic endometrium and endometrial polyps. SEIC and serous carcinoma less than 1 cm in maximum size, without myometrial and vascular invasion or extrauterine metastases, have a favorable prognosis [36–38]. Unlike EEC, there is a potential for

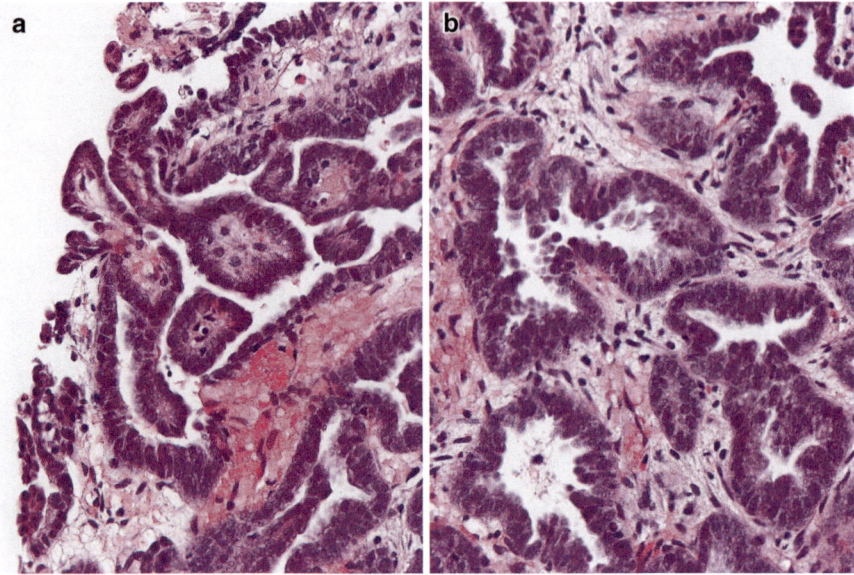

Fig. 11.27 SC (tiny lesion). Same case as Fig. 11.26. (a): showing small papillary structures with fibrovascular stroma. (b) tumor cells show enlarged and pleomorphic nuclei. (HE stain, original magnification a and b: 20×)

extrauterine metastasis to the abdominal cavity. In the fifth edition of the WHO classification in 2020, SEIC is included in the SC group. SEIC is synonymous with SC; and should therefore be used as a descriptive, not diagnostic term [39]. Endometrial cytology plays an important role in diagnosing SEIC, which is often asymptomatic and has a small size.

In SEIC, tumor cells replace the normal endometrial lining (refer to Chap. 12). In addition to showing a tubular structure that retains the original glandular shape, small papillary and sieve-like structures are also seen. There may be a distinctive front at the non-neoplastic endometrial glandular epithelium. Tumor cells are polygonal, hobnail-like, and columnar. Nuclear atypia is marked, similar to that of SC, and the N/C ratio is high. Neoplastic nuclei are 4–5 times larger than atrophic endometrial glandular nuclei in the background.

The cytologic findings in SEIC are almost the same as those of SC described above, except that the background is clear and the degree of nuclear overlapping is often one or two layers [40, 41] (Figs. 11.28, 11.29, 11.30 and 11.31).

11.2.4 Explanatory Note

Zheng et al. reported that approximately 90% of SCs show mutation-pattern overexpression of *p53* protein, with a frequency of *TP53* gene mutations of 96%. The estrogen receptor (ER) is expressed in less than 30% of cases, and insulin-like

Fig. 11.28
SEIC. Medium-sized clusters derived from SEIC, show nuclear overlapping of more than three layers, in contrast to an atrophic endometrial cell cluster (lower left). (Papanicolaou stain, original magnification 40×)

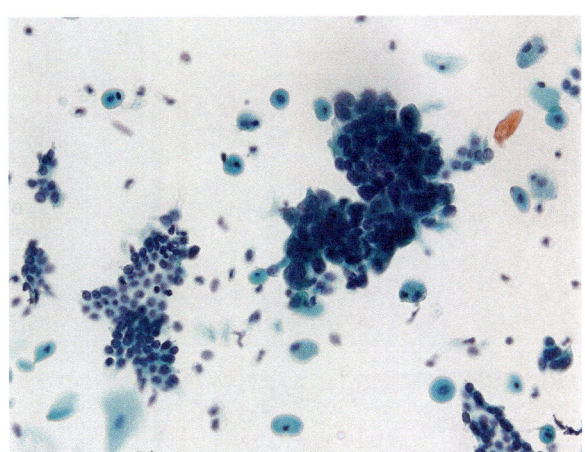

Fig. 11.29
SEIC. Medium-sized irregular cluster of tumor cells from SEIC shows enlarged and pleomorphic nuclei. (Papanicolaou stain, original magnification 40×)

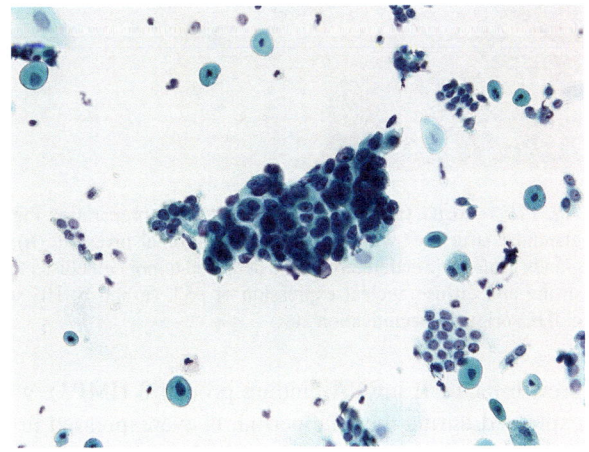

Fig. 11.30 SEIC. Large and eosinophilic nucleoli are seen in many tumor cells of SEIC. (Papanicolaou stain, original magnification 60×)

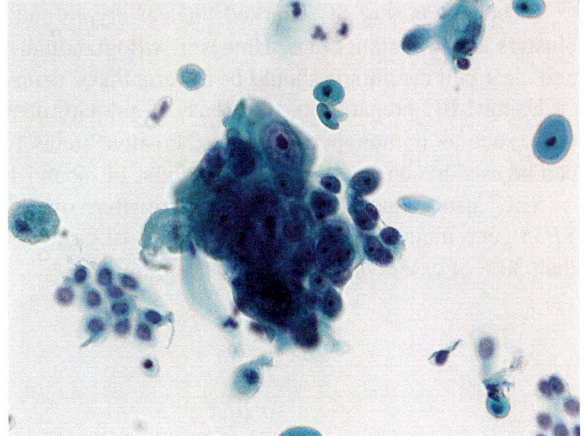

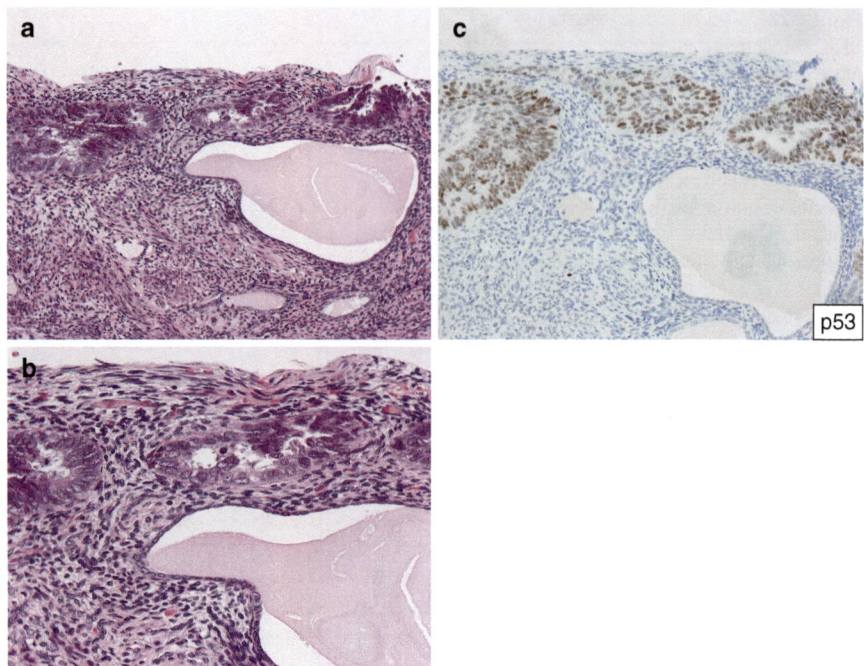

Fig. 11.31 SEIC. (**a**): corresponding histologic preparation Figs. 11.10, 11.11 and 11.12, shows glandular structures with no evidence of stromal invasion. (**b**): tumor cells are confined to the glands in atrophic endometrium and detached tumor cell clusters. (**c**): almost all tumor cells exhibit strong and diffuse nuclear expression of *p*53. (**a** and **b**: HE stain, original magnification 20×, **c**: IHC, original magnification 20×)

growth factor II mRNA-binding protein 3 (IMP3), which is an oncofetal protein expressed during the fetal period, is overexpressed in 91% of cases. Furthermore, the labeling index of Ki-67 is as high as 30–50% or more, and p16 expression is observed in more than 90% of cases [39, 42, 43].

When diagnosing SC, marked nuclear atypia and irregular-shaped tumor cell clusters are important clues. However, villoglandular-type EEC, high-grade EEC, and clear cell carcinoma should be differentiated from SC.

Using LBC preparations, it is easy to prepare unstained samples for ancillary tests, such as immunocytochemistry. Positive stains for *p*53, p16, ER, and IMP3 can be used to support the diagnosis (Figs. 11.32 and 11.33).

SEIC also frequently shows mutation-pattern overexpression of *p*53 protein, and *TP53* gene mutations are seen in 63–72% of cases. ER are also expressed in less than 30% of cases, similar to serous carcinoma.

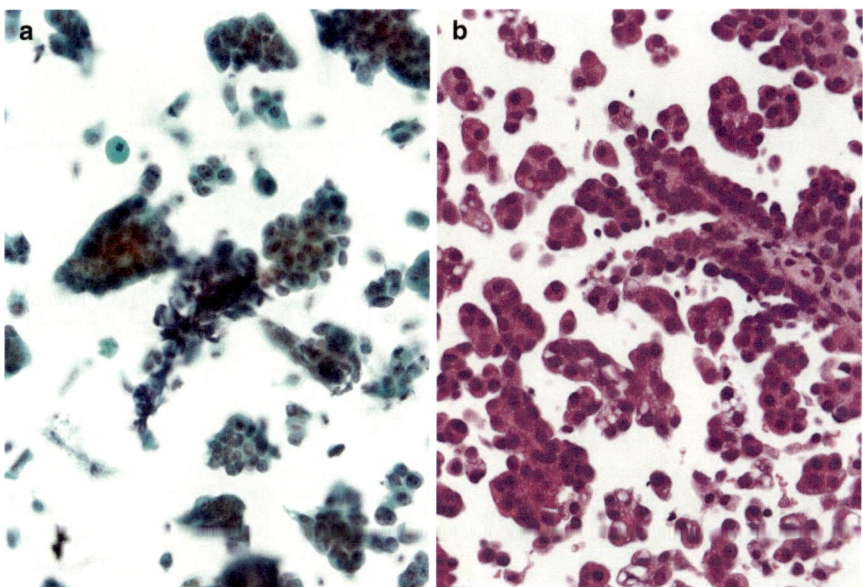

Fig. 11.32 SC. (**a**): small to medium-sized irregular clusters of tumor cells show enlarged, pleomorphic nuclei, with conspicuous nucleoli, and overlapping of more than three layers. (**b**): corresponding histologic preparation shows papillary structures and detached tumor cells clusters. (**a**: Papanicolaou stain, original magnification 40×, **b**: HE stain, original magnification 20×)

Endometrial glandular dysplasia (EmGD), a precancerous lesion of endometrial serous cancer, has been proposed to be a possible precursor of serous cancer (both SEIC and SC) [39, 44], judging from the occurrence of $p53$ abnormalities in the resting atrophic endometrium (so-called "$p53$ signature") [45]. This condition shows coexistence and transition from the surrounding atrophic endometrial glands or SEIC. The histopathologic features of EmGD consist of nuclear hyperchromasia with inconspicuous nucleoli and no atypical mitoses. The size of the lesion may be as small as 1 mm or less. Many of them show a mutation-pattern overexpression of $p53$ protein, and the frequency of $TP53$ gene mutations is 43%. ER and PgR are expressed in 70–95%, 60–90% of cases, respectively. Cytological examination plays an important role in the detection of this state and may assist appropriate clinical management in order to prevent the development of endometrial serous cancer (Figs. 11.34, 11.35 and 11.36).

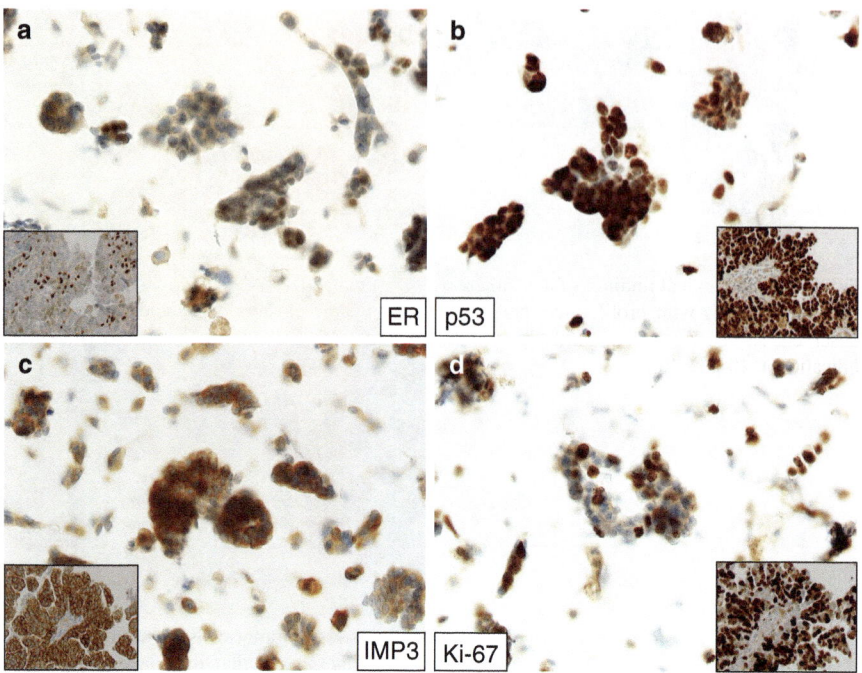

Fig. 11.33 SC (same cases as Fig. 11.32). Immunocytochemical staining (**a**): tumor cells do not express or show reduced expression of ER (inset; IHC staining) (**b**): almost all tumor cells exhibit strong and diffuse nuclear expression of *p53* (inset; IHC staining) (**c**): almost all tumor cells show cytoplasmic expression of IMP3 (inset; IHC staining). (**d**): increased ratio of Ki-67 labeled tumor cells in cluster (inset; IHC staining, original magnification 20×) (ICC, original magnification **a–d**: 40×)

Fig. 11.34 A sheet-like epithelial cell cluster is seen. Epithelial cells show increased nuclear size with anisonucleosis and hyperchromasia, suggesting neoplastic nature. (Papanicolaou stain, original magnification 40×)

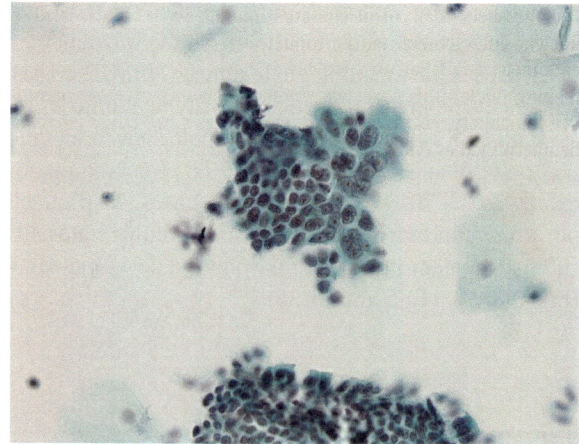

Fig. 11.35 Same cases as Fig. 11.14; showing mild nuclear overlapping. (Papanicolaou stain, original magnification 40×)

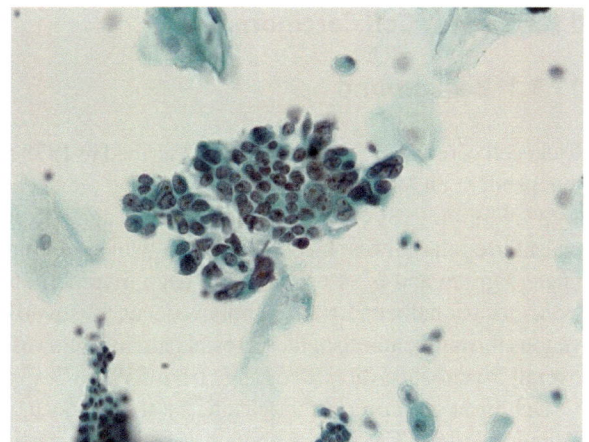

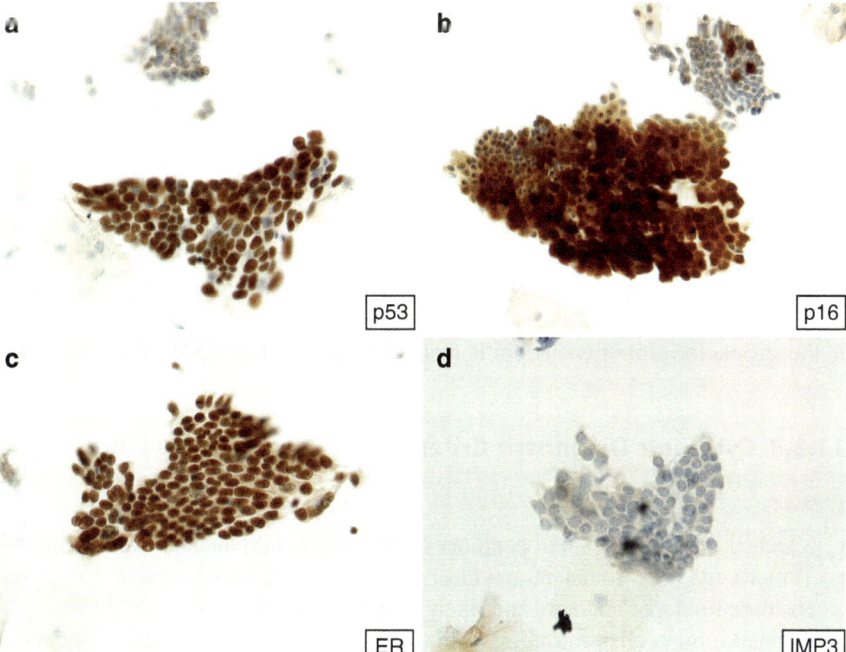

Fig. 11.36 Cytologic preparation from the same sample of Figs. 11.34 and 11.35, (**a**): ER is expressed in almost all cells. (**b**): almost all tumor cells exhibit strong and diffuse expression of p16. (**c**): almost all tumor cells exhibit strong and diffuse nuclear expression of *p*53. (**d**): IMP3 is not expressed. (ICC, original magnification **a–d**: 40×)

11.3 Clear Cell Carcinoma

11.3.1 Background

Clear cell carcinoma (CCC) was first described in 1973 and classified as an estrogen-independent endometrial carcinoma [46]. The prevalence of CCC is approximately 1–6%. Similar to SC, CCC occurs in patients aged 65 years or older, and postmenopausal irregular uterine bleeding is a frequent symptom. CCC tends to show a high nuclear grade and is associated with deep myometrial invasion and vascular invasion. Occasionally endometrial polyps occur. It is worth mentioning that the risk of venous thromboembolism increases in patients with CCC. Studies have reported the overall 5-year survival rate to range from 55% to 78% [47–49].

DeLair et al. reported that genetic mutations occur in *POLE*, *MMR-D*, and *p53* in endometrial CCC [50]. Although it had been considered a Type 2 endometrial carcinoma, its genomic profile shows that endometrial CCC can be regarded as a tumor with intermediate features between EEC and SC.

11.3.2 Definition

In the fifth edition of the WHO classification of 2020, CCC is defined as a carcinoma with a papillary, tubulocystic, and/or solid architectural pattern and variably pleomorphic, cuboidal, flat, or hobnail cells with clear or eosinophilic cytoplasm [49]. Nuclear atypia is generally moderate to severe, with anisonucleosis and distinct eosinophilic large nucleoli. Atypical mitoses are rarely seen. Deposits of basement membrane-like substances, including type IV collagen and laminin, are found in the stroma in form of eosinophilic hyalinized material [51, 52].

11.3.3 Cytologic Diagnostic Criteria (Figs. 11.37, 11.38, 11.39, 11.40 and 11.41)

- Sheet-like clusters or small papillary clusters with mild nuclear overlapping.
- Tumor cells have abundant and clear cytoplasm with oval to round nuclei with eosinophilic large nucleoli, and finely granular chromatin.
- Hobnail tumor cells protruding from the margin of clusters and a low N/C ratio.

11.3.4 Explanatory Note

EEC with clear cell areas secondary to secretory changes and squamous differentiation should be differentiated from CCC.

Immunohistochemically, endometrial CCC shows usually a negative or reduced expression of estrogen receptor (ER) and progesterone receptor (PgR), whereas it is frequently positive for hepatocyte nuclear factor-1 beta (HNF-1β) and Napsin A;

Fig. 11.37
CCC. Irregularly shaped
sheet-like cluster of tumor
cells is seen. Almost all
tumor cells have abundant
clear cytoplasm.
(Papanicolaou stain,
original magnification 40×)

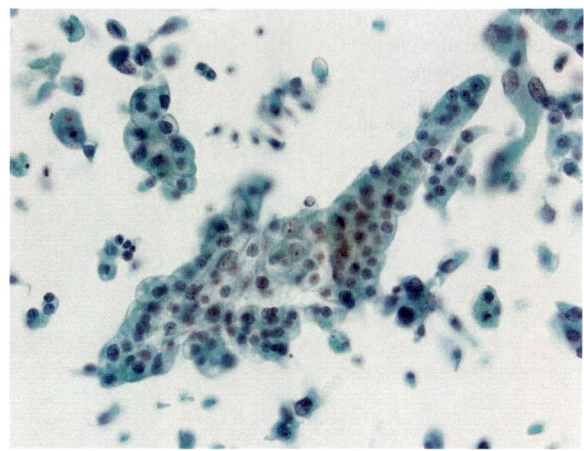

Fig. 11.38
CCC. Corresponding
histologic preparation
shows a complex papillary
pattern and tubular
structures. (HE stain,
original magnification **a**
and **b**: 20×)

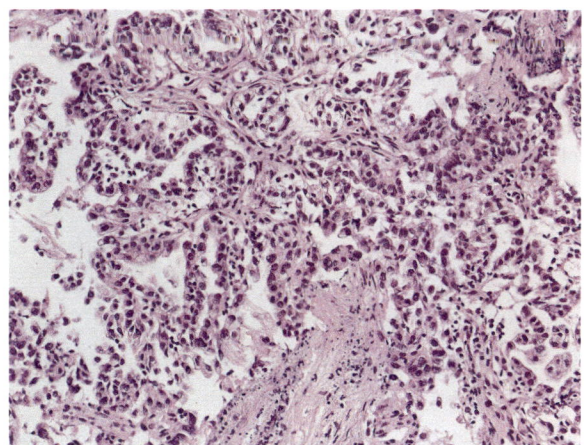

Fig. 11.39 CCC. Tumor
cells have abundant clear
or pale eosinophilic
cytoplasm, and also show
increased nuclear size
with, anisonucleosis.
(Papanicolaou stain,
original magnification 40×)

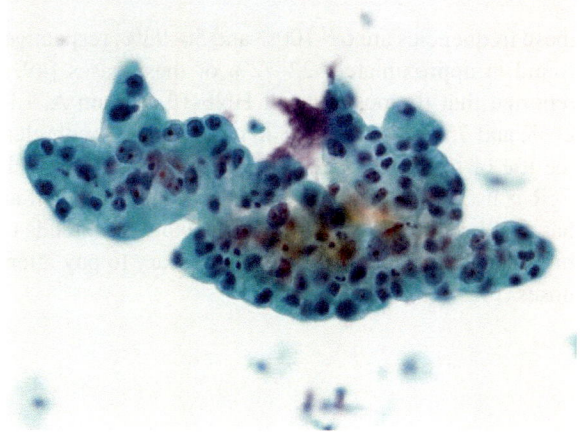

Fig. 11.40 CCC. Tumor
cell cluster shows mild
nuclear overlapping.
(Papanicolaou stain,
original magnification 40×)

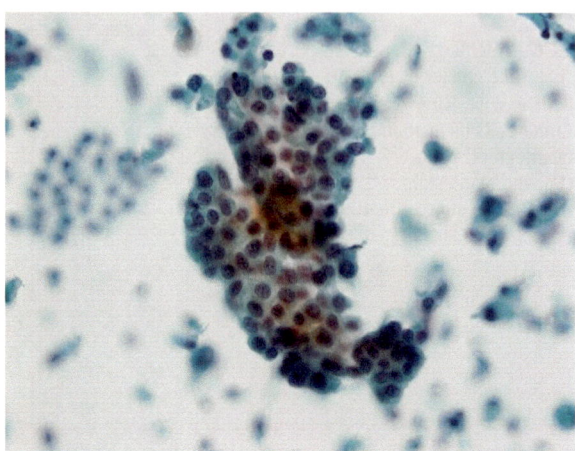

Fig. 11.41 CCC. Tumor
cell clusters show mild
nuclear overlapping.
Tumor cells show.
Increased nuclear size,
pleomorphism, fine
granular chromatin, and
conspicuous nucleoli.
(Papanicolaou stain,
original magnification 40×)

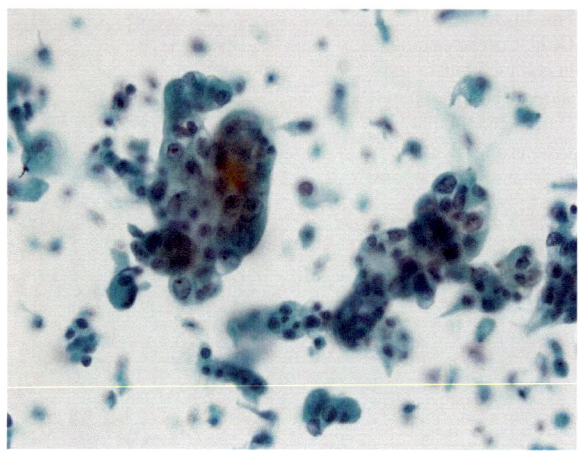

these frequencies are 67–100% and 56–93%, respectively. Overexpression of *p*53 is
found in approximately 22–72% of these cases [49, 53]. A study by Lim et al.
reported that the positivity of HNF-1β, Napsin A, ER, and PgR was 43%, 14%,
86%, and 75%, respectively, in cases of EEC with clear cell areas [54]. Therefore,
the use of immunocytochemical panels composed of HNF-1β, Napsin A, ER, and
PgR is useful for distinguishing EEC with clear cell areas from CCC. However, it
has also been reported that HNF-1β expression tends to be also frequent in SC and
high-grade EEC, and it is hence necessary to pay attention to the differential diagnoses (Figs. 11.42 and 11.43).

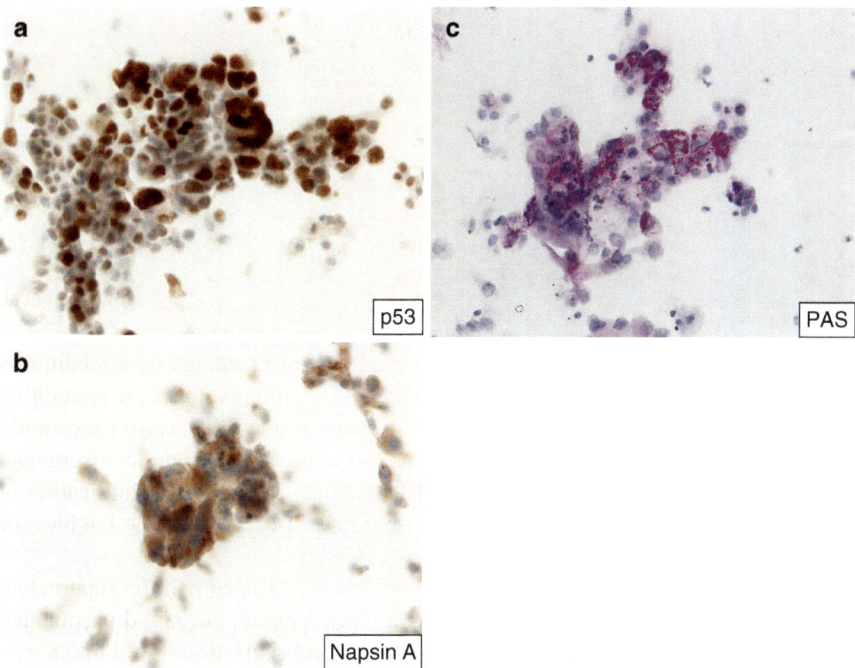

Fig. 11.42 CCC. Cytologic preparation from the same sample of Figs. 11.39, 11.40 and 11.41, (**a**): approximately half of tumor cells exhibit strong nuclear expression of *p53*. (**b**): tumor cells are stained for Napsin A. (**c**): tumor cells have PAS-positive glycogen in cytoplasm. (**a** and **b**: ICC, original magnification 40×, **c**: PAS reaction, original magnification 40×)

The Arias-Stella reaction (ASR) and metaplastic changes due to hormonal or irritative stimulation are also difficult to differentiate from CCC. Because these are benign lesions, overdiagnosis should be avoided. In ASR, epithelial cell clusters are composed of cells with clear or vacuolated abundant cytoplasm containing glycogen. The nuclei show some degree of atypia, with an irregular shape, anisonucleosis, relative hyperchromasia, and presence of intranuclear cytoplasmic inclusions [55]. Philip et al. reported that HNF-1β and Napsin A are highly expressed in ASR (100% and 96%, respectively). Expression of the ER and PgR is also reduced or absent [56]. Because of the overlapping IHC profile of ASR, immunohistochemical studies for differentiated CCC are limited. Clinical information, such as the presence or absence of pregnancy or hormonal drug use, is important. On the other hand, metaplastic changes with large nucleoli mimicking CCC are positive for ER, PgR, and negative for Napsin A and HNF-1β. This expression pattern is a useful ancillary finding for distinguishing CCC (Figs. 11.44 and 11.45).

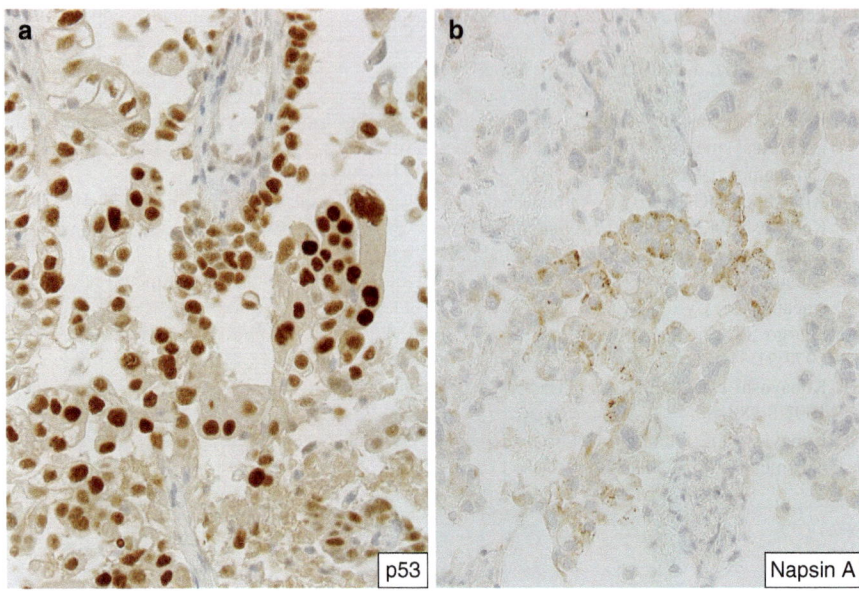

Fig. 11.43 CCC. Corresponding histologic preparation Figs. 11.39, 11.40 and 11.41, (**a**): approximately almost tumor cells exhibit strong nuclear expression of *p*53. (**b**): tumor cells are stained for Napsin A. (IHC, original magnification **a** and **b**: 20×)

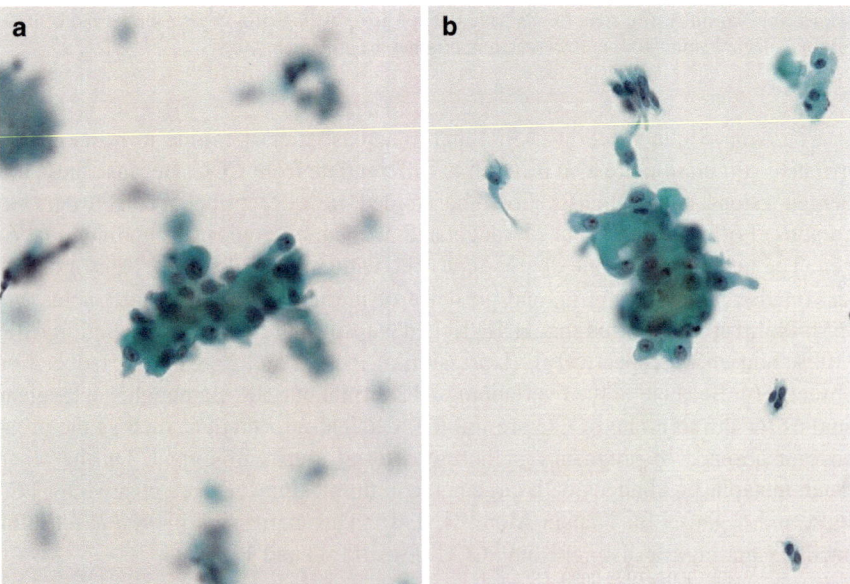

Fig. 11.44 (**a**) Metaplastic epithelial cells with abundant pale eosinophilic cytoplasm should be differentiated from CCC. (**b**) These epithelial cells show nuclear enlargement with prominent nucleoli. (Papanicolaou stain, original magnification **a** and **b**: 40×)

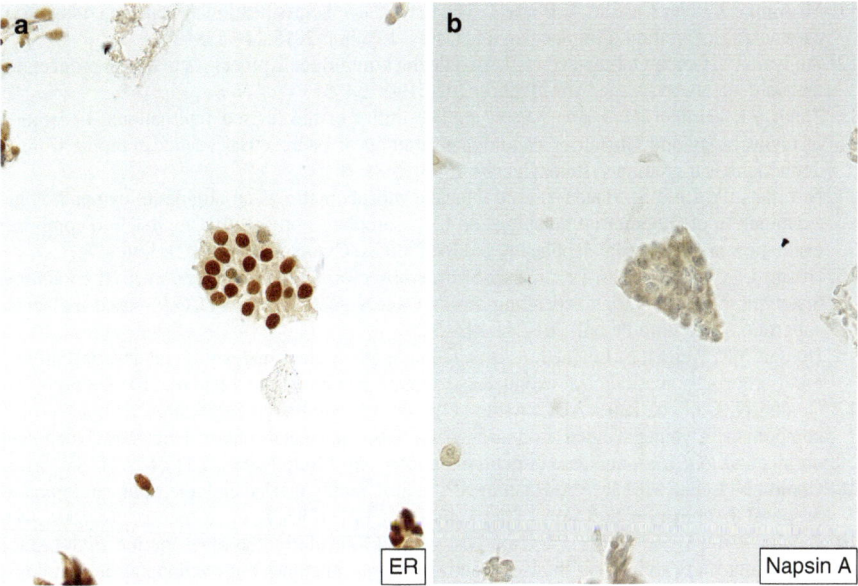

Fig. 11.45 Cytologic preparation from the same sample of Fig. 11.44, (**a**): approximately almost all epithelial cells show expression of ER. (**b**): Napsin A is not expressed. (ICC, original magnification **a** and **b**: 40×)

References

1. Bosse T, Lortet-Tieulent J, Davidson B, et al. Endometrioid carcinoma of the uterine corpus. In: WHO Classification of Tumor Editorial Board, editor. WHO classification of tumor of female reproductive organs. 5th ed. Lyon: IARC Press; 2020. p. 252–5.
2. Ellenson LH, Ronnett BM, Soslow RA, et al. Endometrial carcinoma. In: Kurman RJ, Ellenson LH, Ronnett BM, editors. Blaustein's pathology of the female genital tract. 7th ed. New York: Springer-Verlag; 2019. p. 473–533.
3. Bokhman JV. Two pathogenetic types of endometrial carcinoma. Gynecol Oncol. 1983;15:10–7.
4. Murali R, Soslow RA, Weigelt B. Classification of endometrial carcinoma: more than two types. Lancet Oncol. 2014;15:e268–78.
5. Kandoth C, Schultz N, Cherniack AD, et al. Integrated genomic characterisaion of endometrial carcinoma. Nature. 2013;497:57–73.
6. Norimatsu Y, Yanoh K, Kobayashi TK. The role of liquid-based preparation in the evaluation of endometrial cytology. Acta Cytol. 2013;57:423–35.
7. Yanoh K, Norimatsu Y, Munakata S, et al. Evaluation of endometrial cytology prepared with the Becton Dickinson SurePath™method:a pilot study by the Osaki study group. Acta Cytol. 2014;58:153–61.
8. Fulciniti F, Yanoh K, Karakitsos P, et al. The Yokohama system for reporting directly sampled endometrial cytology: the quest to develop a standardized terminology. Diagn Cytopathol. 2018;46:400–12.
9. Norimatsu Y, Kouda H, Kobayashi TK, et al. Utility of liquid-based cytology in endometrial pathology: diagnosis of endometrial carcinoma. Cytopathology. 2009;20:395–402.
10. Norimatsu Y, Yanoh K, Hirai Y, et al. A diagnostic approach to endometrial cytology by means of liquid-based preparations. Acta Cytol. 2020;64:195–207.

11. McAlpine J, Leon-Castillo A, Bosse T. The rise of novel classification system for endometrial carcinoma; integration of molecular subclasses. J Pathol. 2018;244:538–49.
12. Sugiyama Y, Gotoh O, Fukui N, et al. Two distinct tumorigenic processes in endometrial endometrioid adenocarcinoma. Am J Pathol. 2020;190:234–51.
13. Zaino RJ, Kurman RJ, Diana KL, et al. The utility of the revised International Federation of Gynecology and Obstetrics histological grading of endometrial adenocarcinoma using a defined nuclear grading system. Cancer. 1995;75:81–6.
14. Norimatsu Y, Irino S, Maeda Y, et al. Nuclear morphometry as an adjunct to cytopathologic examination of endometrial brushings on LBC samples: a prospective approach to combined evaluation in endometrial neoplasms and look alikes. Cytopathology. 2021;32:65–74.
15. Hoang LN, Kinloch MA, Leo JM, et al. Interobserver agreement in endometrial carcinoma histotype diagnosis varies depending on the Cancer genome atlas (TCGA)-based molecular subgroup. Am J Sure Pathol. 2017;41:245–52.
16. Hussein YR, Weigelt B, Levine DA, et al. Clinicopathological analysis of endometrial carcinomas harboring somatic *POLE* exonuclease domain mutations. Mod Pathol. 2015;28:505–14.
17. Conlon N, Da Cruz Paula ADC, Ashley CW, et al. Endometrial carcinomas with a "serous" component in young women are enriched for DNA mismatch repair deficiency, lynch syndrome, and *POLE* exonuclease domain mutations. Am J Surg Pathol. 2020;44:641–8.
18. Conlon N, Leitao MM Jr, Abu-Rustum NR, et al. Grading uterine endometrioid carcinoma. A proposal that binary is best. Am J Surg Pathol. 2014;38:1583–7.
19. Joehlin-Price A, Van Ziffle JV, Hills NK, et al. Molecularly classified uterine FIGO grade 3 endometrioid carcinomas show distinctive clinical outcomes but overlapping morphologic features. Am J Surg Pathol. 2021;45:421–9.
20. Bosse T, Nout RA, McAlpine JN, et al. Molecular classification of grade 3 endometrioid endometrial cancers identifies distinctive prognostic subgroups. Am J Surg Pathol. 2018;42:561–8.
21. Soslow RA, Tornos C, Park KJ, et al. Endometrial carcinoma diagnosis: use of FIGO grading and genomic subcategories in clinical practice: recommendations of the international society of gynecological pathologists. Int J Gynecol Pathol. 2019;38(Suppl 1):S64–S74.
22. Norimatsu Y, Miyamoto M, Kobayashi TK, et al. Diagnostic utility of phosphatase and tensin homolog, β-catenin, and p53 of endometrial carcinoma by thin-layer endometrial preparations. Cancer. 2008;114:155–64.
23. Di Lorito AD, Rosini S, Falo E, et al. Molecular alterations in endometrial archived liquid-based cytology. Diagn Cytopathol. 2013;41:492–6.
24. Di Lorito AD, Zappacosta R, Capanna S, et al. Expression of PTEN in endometrial liquid-based cytology. Acta Cytol. 2014;58:495–500.
25. Lu Li YW, Douville C, Cohen JD, et al. Evaluation of liquid from the Papanicolaou test and other liquid biopsies for detection of endometrial and ovaria cancers. Sci Transl Med. 2018;10:eaap8793. https://doi.org/10.1126/scitranslemd.aap8793.
26. Bokhman JV. Two pathogenetic types of endometrial carcinoma. Gynecol Oncol. 1983;15:10–7.
27. Hendrickson M, Ross J. Eifel, et al. Uterine papillary serous carcinoma. A highly malignant form of endometrial adenocarcinoma. Am J Surg Pathol. 1982;6:93–108.
28. Sherman ME, Rosenshein NB, Kurman RJ, et al. Uterine serous carcinoma. A morphologically diverse neoplasm with unifying clinicopathological features. Am J Surg Pathol. 1992;16:600–10.
29. Ueda SM, Kapp DS, Cheung MK, et al. Trends in demographic and clinical characteristics in women diagnosed with corpus cancer and their potential impact on the increasing number of deaths. Am J Obstet Gynecol. 2008;198:218.e1–6.
30. Moll UM, Chalas E, Auguste M, et al. Uterine papillary serous carcinoma evolves via a p53-driven pathway. Hum Pathol. 1996;27:1295–300.
31. Tashiro H, Isacson C, Levine R, et al. p53 gene mutation in uterine serous carcinoma and occurs early in their pathogenesis. Am J Pathol. 1997;150:177–85.

32. Kandoth C, Schultz N, Cherniack AD, et al. Integrated genomic characterizaion of endometrial carcinoma. Nature. 2013;497:57–73.
33. Ellenson LH, Parkash V, Stewart CJR. Serous carcinoma of the uterine corpus. In: WHO Classification of Tumor Editorial Board, editor. WHO classification of tumor of female reproductive organs. 5th ed. Lyon: IARC Press; 2020. p. 256–7.
34. Ellenson LH, Ronnett BM, Soslow RA, et al. Endometrial carcinoma. In: Kurman RJ, Ellenson LH, Ronnett BM, editors. Blaustein's pathology of the female genital tract. 7th ed. New York: Springer-Verlag; 2019. p. 473–533.
35. Zaino R, Carinelli SG, Ellenson LH, et al. Epithelial tumours and precursors. In: Kurman RJ, Carangiu ML, Herrington CS, et al., editors. WHO classification of tumor of female reproductive organs. 4th ed. Lyon: IARC Press; 2014. p. 125–35.
36. Carcangiu ML, Tan LK, Chambers JT. Stage IA uterine serous carcinoma: A study of 13 cases. Am J Surg Pathol. 1997;21:1507–14.
37. Wheeler DT, Bell KA, Kurman RJ, et al. Minimal uterine serous carcinoma. Diagnosis and clinicopathologic correlation. Am J Surg Pathol. 2000;24:797–806.
38. Hui P, Kelly M, O'Malley DM, et al. Minimal uterine serous carcinoma: a clinicopathologic study of 40 cases. Mod Pathol. 2005;18:75–82.
39. Zheng W, Xiang L, Fadare O, et al. A proposed model of endometrial serous carcinogenisis. Am J Surg Pathol. 2011;35:e1–e14.
40. Yasuda M, Katoh T, Hori S, et al. Endometrial intraepithelial carcinoma in association with polyp: review of eight cases. Diagn Pathol. 2013;8:25–31.
41. Umezawa T, Nomura K, Tsuchiya S, et al. Serous endometrial intraepithelial carcinoma. J Jpn Soc Clin Cytol. 2012;51:188–91. (in Japanese with English abstract).
42. Zheng W, Yi X, Fadare O, et al. The Oncofetal protein, IMP3. a novel biomarker for endometrial serous carcinoma. Am J Surg Pathol. 2008;32:304–15.
43. Yomelyanova A, Ji H, Shih IM, et al. Utility of p16 expression for distinction of uterine serous carcinomas from endometrial endometrioid and endocervical adenocarcinomas. Immunohistochemical analysis of 201 cases. Am J Surg Pathol. 2009;33:1504–14.
44. Fadare O, Zheng W. Endometrial glandular dysplasia (EmGD):morphological and biologically distinctive putative precursor lesion of Type II endometrial cancers. Diagn Pathol. 2008;3:6.
45. Jarboe EA, Miron A, Monte N, et al. Evidence for a latent precursor (p53 signature) that may precede serous endometrial intraepithelial carcinoma. Mod Pathol. 2009;22:345–50.
46. Silverberg SG, De Giorgi LS. Clear cell carcinoma of the endometrium. Clinical, pathologic, and ultrastructural findings. Cancer. 1973;31:1127–40.
47. Ellenson LH, Ronnett BM, Soslow RA, et al. Endometrial carcinoma. In: Kurman RJ, Ellenson LH, Ronnett BM, editors. Blaustein's pathology of the female genital tract. 7th ed. New York: Springer-Verlag; 2019. p. 473–533.
48. Abeler VM, Vergote IB, Kjorstad KE, et al. Clear cell carcinoma of the endometrium. Prognosis and metastatic pattern. Cancer. 1996;78:1740–7.
49. Fadare O, Stewart CJR. Clear cell carcinoma of the uterine corpus. In: WHO Classification of Tumor Editorial Board, editor. WHO classification of tumor of female reproductive organs. 5th ed. Lyon: IARC Press; 2020. p. 258–9.
50. DeLair DF, Burke KA, Selenica P, et al. The genetic landscape of endometrial clear cell carcinomas. J Pathol. 2017;243:230–41.
51. Ellenson LH, Ronnett BM, Soslow RA, et al. Endometrial carcinoma. In: Kurman RJ, Ellenson LH, Ronnett BM, editors. Blaustein's pathology of the female genital tract. 7th ed. New York: Springer-Verlag; 2019. p. 473–533.
52. Wakui K, Matsui N, Yasuda M, et al. Cytological study of endometrial clear cell carcinoma. Analysis of structural pattern of tumor cells. J Jpn Clin Cytol. 2008;47:269–74. (in Japanese with English abstract).
53. Fadare O, Desouki MM, Gwin K, et al. Frequent expression of Napsin A in clear cell carcinoma of the endometrium: potential diagnostic utility. Am J Surg Pathol. 2014;38:189–96.

54. Lim D, Philip PC, Cheung ANY, et al. Immunohistochemical comparison of ovarian and uterine endometrioid carcinoma, endometrioid carcinoma with clear cell change, and clear cell carcinoma. Am J Surg Pathol. 2015;39:1061–9.
55. Ip PPC, Djordjevic B. Arias-Stella reaction of the uterine corpus. In: WHO Classification of Tumours Editorial Board, editor. WHO Classification of tumor of female reproductive organs. 5th ed. Lyon: IARC Press; 2020. p. 271.
56. Philip PC, Wang SY, Wong OGW, et al. Napsin A, Hepatocyte Nuclear Factor −1-Beta (HNF-1β), Estrogen and Progesteron receptors expression in Arias-Stella raction. Am J Surg Pathol. 2019;43:325–33.

Endometrial Glandular and Stromal Breakdown (EGBD) as Benign Mimics of Malignancy

12

Yoshiaki Norimatsu, Tadao K. Kobayashi, Yasuo Hirai, and Franco Fulciniti

Abbreviations

CAM 5.2	cytokeratin CAM 5.2
CCC	clusters of cancer cells
CSC	condensed stromal clusters
DUB	dysfunctional uterine bleeding
EC	endometrial carcinoma
EGBD	endometrial glandular and stromal breakdown
EH	endometrial hyperplasia
EIC	endometrial intraepithelial carcinoma
ESC	endometrial serous carcinoma
G1-ECC	grade 1 endometrial endometrioid carcinoma
G3-ECC	grade 3 endometrial endometrioid carcinoma
ICC	immunocytochemistry
IHC	immunohistochemistry

Y. Norimatsu (✉)
Department of Medical Technology, Faculty of Health Sciences, Ehime Prefectural University of Health Sciences, Tobe-cho, Iyo-gun, Ehime, Japan
e-mail: ynorimatsu@epu.ac.jp

T. K. Kobayashi
Division of Health Sciences, Cancer Education and Research Center, Osaka University Graduate School of Medicine, Osaka, Japan

Y. Hirai
Department of Obstetrics and Gynecology, Faculty of Medicine, Dokkyo Medical University, Tochigi, Japan

PCL Japan Pathology and Cytology Center, PCL Inc., Saitama, Japan

F. Fulciniti
Clinical Cytology Service, Istituto Cantonale dì Patologia, Ente Ospedaliero Cantonale, Locarno, Switzerland

© The Author(s), under exclusive license to Springer Nature Singapore Pte Ltd. 2022
Y. Hirai, F. Fulciniti (eds.), *The Yokohama System for Reporting Endometrial Cytology*, https://doi.org/10.1007/978-981-16-5011-6_12

IMP3	Insulin-like growth factor-II mRNA-binding protein 3
LBC	liquid-based cytology
LGB	light green body
MCIP	metaplastic clusters with irregular protrusion
SPSC	Surface Papillary Syncytial Change
WT-1	Wilms' tumor 1 protein

12.1 Background

Uterine bleeding that occurs at irregular intervals (which can be prolonged and excessive, or scanty and prolonged) is considered dysfunctional uterine bleeding (DUB) if there is no easily assignable cause. Anovulatory cycles may lead to DUB with changes in the endometrium; in these cases, DUB occurs frequently and many patients are referred to the gynecology outpatient clinics for evaluation [1]. At the time of menopause, anovulatory cycles frequently cause changes in simple atrophic endometrium leading to abnormal bleeding in many cases.

Anovulatory DUB due to raised estrogen levels is more often associated histologically with persistent Proliferative Endometrium (PE) and Endometrial Hyperplasia (EH); they may both lead to breakthrough bleeding [2]. The origin of breakthrough bleeding in PE and EH can be traced to stromal breakdown, which is associated with pools of extravasated erythrocytes following the disruption of the capillaries plugged by platelet/fibrin thrombi [3, 4]. In this aspect, cytologic findings of anovulatory DUB may simulate EH, leading to difficulty in interpretation of directly sampled endometrial mucosa. Hence, correct recognition of the cytologic findings in anovulatory endometrium is also required to rule out EH.

The appearance of atypical glandular cells in cervicovaginal smears was reported in 1975 by Ehrmann [5] in 2 cases with false-positive cytology, which were retrospectively thought to be associated with endometrial stromal breakdown. Since then, there has been no description of atypical cells in the anovulatory endometrium.

12.2 Definition

The cytological features of the detached endometrial fragments that reflect the histological architecture of EGBD are described below.

12.2.1 Condensed Stromal Clusters (CSC)

There are various references to the histological features of DUB [1–4]. The changes associated with anovulatory bleeding, which are referred to as EGBD, are characterized by extensive fragmentation of PE; as the ground substance undergoes dissolution, the normal architecture collapses; isolated, fragmented glands come to lie haphazardly, without any surrounding stroma (Figs. 12.1, 12.2, 12.3 and 12.4). These degenerative stromal cells condense and form compact nests of cells with hyperchromatic nuclei and little or no cytoplasm (Figs. 12.5 and 12.6). The cytoarchitectural features reflecting these histological changes are defined as "condensed stromal clusters (CSC)" (Figs. 12.7 and 12.8) [6–8]. This characteristic cellular finding is significant for the cytologic diagnosis of EGBD and is found rather commonly.

12.2.2 Metaplastic Clusters with Irregular Protrusion (MCIP)

This definition comprises various types of hormonally induced metaplastic changes in the endometrium observed both in benign and premalignant or malignant conditions [9–11]. Frequently, in histological samples of EGBD, metaplastic changes

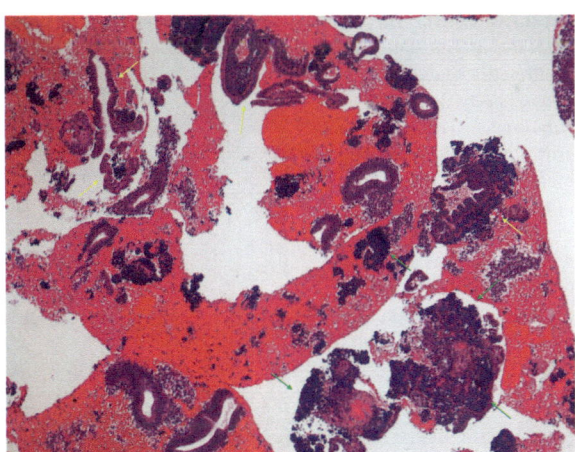

Fig. 12.1 EGBD. There is extensive fragmentation of endometrial glands. Strips of surface epithelium showing SPSC are present (yellow arrows). The stromal cells are condensed into tight clusters (green arrows). These findings are typically found in association with anovulatory cycles. (H&E stain, original magnification 4×)

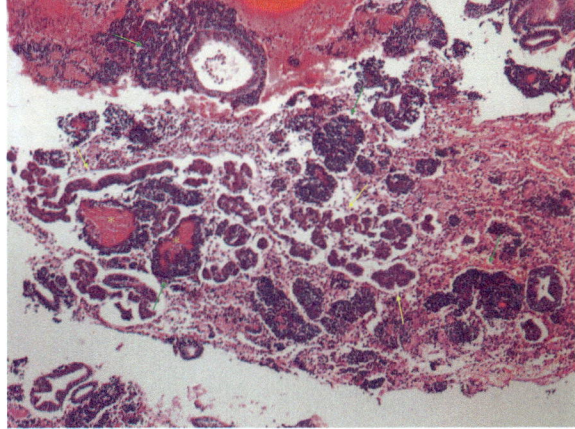

Fig. 12.2 EGBD. There is extensive fragmentation of endometrial glands. Strips of surface epithelium showing SPSC are present (yellow arrows). The stromal cells are condensed into tight clusters (green arrows). Thin-walled ectatic venules may contain prominent fibrin thrombi (*). (H&E stain, original magnification 4×)

Fig. 12.3 EGBD. There is extensive fragmentation of proliferative glands, stromal necrosis, and hemorrhage. (H&E stain, original magnification 4×)

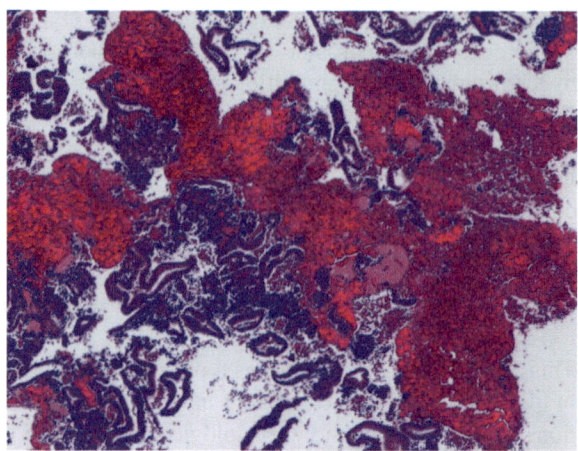

Fig. 12.4 EGBD. There is extensive fragmentation of proliferative glands, stromal necrosis, and hemorrhage. The glandular crowding results from the breakdown and dissolution of the intervening stroma; this should not be confused with crowding as a result of hyperplasia. (H&E stain, original magnification 4×)

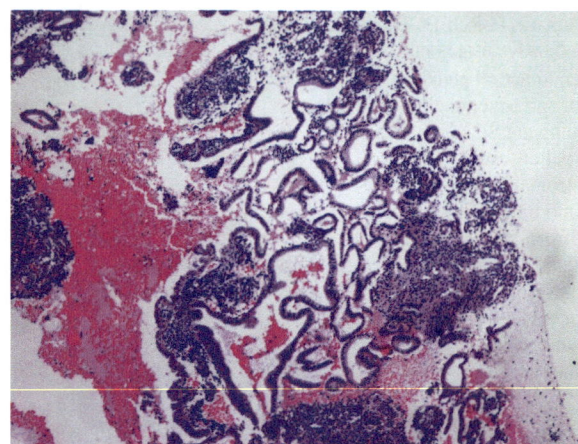

Fig. 12.5 EGBD. Frequently, in the histological samples of EGBD, these degenerative stromal cells condense and form compact cell balls. (H&E stain, original magnification 40×)

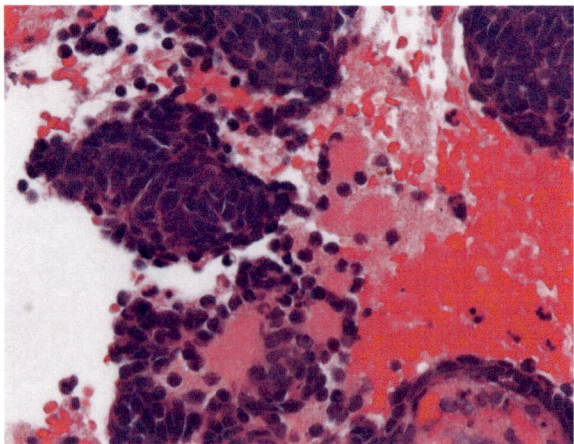

Fig. 12.6 EGBD. The condensed stromal cells show hyperchromatic nuclei with little or no cytoplasm. (H&E stain, original magnification 40×)

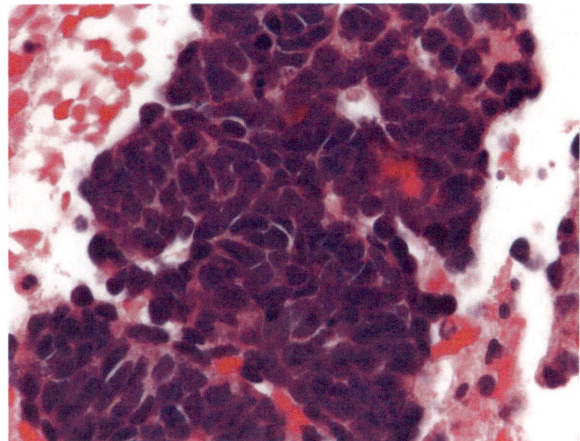

Fig. 12.7 EGBD-condensed stromal clusters (CSC). The cytoarchitectural features reflecting histological findings of Figs. 12.5 and 12.6 are defined as CSC. This characteristic cellular finding is significant for the cytologic diagnosis of EGBD. (Papanicolaou stain, original magnification 40×)

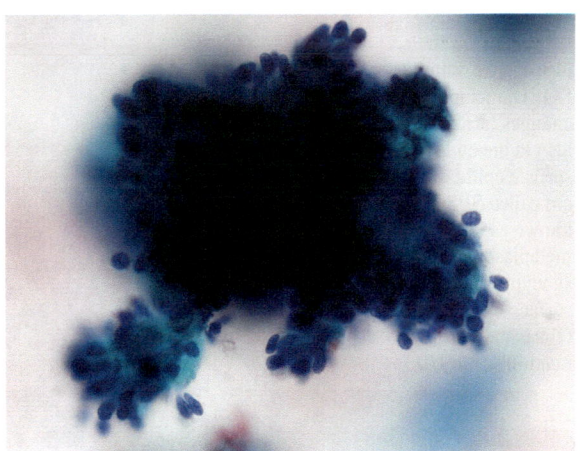

Fig. 12.8 EGBD-CSC. In cytologic samples, irregular protrusion pattern and nuclear overlapping with more than 3 layers are recognized; additionally reniform or spindle nuclei with little or no cytoplasm are observed. The arrow indicates a light green body (LGB). (Papanicolaou stain, original magnification 40×)

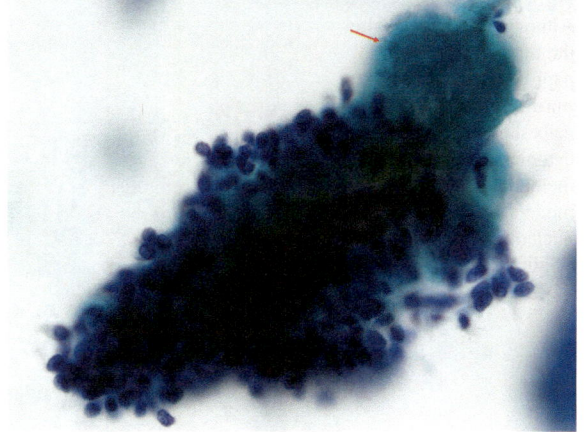

occur also in the endometrial surface epithelium (Figs. 12.9 and 12.10) [1, 10, 11]. These latter are collectively referred to as Surface Papillary Syncytial Change (SPSC), also known as papillary metaplasia, eosinophilic syncytial change, and surface syncytial change.

Although in SPSC the formation of true papillae is lacking (due to the absence of a fibrovascular core), the cells show eosinophilic cytoplasm, indistinct cell membranes, and moderately prominent nucleoli. Occasionally, the nuclei can be enlarged and contain a single prominent nucleolus. Hence SPSC in EGBD may mimic some of the atypical cytologic features found in endometrial intraepithelial carcinoma (EIC) or endometrial atypical hyperplasia.

In a previous study, Norimatsu et al. [7, 12] have found that a higher frequency of metaplastic changes in cellular samples was found in EGBD. In cytologic preparations, these latter consist of glandular cells with thick cytoplasm and spindled or round nuclei with slight to moderate swelling; moderately enlarged nucleoli may be

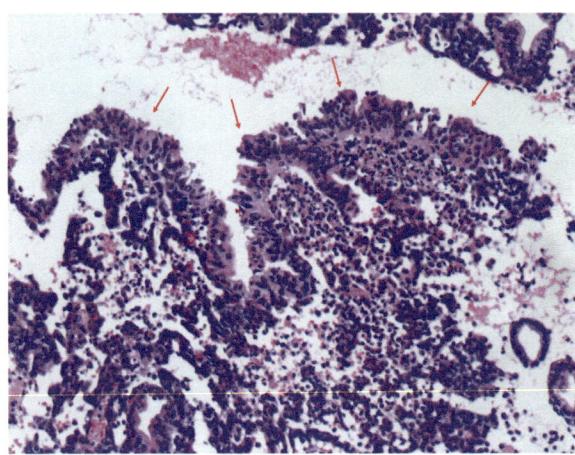

Fig. 12.9 EGBD. Frequently, in the histological samples of EGBD, metaplastic changes (arrows) occur also in the endometrial surface epithelium. They are called SPSC, also known as papillary metaplasia, eosinophilic syncytial change, and surface syncytial change. (H&E stain, original magnification 10×)

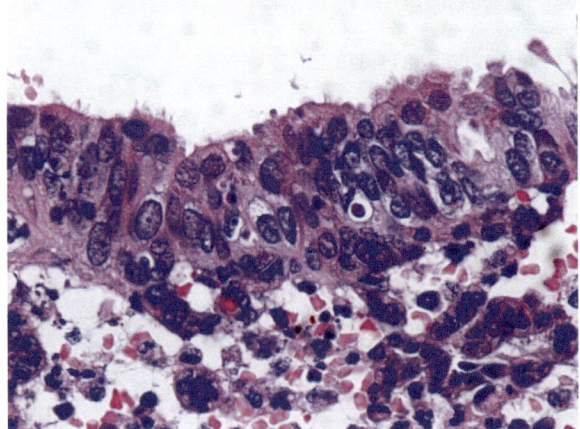

Fig. 12.10 EGBD. Although SPSC is lacking the formation of true papillae (since they lack a fibrovascular core), the cells show eosinophilic cytoplasm, indistinct cell membranes, and moderately prominent nucleoli. (H&E stain, original magnification 60×)

found as sometimes associated with some irregular small epithelial projections which can be seen bulging from the edges of the cell clusters. These findings were defined as "Metaplastic Clusters with Irregular Protrusion (MCIP)" (Figs. 12.11 and 12.12).

In EGBD cases, MCIP appears to be common, occurring in 90.6% of our cases, and they are rather characteristic in comparison with other lesions [7, 12]. Sherman et al. [1] noted that SPSC is an epithelial change that is associated with stromal breakdown and that these lesions typically involve the surface epithelium. Histopathological findings of acute endometrial breakdown, as previously shown by Zaman and Mazur [11] who reported such changes, were intimately admixed with foci of SPSC along the surface epithelium. Due to the similarity between SPSC in histopathology and MCIP, these latter were considered as corresponding to SPSC in cytologic preparations.

Fig. 12.11 EGBD-metaplastic clusters with irregular protrusion (MCIP). The cytoarchitectural features reflecting histological findings of Figs. 12.9 and 12.10 are defined as MCIP. This characteristic cellular finding is significant for the cytologic diagnosis of EGBD. (Papanicolaou stain, original magnification 40×)

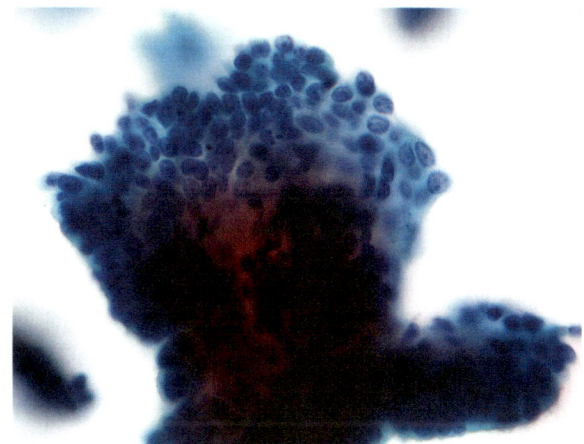

Fig. 12.12 EGBD-MCIP. In cytologic samples, irregular protrusion pattern and nuclear overlapping with more than 3 layers are recognized, additionally also spindled or round nuclei with wide and thick cytoplasm are observed. (Papanicolaou stain, original magnification 40×)

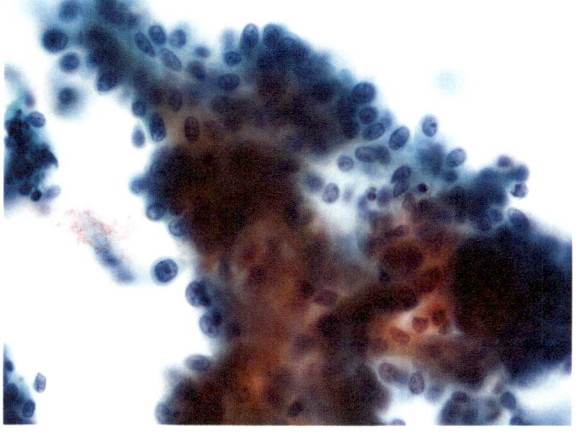

There still remains the problem of understanding the pathogenesis of these changes. Several studies [11, 13, 14] reported that "condensed stromal cells" were contained within SPSCs in histopathological samples of EGBD (Figs. 12.13 and 12.14). Sherman et al. described that SPSC forms microscopic protrusions on the endometrial surface overlying condensed stromal cells or envelop clusters of stromal cells [1]. Zaman and Mazur [11] have shown that the presence of dense clusters of endometrial stroma often with a cap of epithelium is the defining feature of the process of acute endometrial breakdown. Lehman and Hart [14] noted that the presence of rounded clusters of endometrial stromal cells associated with nuclear debris and neutrophils is a characteristic feature. As it concerns the nature of these lesions, it seems that, however, they are neither pathological nor metaplastic, and rather represent a tissue repair response/process following bleeding endometrial breakdown.

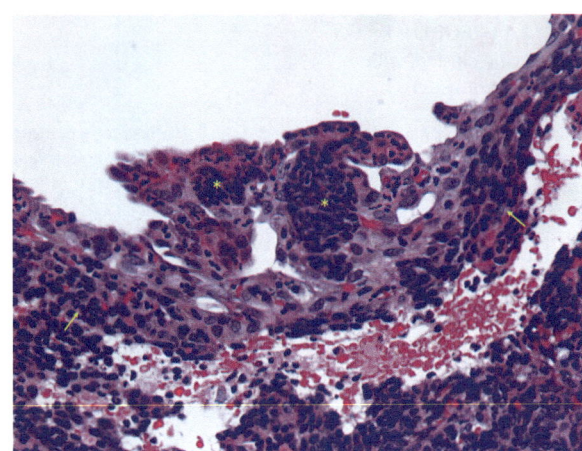

Fig. 12.13 EGBD. Frequently, in the histological samples of EGBD, SPSC form microscopic projections on the endometrial surface overlying condensed stromal cells (yellow arrows) or enveloping cluster of stromal cells (*). (H&E stain, original magnification 20×)

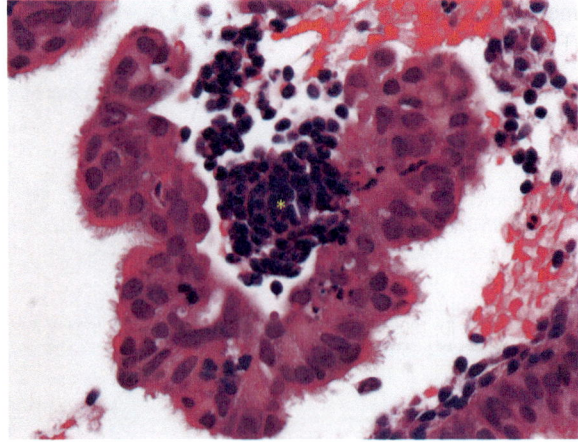

Fig. 12.14 EGBD. The dense clusters of endometrial stroma (*) often with a cap of SPSC epithelium are the defining features of the process of EGBD. (H&E stain, original magnification 40×)

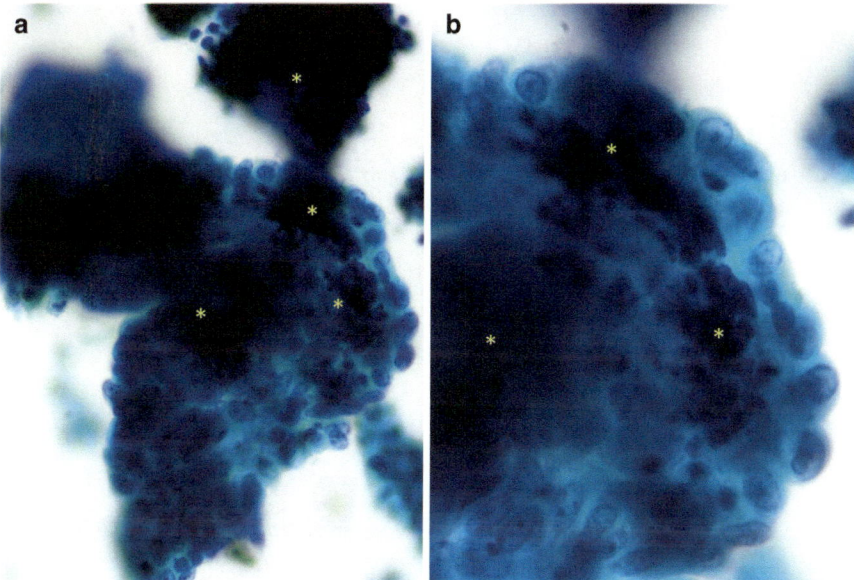

Fig. 12.15 EGBD-MCIP with CSC. When, in cytologic samples, an irregular protrusion pattern with nuclear overlapping >3 layers and nuclear atypia (enlarged nuclei and a single prominent nucleolus) are observed, differential diagnosis with malignancy is necessary. The recognition of an attachment or inclusion of CSC (*) on MCIP can be of great help in confirming the cytologic diagnosis of EGBD. (Papanicolaou stain, original magnification **a**: 40×, **b**: 60×)

Association or co-existence of CSC with MCIP was recognized in 93.1% of EGBD cases [7, 12], and it was a statistically highly significant feature compared to other lesions. Therefore, it is believed that "MCIP with CSC" (Figs. 12.15 and 12.16) originate from SPSC on the endometrial surface epithelium as a response tissue repair following bleeding endometrial breakdown. Their joint finding can hence be of great help in confirming the suggestion of EGBD [7, 12].

12.2.3 Light Green Body (LGB)

In the CSC or background, "Light Green Bodies (LGB)", defined as light green amorphous substance blocks with patchy granular or fibrillary staining pattern, may also be observed [15] (Fig. 12.17). As for the histopathological features of EGBD, it was reported that the endometrium often displays thin-walled ectatic venules that may contain prominent fibrin thrombi, a feature seldom encountered in normal menstrual mucosa (Fig. 12.18) [1–4].

Likewise, Maksem et al. [16] showed that fibrin thrombi are often intimately associated with discrete areas surrounding stromal breakdown in histopathologic preparations, and fibrin thrombi decorated by adherent stromal cells were also identified in cytologic preparations. We speculated that those fibrin thrombi and LGB

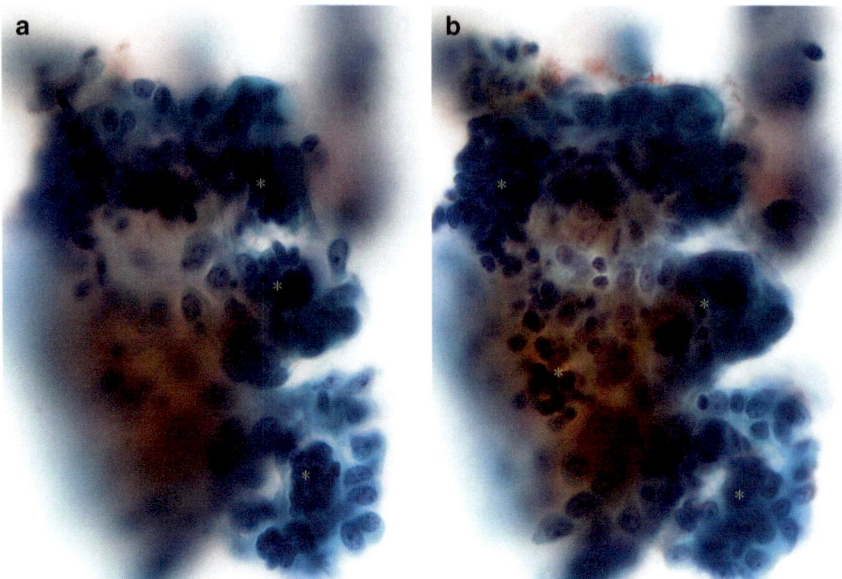

Fig. 12.16 EGBD-MCIP with CSC. The recognition of attachment or inclusion of CSC (*) on MCIP can be of great help in confirming the cytologic diagnosis of EGBD endometrium. (Papanicolaou stain, original magnification **a** and **b**: 40×)

Fig. 12.17 LGB of EGBD. In the CSC or in the background of cytologic samples, LGB (light green amorphous substance blocks with patchy granular or fibrillary staining pattern) may also be observed. Identification of LGB may be another useful diagnostic cytological criterion of EGBD. (Papanicolaou stain, original magnification 40×)

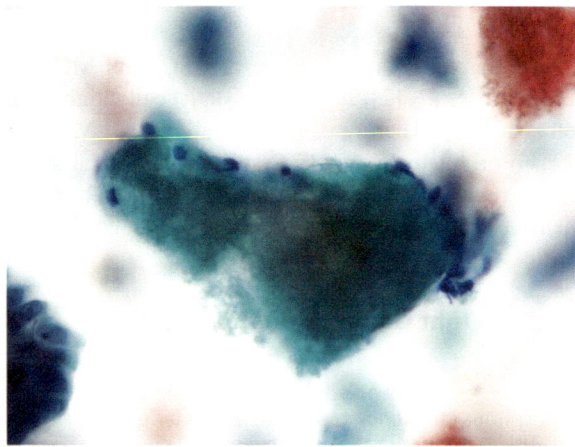

were identical[*1]. The occurrence of CSC including an LGB was 44.8% and an average of 2 LGBs was found per case. In addition, the occurrence of LGBs in the background was 91.4% and an average of 4 LGBs was found per case. Identification of LGB seems to be another useful diagnostic cytological criterion of EGBD.

Fig. 12.18 EGBD. In the histological samples of EGBD, thin-walled ectatic venules that may contain fibrin thrombi can often be found, reflecting the LBG of Fig. 12.17. (H&E stain, original magnification 40×)

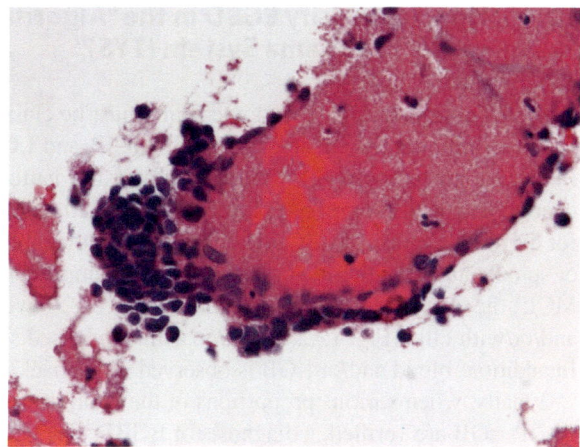

12.3   Criteria

12.3.1 Condensed Stromal Clusters (CSC)

- Degenerative stromal cells condense and form compact nests.
- High diagnostic value in EGBD.

12.3.2 Metaplastic Clusters with Irregular Protrusion (MCIP)

- Metaplastic changes are recognized as structural atypia in cytological preparations.
- MCIP may mimic cell clusters of endometrial carcinoma.

12.3.3 MCIP with Condensed Stromal Clusters (CSC)

- Response of tissue repair following bleeding endometrial breakdown.
- High diagnostic value in EGBD.

12.3.4 Light Green Body (LGB)

- Thrombi mainly formed by clustered platelets.
- A patchy granular or fibrillary staining pattern is observed.

12.4 How to Classify EGBD in the "Algorithmic Approach to the Yokohama System (TYS)"

EGBD, a non-pathological endometrium, must be classified as TYS1. For that reason, it is important to identify the CSC, MCIP, and LGB. As for CSC or MCIP in EGBD, microscopically, an irregular protrusion pattern (Fig. 12.19) and nuclear overlapping with more than 3 layers (Figs. 12.20 and 12.21) may be observed. As for the identification of CSC, the observation of the reniform (Figs. 12.22 and 12.23) or spindle nuclei with little or no cytoplasm is important. As for the identification of MCIP, the observation of spindle nuclei (Fig. 12.24) with wide and thick cytoplasm and/or/with cilia (Fig. 12.25), and/or/with condensed stromal clusters is important. In addition, blood and/or LGB is observed in the background.

Finally, when various proportions of the main three cytological criteria of CSC, MCIP, LGB are verified, a diagnosis of EGBD may be postulated. However, operators unfamiliar with endometrial cytology and the TYS classification may actually classify the CSC or MCIP as TYS4 or TYS5/6 because of an irregular protrusion pattern and nuclear overlapping with more than 3 layers.

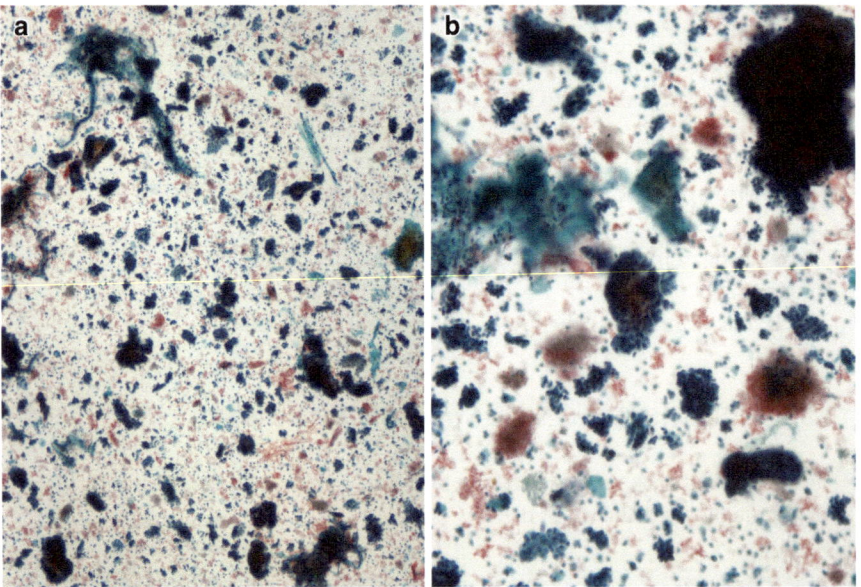

Fig. 12.19 EGBD. At low power microscopic evaluation, an irregular protrusion pattern is recognized. (Papanicolaou stain, original magnification **a**: 4×, **b**: 10×)

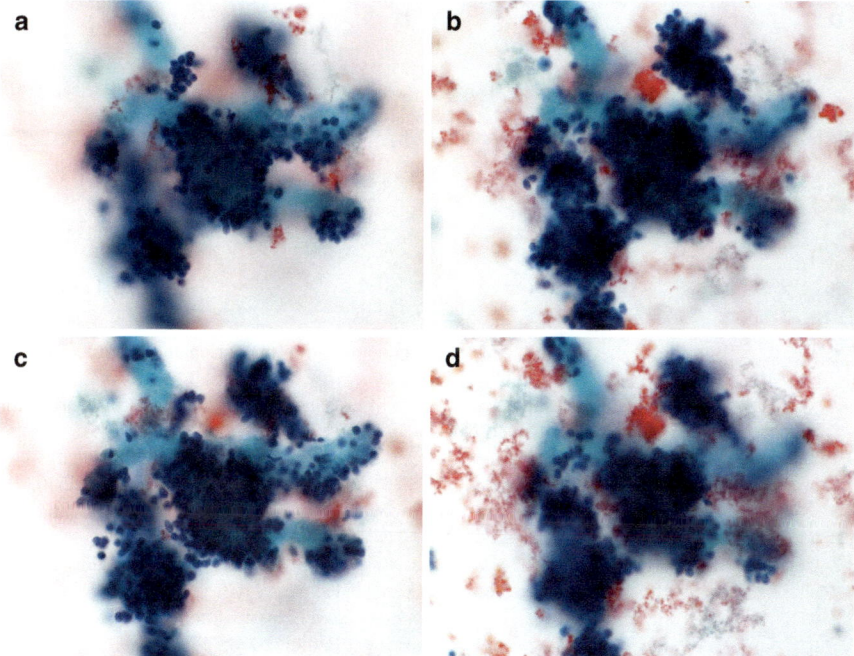

Fig. 12.20 CSC of EGBD. At intermediate microscopic magnification, by fine focusing, nuclear overlapping amounts to more than 3 layers. (Papanicolaou stain, original magnification **a–d**: 40×)

12.5   Differential Diagnosis

As stated before, CSC and MCIP in EGBD cases may simulate the clusters of cancer cells (CCC) in endometrial carcinoma (EC), leading to difficulty in cytological interpretation [10, 12, 17, 18]. Criteria based on the nuclear findings and molecular biology in addition to the cytoarchitectural criteria are hence necessary for a further improvement of the diagnostic accuracy.

12.5.1 Nuclear Findings in LBC: Discrimination Between EGBD and G1-EEC

In order to accurately perform this differential diagnosis, several detailed nuclear criteria must be found in liquid-based cytology (LBC) samples in addition to the cytoarchitectural criteria for EGBD and grade 1-endometrial endometrioid

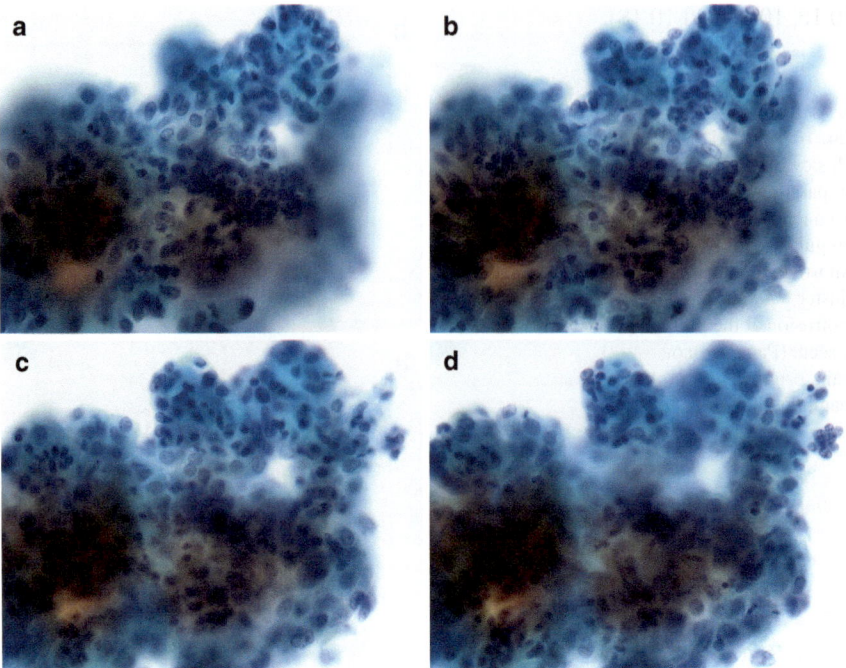

Fig. 12.21 MCIP of EGBD. An irregular protrusion pattern with nuclear overlapping with more than 3 layers is recognized. (Papanicolaou stain, original magnification **a–d**: 40×)

carcinoma (G1-EEC). The following discriminating nuclear findings[*2] should be highlighted [17, 18]: (1) in CSC of EGBD, the nuclei are characteristically small, of reniform or spindle shape, nuclear chromatin tends to be hyperchromatic (Figs. 12.22 and 12.23); (2) in MCIP of EGBD, larger spindle-shaped nuclei are characteristic (Fig. 12.24); (3) as it concerns CCC in G1-EEC, round-oval nuclei appear to predominate.

12.5.2 Immunocytochemical Findings in LBC: Discrimination Between EGBD and EEC

12.5.2.1 CSC of EGBD versus CCC of EEC

CD10 is stated to be a reliable and sensitive immunohistochemical (IHC) marker of normal endometrial stroma and both CD10 [19] and Wilms' tumor 1 protein (WT-1) [20] expression may be of value in diagnosis. Cytokeratin CAM 5.2 (CAM 5.2) may be used as a negative marker in the discrimination between endometrial stromal and smooth muscle tumors of the uterus [21]. Norimatsu et al. [17] studied the immuno-cytochemical (ICC) expression of these markers and found that CSC of EGBD was positive for CD10 and WT-1, while CCC of EEC was positive for CAM5.2. In this

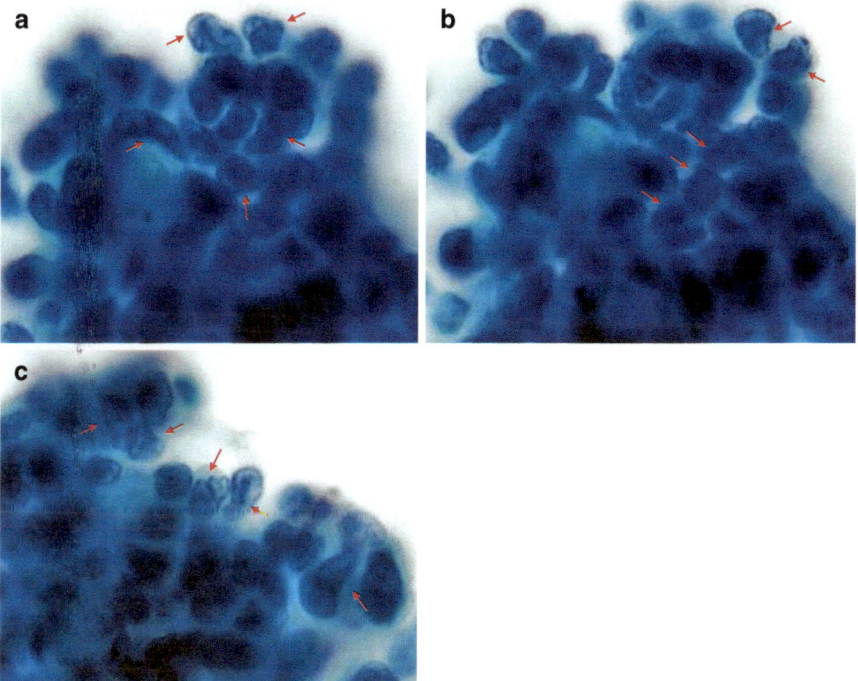

Fig. 12.22 EGBD. At high microscopic magnification, "reniform (arrows) or spindle nuclei with little or no cytoplasm" are observed and a presumptive diagnosis of CSC is made. (Papanicolaou stain, original magnification **a–c**: 100×)

way, a combination of WT-1, CD10 (Fig. 12.26), and CAM5.2 (Fig. 12.27) is useful for the discrimination between CSC of EGBD and CCC of EEC.

12.5.2.2   SPSC (MCIP) of EGBD versus CCC of EEC/ESC

Since MCIP (Figs. 12.28b and 12.29b) of EGBD may occasionally display papillary-like arrangement and some nuclear atypia, resembling endometrial serous carcinoma (ESC) (Fig. 12.28a) or EEC (Fig. 12.29a), it may be misdiagnosed as malignancy [22, 23]. To this concern, p53 overexpression occurs more frequently in non-endometrioid type or poorly differentiated ECs with p53 mutation, so this stain might usefully be exploited for the identification of high-grade ECs [24, 25].

In addition, invasive and noninvasive ESC showed significant overexpression of p53 and Ki-67. Cyclin A expression was involved in the progression to malignancy of the endometrium and was correlated with proliferative activity and prognostic features including histological grade [26]. In previous studies, Zaman et al. [11] reported that SPSC has a low Ki67 index on IHC, and p53 shows a weak and heterogeneous pattern. Norimatsu et al. [27–30] studied the ICC expression of p53 (Fig. 12.30) and cyclin A (Fig. 12.31) for the discrimination between MCIP of EGBD and CCC of EEC/ESC. As a result, since MCIP of EGBD showed a

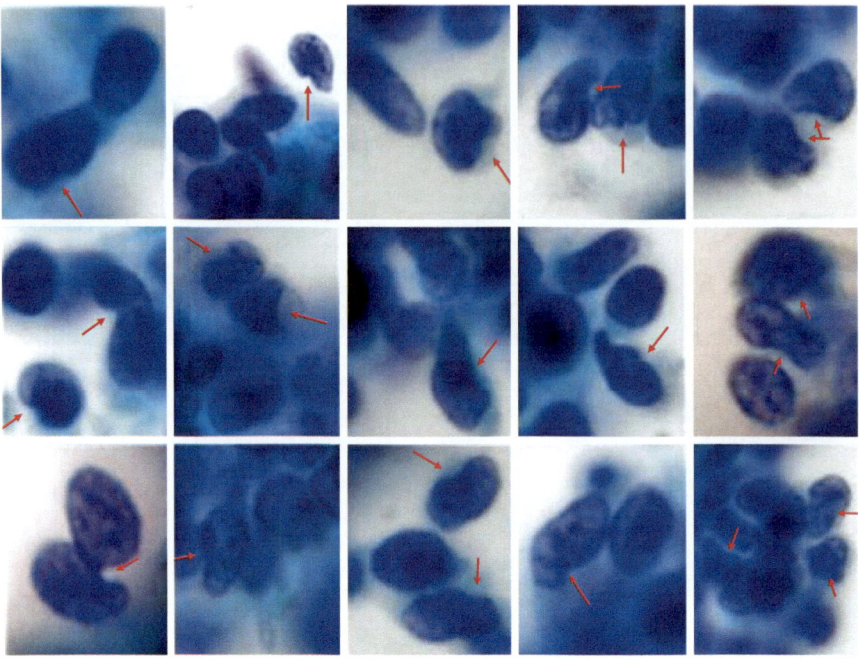

Fig. 12.23 Characteristics of nuclear findings in CSC of EGBD. The adoption or refined nuclear findings is indispensable for accurate LBC differential diagnosis between EGBD and G1-EEC. In CSC of EGBD, the nuclei are characteristically small, of reniform (arrows) or of spindled shape, nuclear chromatin tends to be hyperchromatic. (Papanicolaou stain, original magnification 100×)

significantly lower immunoreactivity for both markers compared with CCC of EEC, they may be helpful for discrimination between EGBD and EEC/ESC.

However, due to the consistent nuclear and cytoplasmic expression of p16 (INK4A) by SPSC, this lesion may be misdiagnosed, in some cases, as surface ESC [31, 32]. Therefore, it is necessary to demonstrate the double negativity of both markers in SPSC. Several authors [33–36] reported that "insulin-like growth factor-II mRNA-binding protein 3 (IMP3)" is a useful ICC marker to distinguish between ESC and EEC (G1, G3). Very recently, Norimatsu et al. [37] also found that IMP3[*3] is one of the helpful ICC markers to distinguish ESC from EEC (G1, G3) or MCIP.

When the score ≥ 3 threshold for IMP3 ICC expression was used, only ESC cases (Fig. 12.32a) showed significantly higher expression than G3- (Fig.12.32b) and G1- (Fig.12.33a) EEC cases, while all of the benign endometrium cases including MCIP showed negative cytoplasmic staining (Fig. 12.33b). This study is the first evaluation in the literature of the ICC expression of IMP3 in a relatively large series of not only EECs but also MCIP. A combination of IMP3 and other biomarkers may be necessary as an ancillary tool in routine endometrial cytology (Figs. 12.34 and 12.35), and this approach seems to represent a promising role in the enhancement of diagnostic reproducibility.

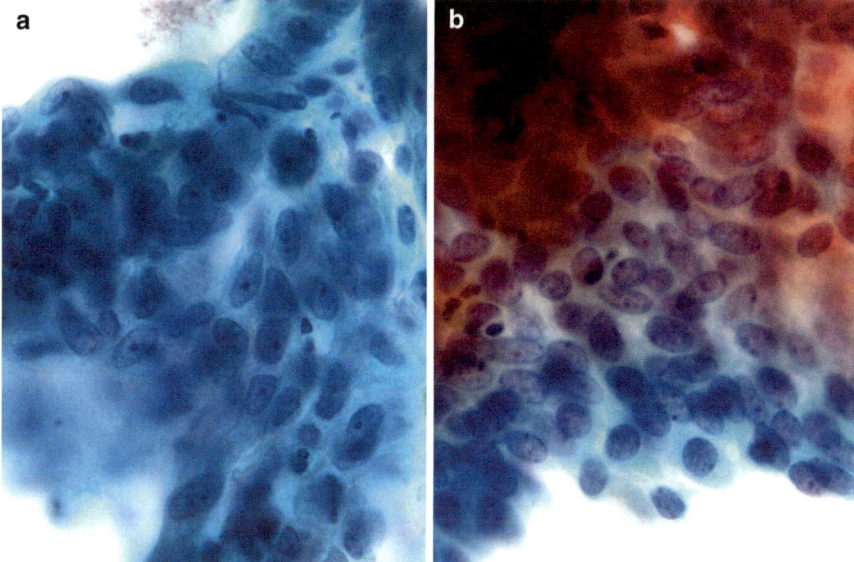

Fig. 12.24 Characteristics of nuclear findings in MCIP of EGBD. Large spindle-shaped nuclei should also be searched and identified in MCIP. (Papanicolaou stain, original magnification **a** and **b**: 60×)

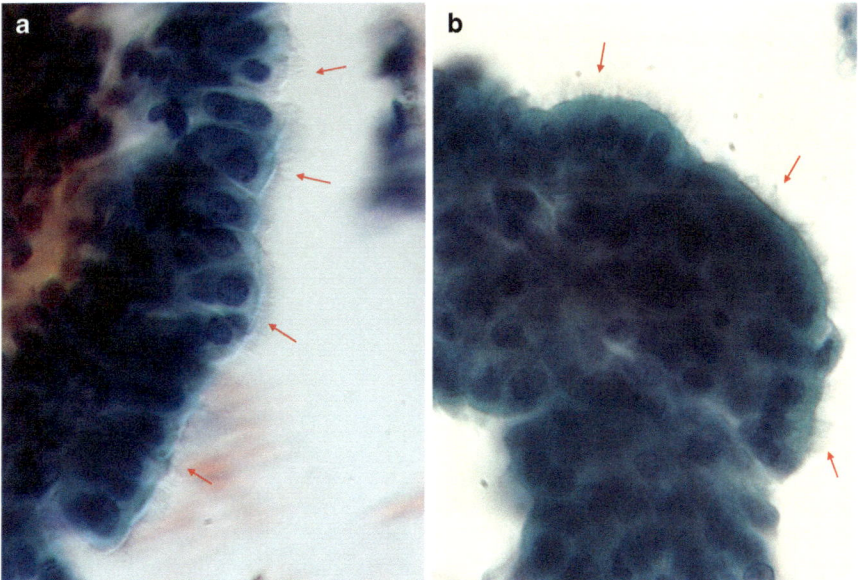

Fig. 12.25 MCIP of EGBD. Ciliated epithelia may also be observed (arrows) is observed in MCIP. (Papanicolaou stain, original magnification **a** and **b**: 60×)

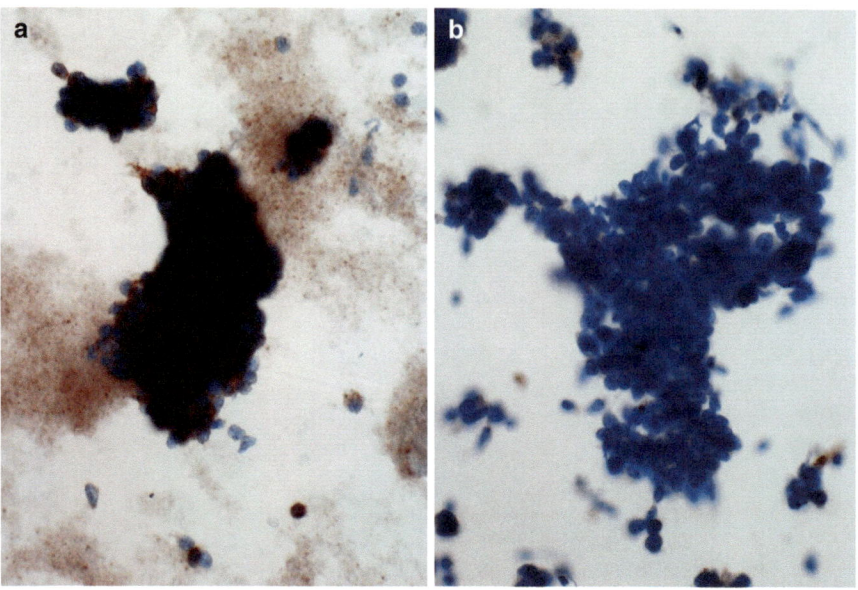

Fig. 12.26 ICC findings (CD10) in CSC of EGBD. As for CD10 immunoreactivity, CSC of EGBD shows brown cytoplasmic staining (**a**), while CCC of G1-EEC shows negative staining (**b**). (ICC stain, original magnification **a** and **b**: 40×)

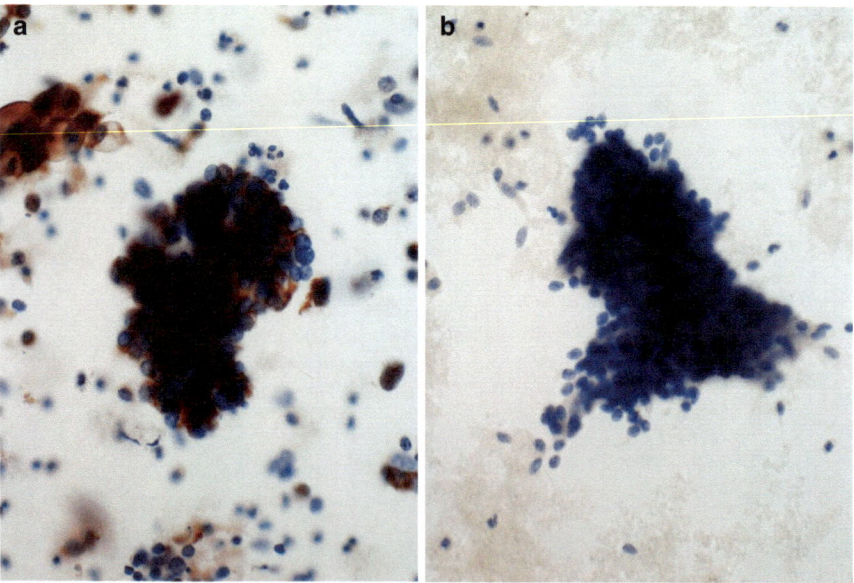

Fig. 12.27 ICC cytokeratin findings in CSC of EGBD. Cytokeratin CAM 5.2 is typically positive for the cytoplasms in CCC of G1-EEC (**a**), while CSC of EGBD shows negative staining (**b**). (ICC stain, original magnification **a** and **b**: 40×)

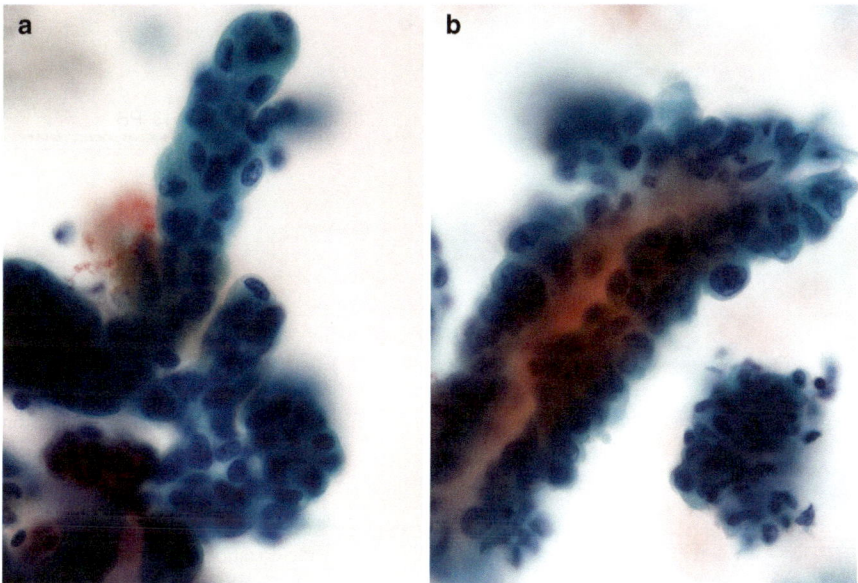

Fig. 12.28 MCIP of EGBD vs CCC of ESC. Since MCIP of EGBD (**b**) may occasionally display papillary-like arrangement and some nuclear atypia, mimicking CCC of ESC (**a**), it may be misdiagnosed as malignancy. (Papanicolaou stain, original magnification **a** and **b**: 40×)

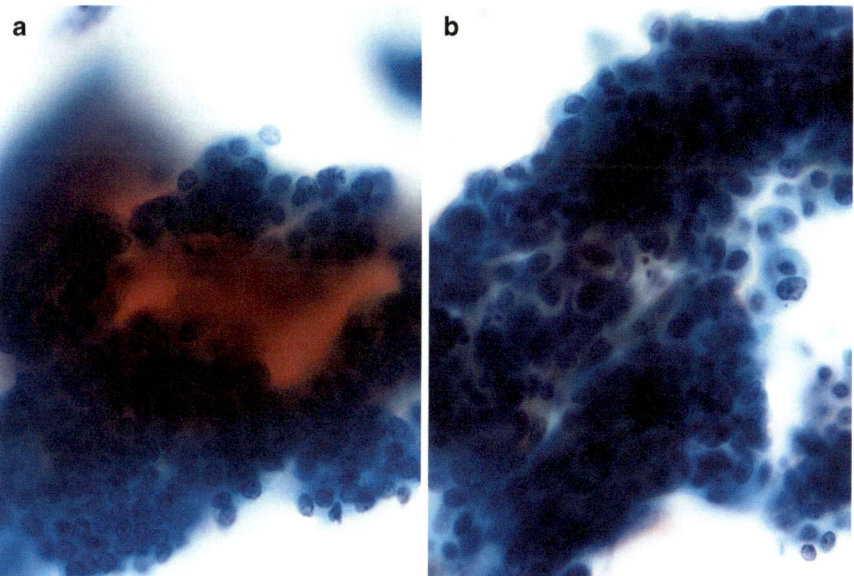

Fig. 12.29 MCIP of EGBD vs CCC of EEC. MCIP of EGBD (**b**) may occasionally display papillary arrangement and some nuclear atypia, resembling CCC of EEC (**a**), hence it may be misdiagnosed as malignancy. (Papanicolaou stain, original magnification **a** and **b**: 40×)

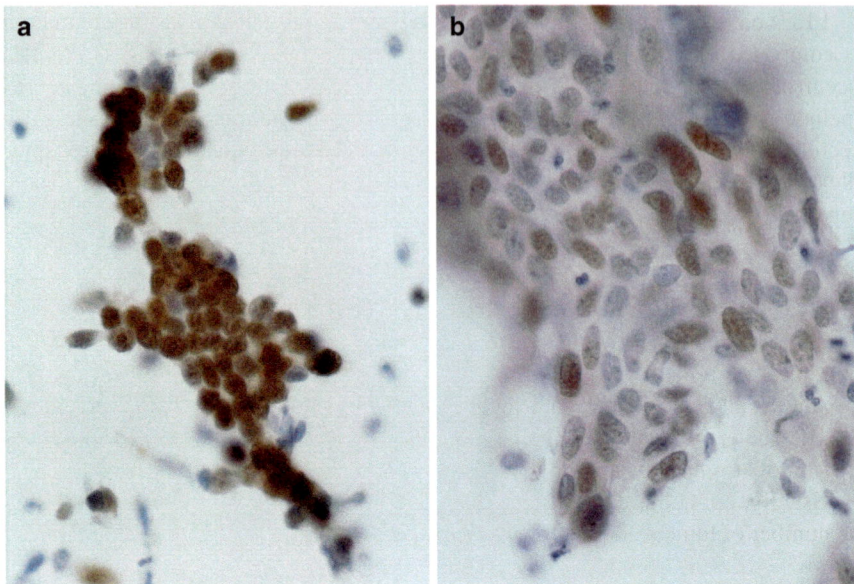

Fig. 12.30 ICC findings (p53) in MCIP of EGBD. As for p53immunoreactivity, MCIP of EGBD (**b**) shows a weak and heterogeneous pattern of nuclei, while CCC of ESC (**a**) shows strong nuclear staining. (ICC stain, original magnification **a** and **b**: 40×)

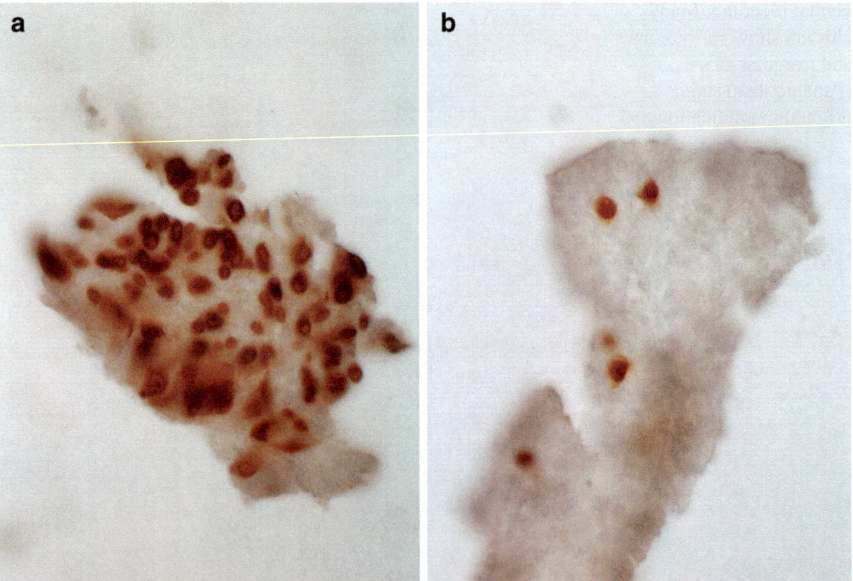

Fig. 12.31 ICC findings (Cyclin A) in MCIP of EGBD. As for cyclin A immunoreactivity, MCIP of EGBD (**b**) shows moderately intense nuclear staining with <5% nuclear labeling index. In contrast, CCC of ESC (**a**) shows moderate to strong nuclear staining with > 50% nuclear labeling index. (ICC stain, original magnification **a** and **b**: 40×)

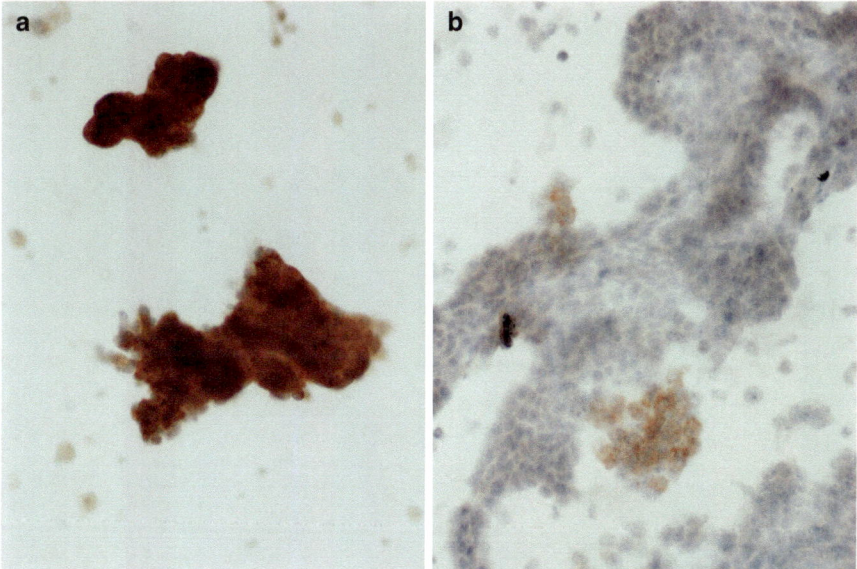

Fig. 12.32 IMP3 ICC findings in MCIP of EGBD. In ESC cases (**a**), tumor cells show diffuse and intense positive cytoplasmic staining, while in G3-EEC (**b**), tumor cells show low-frequency and intensity of the nuclear stain. (ICC stain, original magnification **a** and **b**: 20×)

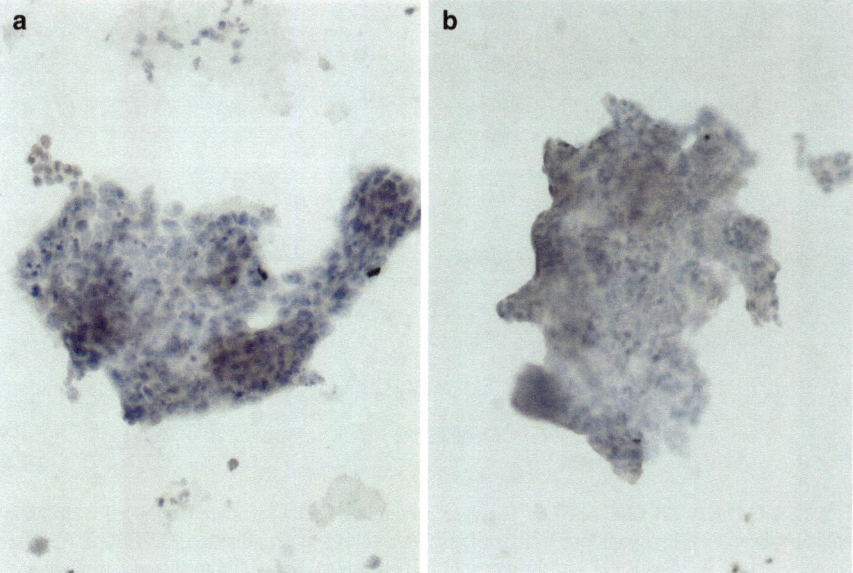

Fig. 12.33 IMP3 ICC findings in MCIP of EGBD. In G1-EEC (**a**) and MCIP of EGBD (**b**), tumor cells show negative staining. (ICC stain, original magnification **a** and **b**: 20×)

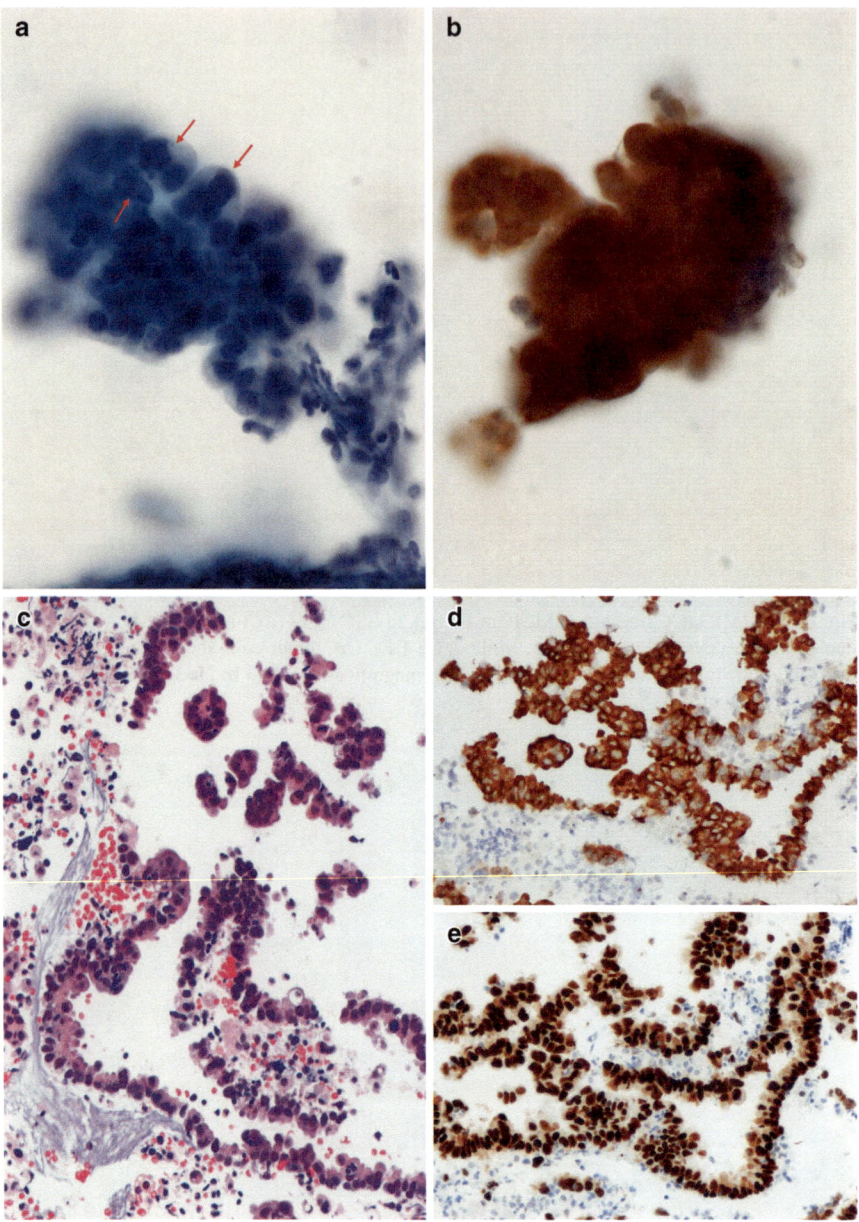

Fig. 12.34 Query G3-EEC or SEC. The case concerns a 68 Y-O woman diagnosed with malignant neoplasm (TYS6) by endometrial cytology. On cytological examination, G3-EEC or SEC was suspected since neoplastic cells in clusters enlarged nuclei (**a**). ICC staining showed strong expression of IMP3 by the neoplastic cells (**b**), and a diagnosis of SEC was strongly favored (Papanicolaou stain (**a**), ICC stain (**b**), original magnification **a** and **b**: 40×).

IHC examination on tissue biopsy showed diffuse expression of the neoplastic cells for IMP3 (**d**) and p53 (**e**), partial expression for estrogen receptor, and no expression for vimentin. ESC was suggested (**c**). The subsequent radical hysterectomy sample showed tubo-ovarian ESC. (H&E stain (**c**), IHC stain (**d** and **e**), original magnification **c–e**: ×20)

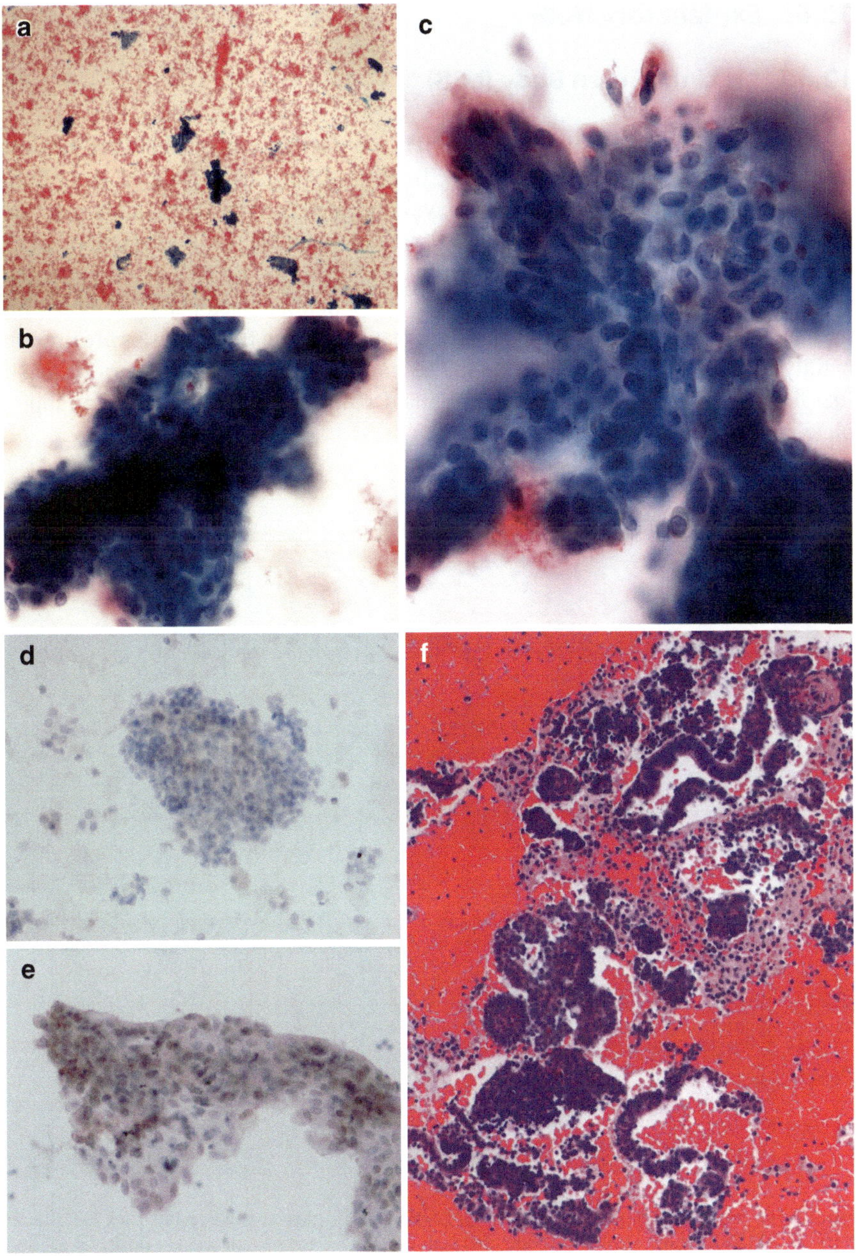

Fig. 12.35 Query MCIP in EGBD. The case concerns a 55 Y-O woman. The endometrial cytology sample shows a few clusters with irregular protrusion patterns (**a**), and nuclear overlapping with > 3 layers in a hemorrhagic background (**b**). Since the spindle-shaped nuclei (**c**) were observed, MCIP of EGBD was suspected. However, the diagnosis of EGBD could not be confirmed because CSC was not found. (Papanicolaou stain (**a–c**), original magnification **a**: ×10; **b** and **c**: ×40).

In the ICC study, IMP3 staining (**d**) was negative and p53 (**e**) showed a weak and heterogeneous pattern. The cytological sample was diagnosed as consistent with MCIP, and subsequent biopsies confirmed the diagnosis of EGBD (**f**). (ICC stain (**d** and **e**), H&E stain (**f**), original magnification **d** and **e**: ×40, **f**: ×20)

12.6   Explanatory Note

12.6.1 *1: Light Green Body (LGB)

As for the nature of LGB (Fig. 12.36a), IHC on paraffin-embedded tissue sections gave positive results for CD31, factor VIII (Fig. 12.36b) and CD42b (Fig. 12.36c), but fibrinogen (Fig. 12.36d) staining gave a negative or weakly positive reaction. The results of ICC on LBC preparations were comparable: LGB stained positively for CD31, factor VIII (Fig. 12.37a, b) and CD42b (Fig. 12.37c, d). In particular, CD42b acts as an adhesion receptor for the von Willebrand factor and thrombin [38].

These above-mentioned results indicate that LGB in an EGBD case were thrombi mainly formed by clustered platelets. The non-uniform granular pattern or mesh-like fiber pattern in LGB could be explained with mesh-like fibrin net entrapping platelets.

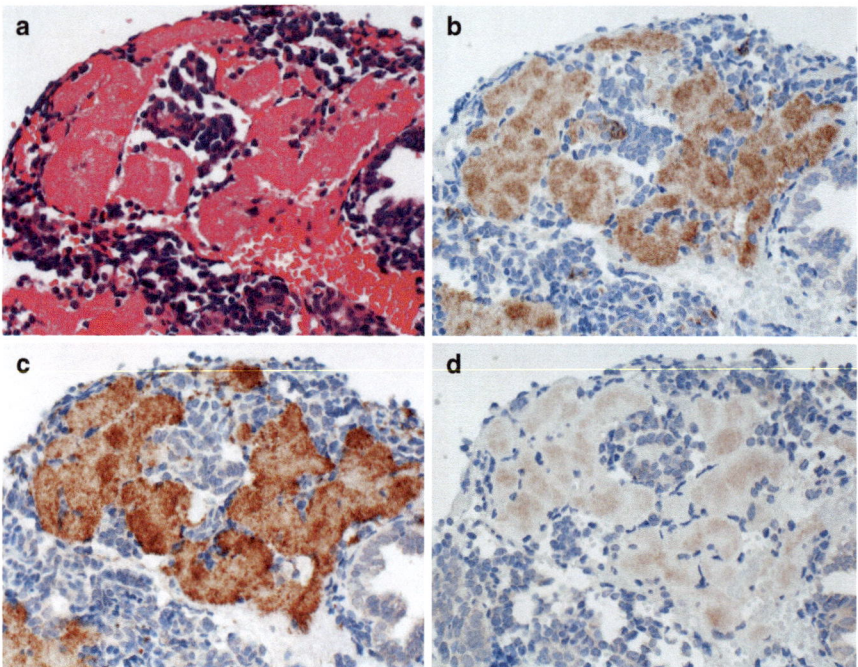

Fig. 12.36 EGBD-LGB. As for the nature of LGB (**a**), IHC gave positive results for CD31, factor VIII (**b**) and CD42b (**c**), while fibrinogen (**d**) staining gave a negative or weakly positive reaction. (H&E stain (**a**), IHC stain (**b–d**), original magnification **a–d**: 40×)

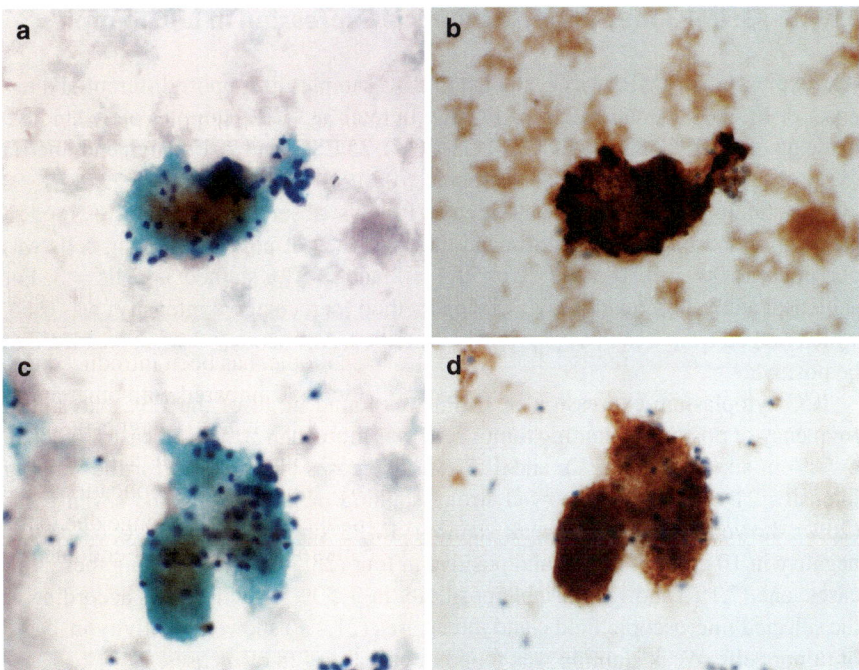

Fig. 12.37 EGBD-LGB. The results of ICC on LBC are comparable: LGB stained positively for factor VIII (**a** and **b**) and CD42b (**c** and **d**). (Papanicolaou stain (**a** and **c**), ICC stain (**b** and **d**), original magnification **a–d**: 20×)

12.6.2 *2: Comparison of the Frequency of Nuclear Shape

As for the 30 cases in each group in the quoted study, five clusters were selected in every case for each group (PE, CSC of EGBD, MCIP of EGBD, and CCC of G1-EEC) at random. With respect to nuclear shape, 50 cells per cell cluster were classified into three subtypes: (1) round-oval, (2) spindle, and (3) reniform, then 5 cell clusters in each case were examined, and the frequency of nuclear shape of 250 cells was calculated.

As for the round-oval shape parameter, CCC of G1-EEC (95.9%) had a significantly higher value in comparison with CSC of EGBD (70.5%) and MCIP of EGBD (82.9%), respectively. As for the spindle shape parameter, its frequency in MCIP of EGBD (17.1%) was significantly higher in comparison with other groups, respectively. As for the reniform shape parameter, its frequency in CSC of EGBD (20.6%) was significantly higher in comparison with other groups, respectively.

12.6.3 *3: IMP3 Immunocytochemical Expression in LBC Samples

ICC expression of IMP3 was evaluated in LBC samples to support differential diagnosis of EEC, ESC, and MCIP of EGBD. In total 333 LBC samples were studied, made up of 97 EEC (83 G1-ECC, 14 G3-ECC), 35 ESC, and 201 benign endometria (51 proliferative, 42 secretory, 38 atrophic, 70 MCIP of EGBD).

The staining intensity was classified as follows: absent = 0, weak = 1+, moderate = 2+ and strong = 3+ reaction and the frequency of positive staining cells was classified as 0% = 0, 1–5% = 1, 6–50% = 2 and > 50% stained/% cells = 3. For statistical analysis, a quantitative scoring method (percentage × intensity) was used, which ranged from 0 to 9, as previously described. Any score ≥ 3 was considered to be positive.

ICC cytoplasmic expression of IMP3 was found in all of the ESC cases. The frequency of positively staining tumor cells was more than 50% in 24 cases (68.6%), 6–50% in six cases (17.1%), and 1–5% in five cases (14.3%). As for the staining intensity, 21 cases (60%) showed strong staining of tumor cells and seven cases (20%) showed moderate or weak staining. IMP3 staining in G3-ECC cases was negative in 10 cases (71.5%), and positive in four (28.5%) cases. Among these four cases, one (7.1%) showed variable positivity in 6–50% of tumor cells according to the selected microscopic fields, and three cases (21.4%) showed positivity in 1–5% of tumor cells. Weak staining was found in only four (28.6%) cases.

IMP3 staining in G1-ECC cases was negative in almost all of the cases (81 cases; 97.6%) and positive only in two (2.4%) cases. Of the two positive cases, one (1.2%) had positive staining in 6–50% or 1–5% of tumor cells according to the selected microscopic fields. IMP3 staining was absent in almost all of G1-ECC cases (81 cases; 97.6%) and only two (2.4%) cases had weak staining. All of the 201 benign endometrial cases including MCIP of EGBD cases showed negative ICC expression of IMP3.

As for the distribution of positive IMP3 expression (with score ≥ 3), only ESC cases showed positive ICC expression (28/35 cases; 80%). These findings were present at a significantly higher frequency in comparison with all the other diagnostic categories (p <0.0001).

References

1. Sherman ME, Mazur MT, Kurman RJ. Benign disease of the endometrium. In: Kurman RJ, editor. Blaustein's pathology of the female genital tract, vol. 5. New York: Springer; 2001. p. 421–66.
2. Vakiani M, Vavilis D, Agorastos T, et al. Histopathological findings of the endometrium in patients with dysfunctional uterine bleeding. Clin Exp Obstet Gynecol. 1996;23:236–9.
3. Livingstone M, Fraser IS. Mechanisms of abnormal uterine bleeding. Hum Reprod Update. 2002;8:60–7.
4. Ferenczy A. Pathophysiology of endometrial bleeding. Maturitas. 2003;45:1–14.
5. Ehrmann RL. Atypical endometrial cells and stromal breakdown two case reports. Acta Cytol. 1975;19:463–9.

6. Norimatsu Y, Kouda H, Kobayashi TK, et al. Utility of thin-layer preparations in the endometrial cytology: evaluation of benign endometrial lesions. Ann Diagn Pathol. 2008;12:103–11.

7. Kobayashi TK, Norimatsu Y, Buccoliero AM. Cytology of the body of the uterus. In: Gray W, Kocjan G, editors. Diagnostic cytopathology. 3rd ed. London: Churchill Livingstone; 2010. p. 689–719.

8. Shimizu K, Norimatsu Y, Kobayashi TK, et al. Endometrial glandular and stromal breakdown, part 1: cytomorphological appearance. Diagn Cytopathol. 2006;34:609–13.

9. Andersen WA, Taylor PT Jr, Fechner RE, et al. Endometrial metaplasia associated with endometrial adenocarcinoma. Am J Obstet Gynecol. 1987;157:597–604.

10. Ronnett BM, Kurman RJ. Precursor lesions of endometrial carcinoma. In: Kurman RJ, editor. Blaustein's pathology of the female genital tract. 5th ed. New York: Springer; 2001. p. 467–500.

11. Zaman SS, Mazur MT. Endometrial papillary syncytial change. A nonspecific alteration associated with active breakdown. Am J Clin Pathol. 1993;99:741–5.

12. Norimatsu Y, Shimizu K, Kobayashi TK, et al. Endometrial glandular and stromal breakdown,part 2: cytomorphology of papillary metaplastic changes. Diagn Cytopathol. 2006;34:665–9.

13. Hendrickson MR, Kempson RL. Endometrial epithelial metaplasias: proliferations frequently misdiagnosed as adenocarcinoma: report of 89 cases and proposed classification. Am J Surg Pathol. 1980;4:525–42.

14. Lehman MB, Hart WR. Simple and complex hyperplastic papillary proliferations of the endometrium: a clinicopathologic study of nine cases of apparently localized papillary lesions with fibrovascular stromal cores and epithelial metaplasia. Am J Surg Pathol. 2001;25:1347–54.

15. Norimatsu Y, Kawai M, Kamimori A, et al. Endometrial glandular and stromal breakdown, part 4: cytomorphology of "condensed cluster of stromal cells including a light green body". Diagn Cytopathol. 2012;40:204–9.

16. Maksem JA, Robboy SJ, Bishop JW, et al. Benign endometrial abnormalities. In: Rosenthal DL, editor. Endometrial cytology with tissue Correlatioms, ed 1. New York: Springer Science and Business Media; 2009. p. 97–152.

17. Norimatsu Y, Yuminamochi T, Shigematsu, et al. Endometrial glandular and stromal breakdown, part 3: cytomorphology of 'condensed cluster of stromal cells'. Diagn Cytopathol. 2009;37:891–6.

18. Norimatsu Y, Shigematsu Y, Sakamoto S, et al. Nuclear features in endometrial cytology: comparison of endometrial glandular and stromal breakdown and endometrioid adenocarcinoma grade 1. Diagn Cytopathol. 2012;40:1077–82.

19. McCluggage WG, Sumathi VP, Maxwell P. CD10 is a sensitive and diagnostically useful immunohistochemical marker of normal endometrial stroma and of endometrial stromal neoplasms. Histopathology. 2001;39:273–8.

20. Sumathi VP, Al-Hussaini M, Connolly LE, et al. Endometrial stromal neoplasms are immunoreactive with WT-1 antibody. Int J Gynecol Pathol. 2004;23:241–7.

21. Oliva E, Young RH, Amin MB, et al. An immunohistochemical analysis of endometrial stromal and smooth muscle tumors of the uterus: a study of 54 cases emphasizing the importance of using a panel because of overlap in immunoreactivity for individual antibodies. Am J Surg Pathol. 2002;26:403–12.

22. Nicolae A, Preda O, Nogales FF. Endometrial metaplasias and reactive changes: a spectrum of altered differentiation. J Clin Pathol. 2011;64:97–106.

23. McCluggage WG, McBride HA. Papillary syncytial metaplasia associated with endometrial breakdown exhibits an immunophenotype that overlaps with uterine serous carcinoma. Int J Gynecol Pathol. 2012;31:206–10.

24. Shih HC, Shiozawa T, Kato K, et al. Immunohistochemical expression of cyclins, cyclin-dependent kinases, tumor-suppressor gene products, Ki-67, and sex steroid receptors in endometrial carcinoma: positive staining for cyclin a as a poor prognostic indicator. Hum Pathol. 2003;34:471–8.

25. Kounelis S, Kapranos N, Kouri E, et al. Immunohistochemical profile of endometrial adeno-carcinoma: a study of 61 cases and review of the literature. Mod Pathol. 2000;13:379–88.
26. Kyushima N, Watanabe J, Hata H, et al. Expression of cyclin a in endometrial adenocarcinoma and its correlation with proliferative activity and clinicopathological variables. J Cancer Res Clin Oncol. 2002;128:307–12.
27. Norimatsu Y, Moriya T, Kobayashi TK, et al. Immunohistochemical expression of PTEN and beta-catenin for endometrial intraepithelial neoplasia in Japanese women. Ann Diagn Pathol. 2007;11:103–8.
28. Norimatsu Y, Miyamoto T, Kobayashi TK, et al. Utility of thin-layer preparations in endome-trial cytology: immunocytochemical expression of PTEN, beta-catenin and p53 for benign endometrial lesions. Diagn Cytopathol. 2008;36:216–23.
29. Norimatsu Y, Miyamoto M, Kobayashi TK, et al. Diagnostic utility of phosphatase and tensin homolog, beta-catenin, and p53 for endometrial carcinoma by thin-layer endometrial prepara-tions. Cancer. 2008;114:155–64.
30. Norimatsu Y, Ohsaki H, Yanoh K, et al. Expression of immunoreactivity of nuclear findings by p53 and cyclin a in endometrial cytology: comparison with endometrial glandular and stromal breakdown and endometrioid adenocarcinoma grade 1. Diagn Cytopathol. 2013;41:303–7.
31. Nicolae A, Preda O, Nogales FF. Endometrial metaplasias and reactive changes: a spectrum of altered differentiation. J Clin Pathol. 2011;64:97–106.
32. McCluggage WG, McBride HA. Papillary syncytial metaplasia associated with endometrial breakdown exhibits an immunophenotype that overlaps with uterine serous carcinoma. Int J Gynecol Pathol. 2012;31:206–10.
33. Li C, Zota V, Woda BA, et al. Expression of a novel oncofetal mRNA-binding protein IMP3 in endometrial carcinomas: diagnostic significance and clinicopathologic correlations. Mod Pathol. 2007;20:1263–8.
34. Zheng W, Yi X, Fadare O, et al. The oncofetal protein IMP3: a novel biomarker for endometrial serous carcinoma. Am J Surg Pathol. 2008;32:304–15.
35. Mhawech-Fauceglia P, Herrmann FR, et al. IMP3 distinguishes uterine serous carcinoma from endometrial endometrioid adenocarcinoma. Am J Clin Pathol. 2010;133:899–908.
36. Alkushi A, Köbel M, Kalloger SE, et al. High-grade endometrial carcinoma: serous and grade 3 endometrioid carcinomas have different immunophenotypes and outcomes. Int J Gynecol Pathol. 2010;29:343–50.
37. Norimatsu Y, Yanoh K, Maeda Y, et al. Insulin-like growth factor-II mRNA-binding protein 3 immunocytochemical expression in direct endometrial brushings: possible diagnostic help in endometrial cytology. Cytopathology. 2019;30:215–22.
38. Cranmer SL, Ulsemer P, Cooke BM, et al. Glycoprotein (GP) Ib-IX-transfected cells roll on a von Willebrand factor matrix under flow: importance of the GPIb/actin-binding protein (ABP-280) interaction in maintaining adhesion under high shear. J Biol Chem. 1999;274:6097–106.

Atypical Endometrial Cells (ATEC)

13

Kenji Yanoh

13.1 Background

In the 1988 Bethesda System for reporting cervical/vaginal cytologic diagnoses [1], the Papanicolaou classification was abolished because the Papanicolaou classification had no equivalent in diagnostic histopathologic terminology. In the 1988 system, the new diagnostic terms "low-grade and high-grade squamous intraepithelial lesions" were adopted. Additionally, "Atypia" was defined for use with those cases in which the cytologic findings are of undetermined significance. "Atypia" should not be used as a diagnosis for otherwise defined inflammatory, pre-neoplastic, or neoplastic cellular changes. To assist the referring physician, a report in which cells are described as "atypical" should include a recommendation for further evaluation that may help to determine the significance of the atypical cells. As a result, in the classification of squamous cell abnormalities, "Atypical squamous cells of undetermined significance" (ASCUS), "Squamous intraepithelial lesion" (SIL), and "Squamous cell carcinoma" were proposed. In the 2001 Bethesda system (TBS 2001) [2], the term "atypical squamous cells" was qualified as "of undetermined significance" (ASC-US) or "cannot exclude high-grade

This chapter is based on Yanoh K Atypical endometrial cells (ATEC). In: Hirai Y, Yanoh K, Norimatsu Y, editors. Kijyutsushiki Naimakusaiboushin Houkokuyoushiki ni motoduku Shikyuu Saiboushin Atlas; Tokyo: Igakushoin; 2015. p. 72–97. (in Japanese)

K. Yanoh (✉)
Department of Obstetrics and Gynecology, Suzuka Chuo General Hospital,
Suzuka, Mie, Japan
e-mail: kyanou@sch.miekosei.or.jp

SIL" (ASC-H). Eliminating the "Atypical squamous cells" (ASC) category by classifying every specimen as "negative for intraepithelial lesion/malignancy" (NILM) or "SIL" using light microscopy alone is not feasible. However, misclassification may lead to loss of both sensitivity and positive predictive value. Therefore, in TBS 2001, the category "Atypical squamous cells" (ASC) was retained.

Regarding endometrium, there was an increasing need in Japan for a new reporting format to replace the classical reporting format, which used terms such as "Negative," "Suspicious", and "Positive." In 2012, a study group of The Japanese Society of Clinical Cytology (JSCC) published a descriptive reporting format for endometrial cytology in the Bethesda style [3]. A category designated "Atypical endometrial cells" (ATEC) was newly adopted in this reporting format. The ATEC category has to be applied when, for any reason, a histological diagnosis cannot be assigned, for instance, due to inflammatory changes, metaplastic changes, iatrogenic effects, or any other changes with some cytomorphological impact.

Another descriptive reporting system for endometrial cytology was advocated by a committee in Greece [4]. Both reporting systems employ a "Bethesda-style" format. As a result, these two reporting systems have similar diagnostic algorithms, including the use of the ATEC term. After intensive discussion during and after the ICC 2016 symposium held in Yokohama, Japan, we, attendants of this symposium, have combined the two reporting systems to generate TYS "the Yokohama System for reporting directly sampled endometrial cytology" [5].

According to the TYS, ATEC includes ATEC-US "Atypical endometrial cells, of undetermined significance" and ATEC-AE "Atypical endometrial cells, cannot exclude endometrial atypical hyperplasia/endometrioid intraepithelial neoplasia (EAH/EIN) or malignant conditions". ATEC-US is selected when atypical endometrial cells are observed. However, in ATEC-US, the diagnostic significance of these atypical endometrial cells cannot be determined due to cytomorphological changes that may be associated with inflammation, hormonal imbalance, or other phenomena. ATEC-AE is selected when the possibility of atypical endometrial hyperplasia or malignant tumor is not excluded. However, cytologic changes are qualitatively or quantitatively insufficient to permit definitive diagnosis as EIN or malignant neoplasm. While a more detailed cytological diagnosis can be selected for cases of "Negative for malignancy," "endometrial hyperplasia," "atypical endometrial hyperplasia," or "malignant tumor" with endometrial cytology, for cases evaluated as ATEC/either ATEC-US or ATEC-AE must be selected, without exception.

The ATEC category does not represent a single biological entity. Moreover, this category comprises two subcategories, including ATEC-US and ATEC-AE. In TYS, ATEC-US includes a wide spectrum of conditions, ranging from benign conditions to neoplastic changes. In future work, scientific verification research, like that of the ASCUS/LSIL triage study (ALTS), is expected to render the ATEC-US classification more definite and easier to use.

13.2   Definition

ATEC, which refers to cytologic changes suggestive of abnormal endometrial conditions including endometrial carcinoma, does not represent a single biological entity. Accordingly, ATEC includes a wide spectrum of conditions ranging from benign changes resulting from inflammation, abnormal hormonal circumstances, iatrogenic effects, or intrauterine device (IUD), to neoplastic changes. Pre-neoplastic conditions may be included. This category was devised to designate the cellular interpretation of an entire specimen, not that of individual cells. While rigid clinical triage methods exist for ASC, no clinical triage methods have been proposed for ATEC. As for ATEC-AE, endometrial biopsy or curettage is considered to be necessary, because atypical endometrial hyperplasia or carcinoma might be included. Given that ATEC-US is expected to include a wide spectrum of conditions ranging from a benign endometrium to neoplastic changes, it is difficult to determine a triage method for such cases at the present time. Currently, repeated endometrial cytology during 2 or 3 months or endometrial biopsy is recommended. To avoid abuse of this category, ATEC-US should be used in less than 5% of the total results. In Chap. 13, some microscopic images are shown as a result of the OSG training activities. The OSG (Osaki study group) was established in May 2012 as a non-profit organization in Japan with the purpose of conducting research and training activities to contribute to the dissemination of knowledge and skills related to endometrial cytology. Basically, a set of several images of endometrial cells from LBC preparations was chosen, which had been preliminarily examined by a group of experienced cytopathologists, and classified as most possibly representing the "ATEC" category of the TYS. The images were circulated to 14 OSG volunteers, and diagnosed according to the criteria of "The Yokohama System for Reporting Endometrial Cytology". The only clinical information provided was the patient's age. The results were tabulated using "The Yokohama System for Reporting Endometrial Cytology". Each panel member worked blindly and the individual diagnostic considerations were tabulated before the final cytological interpretation. The percentage in parentheses in the figures in Chap. 13 shows the results of the relative percentage of diagnostic observations.

13.3   ATEC-US

The term ATEC-US (TYS2) is used when atypical endometrial cells are observed but the significance of such cells cannot be determined because of interference by inflammatory, abnormal hormonal environmental, or iatrogenic effects. For cases within the cytological categories mentioned above, ATEC-US (TYS2) should be selected. When referring to the ATEC-US category, the reason for cytological diagnosis should be described. Usually, in ATEC-US, cell clusters consist of fewer than three layers. In these cells, metaplastic change usually is observed in the cytoplasm. Necrotic background or isolated atypical cells usually are not present.

13.4   ATEC-AE

ATEC-AE (TYS4) is applied when endometrial atypical hyperplasia/endometrioid intraepithelial neoplasia (EAH/EIN) or malignant neoplasm is suggested, but the findings are not sufficient to interpret these events as malignant neoplasms (TYS6). The inability to interpret these events may be because the number of atypical cells is too small for cytologic decision, or because the atypia is caused by interference associated with inflammatory, abnormal hormonal environmental, or iatrogenic effects. At low-power microscopic magnification, cell clusters with an irregular protrusion pattern show nuclear overlapping of three or more layers. ATEC-AE should be applied when cell clumps with irregular protrusion patterns composed of metaplastic cells are detected at medium-power microscopic magnification. In this case, typical EGBD findings cannot be observed (Figs. 13.1, 13.2, 13.3, 13.4, 13.5, 13.6, 13.7, 13.8, 13.9, 13.10, 13.11, 13.12 and 13.13).

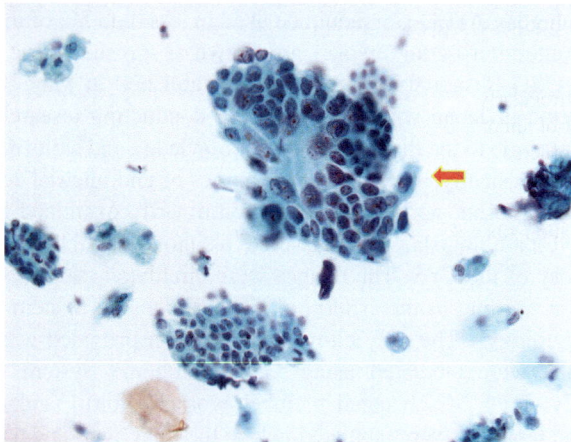

Fig. 13.1   ATEC - US (78.6%). 50s. Small atypical cell clusters of less than three layers. Polygonal cells with large nuclei, distortion of nuclear axis, increased chromatin, and prominent nucleoli are observed compared with adjacent normal cell cluster. Ciliary metaplastic change is also present (red arrow). This metaplastic finding is usually seen in endometrial glandular and stromal breakdown (EGBD) cases. In this case, no typical EGBD finding was seen in any other field except bloody background. Although these findings in this figure are often observed in atrophic endometrium, "Negative for malignancy" should not be applied in this case, because they are occasionally observed in serous EIC too. (Papanicolaou stain, original magnification 40×)

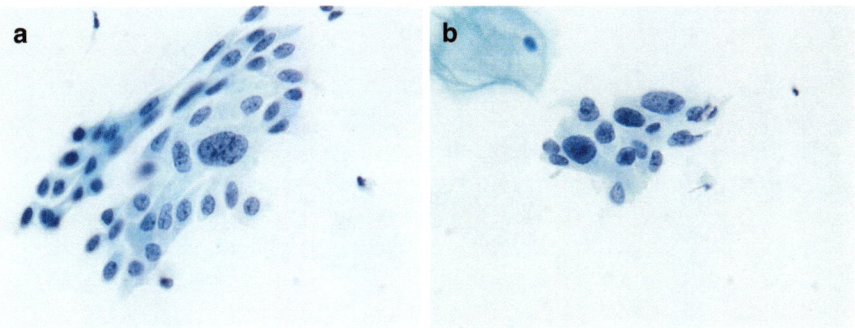

Fig. 13.2 ATEC-US (78.6%). Late 50s. Prominent large nuclei are observed. Cellular overlapping is not observed. A remarkable polygonal cell with hyperchromatic nuclei and metaplastic change can be observed in **a** and **b**. Histology test after endometrial cytology confirmed atrophic endometrium. (Papanicolaou stain, original magnification **a**, **b**: 40×)

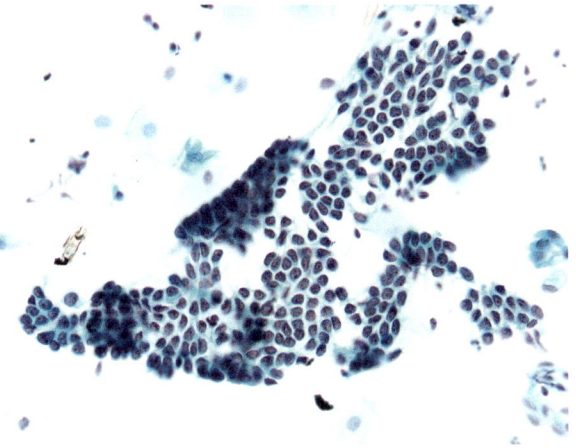

Fig. 13.3 ATEC-US (64.3%). 70s. Endometrial cell cluster less than two layers without endometrial stromal cells is observed. Prominent nucleoli are observed. On the contrary, nuclear findings and size are almost the same. In this case, the effect of abnormal hormonal environment was suspected because intermediate squamous cells of the uterine cervix are accompanied. In this case, "Negative for malignancy; atrophic endometrium" was also suspected. This cellular sample was collected from operative material in which findings of partial p53 positive area were diagnosed. (Papanicolaou stain, original magnification 40×)

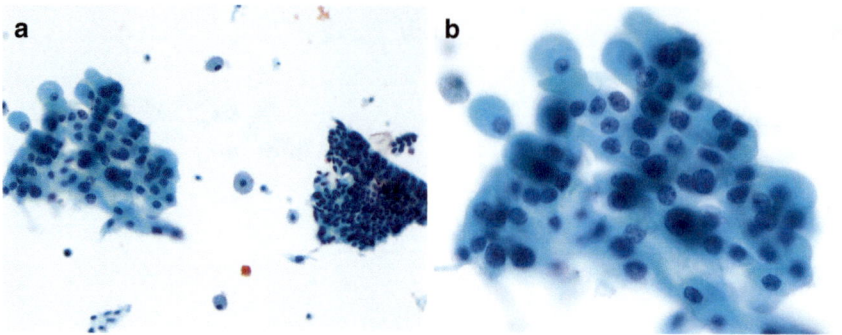

Fig. 13.4 ATEC-US (78.6%). Early 70s. Cell cluster composed with metaplastic cells without cellular overlapping is observed accompanied with typical atrophic cell cluster. Though nucleoli are observed, these are not so prominent as observed in the malignant nucleus. Many OSG members hesitated to interpret it as "Negative for malignancy." In this case, following endometrial biopsy was done. However, histological diagnosis was not able to be done because of an unsatisfactory specimen. In an aged woman, obtained specimen of biopsy is sometimes insufficient. Repeat endometrial cytological test and/or ultrasonographic test will be desired to be considered in such a case. (Papanicolaou stain, original magnification **a** and **c**: 20×, **b**: 40×)

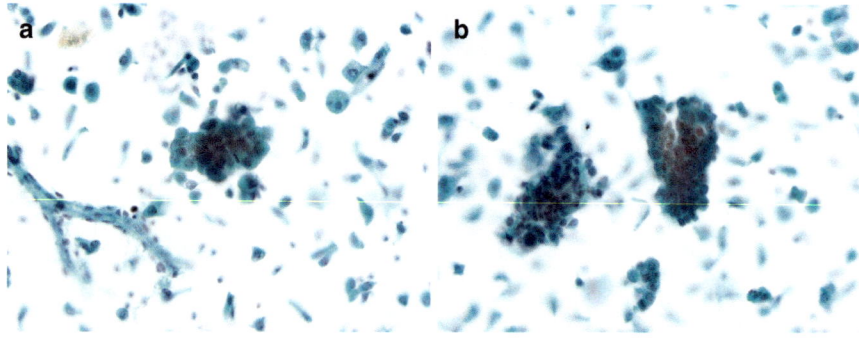

Fig. 13.5 ATEC-US (42.9%), ATEC-AE (35.7%). 50s. Small cell cluster which is consisted of less than three atypical cell layers is observed. Metaplastic changes, anisocytosis of the nucleus, prominent nucleoli are observed. Though the malignant tumor was suspected because of scattered isolated endometrial cells, 42.9% of OSG members interpreted as ATEC-US while 35.7% interpreted as ATEC-AE, because the cell cluster consisted of more than three layers was not observed. In the case of fewer cells in specimen, even though malignant tumor is suspected, ATEC-US rather than ATEC-AE tends to be interpreted. In this case, pathological diagnosis with following endometrial biopsy was endometrioid carcinoma, grade1. (Papanicolaou stain, original magnification **a**, **b**: 40×)

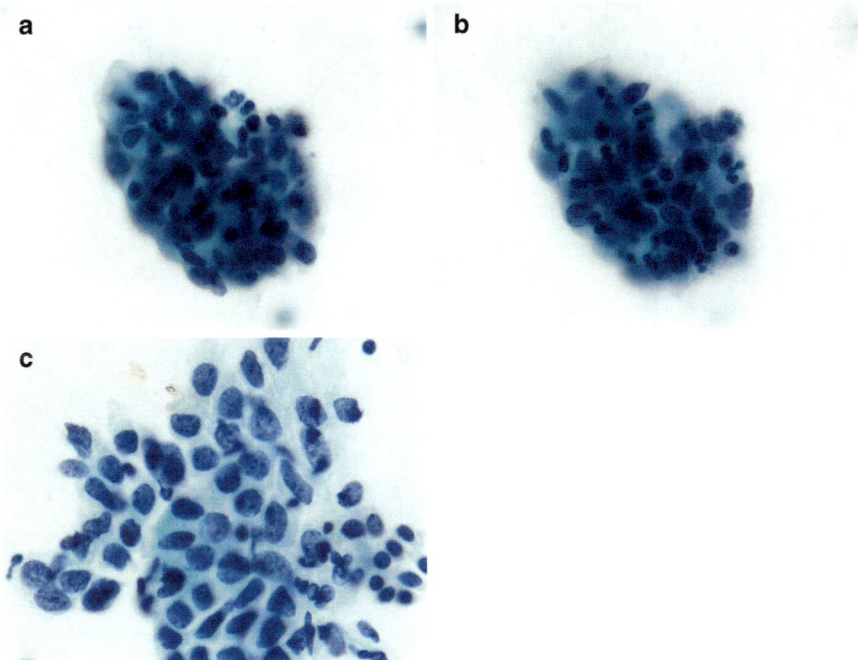

Fig. 13.6 ATEC-US (42.9%), ATEC-AE (42.9%). Late 50s. Cluster of metaplastic cells with anisocytosis of the nucleus, prominent nucleoli, invaded neutrophils, karyotype irregularity. Interpretation was divided ATEC-US and ATEC-AE depend on the recognition that this cell cluster consisted of two cell layers or three. (**a** and **b**) are both the same field of view, on a different focus. Focusing up and down could clarify the interpretation. In this case, normal endometrium that was affected with drugs or abnormal hormonal effect was suspected. Final pathological diagnosis with following endometrial biopsy was atrophic endometrium. (Papanicolaou stain, original magnification **a**, **b**, **c**: 80×)

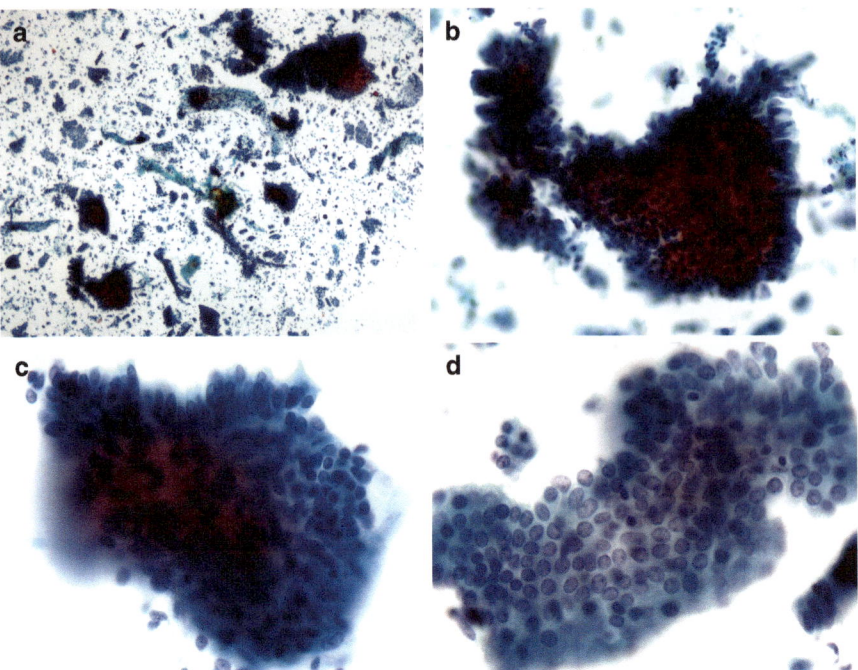

Fig. 13.7 ATEC-AE (35.7%), ATEC-US (28.6%). Early 60s. Attention should be paid to high cellularity in spite of the menopausal case. The effect of abnormal hormonal environment on atrophic endometrium is suspected because of metaplastic change and clinical information. Many irregular-shaped clusters consisted of large cells, anisocytosis of the nucleus, increased chromatin, more than three layers are observed. Final pathological diagnosis with endometrial biopsy was endometrial hyperplasia accompanied by endometrial polyp. (Papanicolaou stain, original magnification **a**: 4×, **b**: 20×, **c** and **d**: 40×)

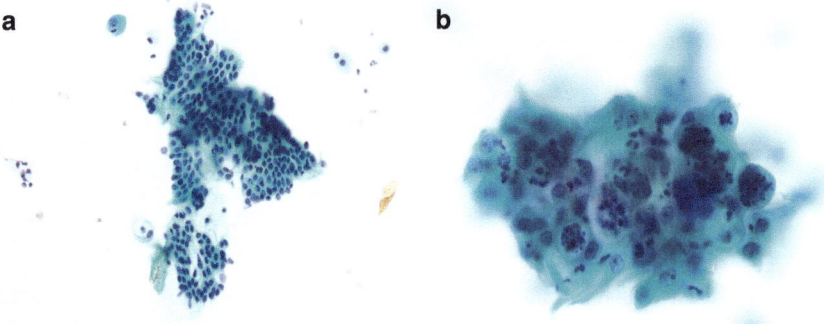

Fig. 13.8 ATEC-AE (64.3%). Early 60s. Abnormal cell cluster with irregular protrusion. Focusing up and down may clarify the interpretation. Metaplastic cytoplasmic change is observed. This change might be caused due to inflammation because of inflammatory reactive changes. While findings of anisocytosis of the nucleus, and prominent nucleoli are observed, nuclear size is almost the same. In this case, typical EGBD findings were not observed any other field. (Papanicolaou stain, original magnification **a**: 10×, **b**: 40×)

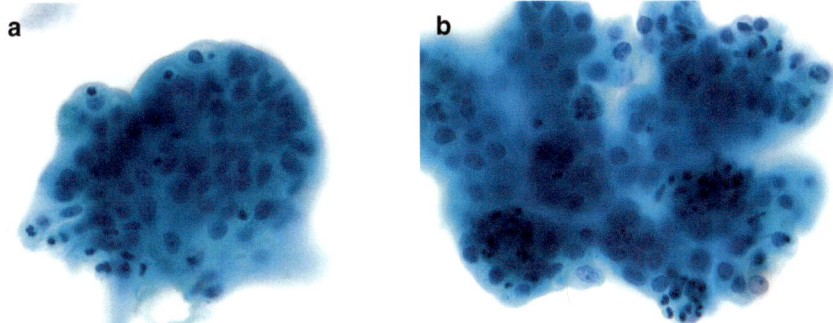

Fig. 13.9 ATEC-AE (71.4%). Early 70s. Medium size cell cluster consisted of cells with a dense metaplastic cytoplasm and prominent nucleoli. This cell cluster shows more than three layers without endometrial stromal cells. Though nuclear axis is not the same, protrusion of cell from cluster or isolated cell is not isolated. Malignant tumor such as endometrioid carcinoma is not difficult to be interpreted because bloody/necrotic background was not observed, and relatively uniform pattern of nuclear findings. Following endometrial biopsy revealed endometrioid carcinoma, grade 1. (Papanicolaou stain, original magnification **a**, **b**: 40×)

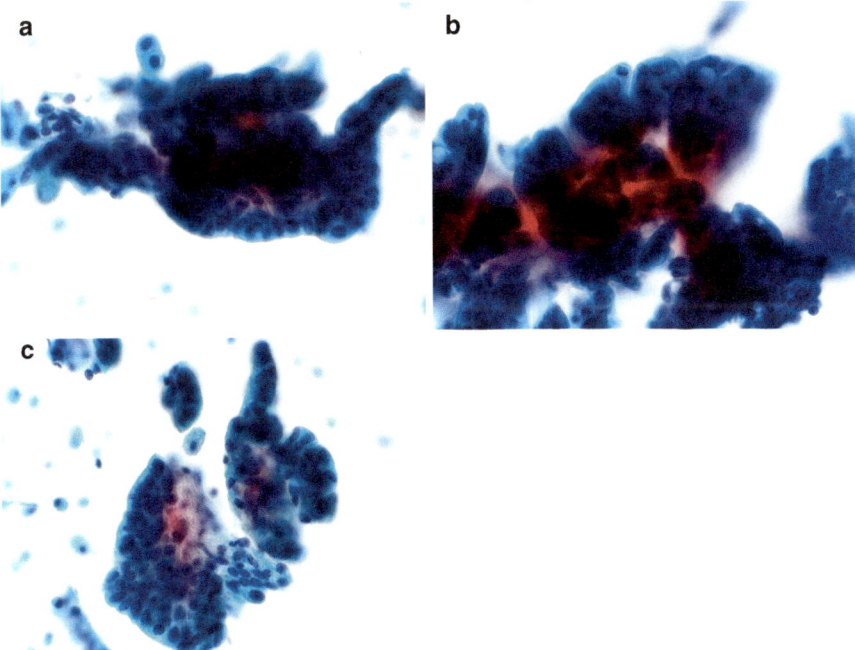

Fig. 13.10 ATEC-AE (71.4%). Late 50s. Abnormal cell cluster showing irregular protrusion without endometrial stromal cells. Possible interpretation is a malignant tumor for more than three layers, irregular cell array. Moreover, there is no typical EGBD finding. Interpretation as atypical endometrial hyperplasia or malignant neoplasm were a few because of underestimation for metaplastic cytoplasm. Following endometrial biopsy and operation revealed atypical endometrial hyperplasia with endometrial hyperplasia without atypia. (Papanicolaou stain, original magnification **a**, **b**, **c**: 40×)

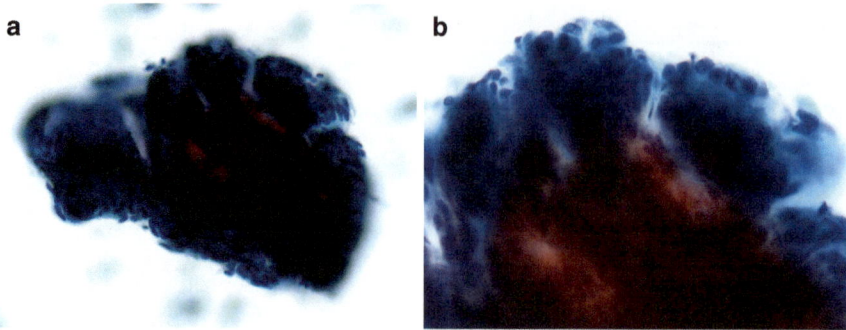

Fig. 13.11 ATEC-AE (71.4%). Early 80s. Abnormal cell cluster showing irregular protrusion and three layers without endometrial stromal cells. Because this cluster consisted of large nuclei, possible interpretation should be atypical endometrial hyperplasia or malignant tumor. In this case, because these abnormal cell clusters were seldom observed, many OSG members interpreted as ATEC-AE. Remaining members were interpreted as malignant tumors. Following histological test was endometrioid carcinoma, grade1. (Papanicolaou stain, original magnification **a**: 10×, **b**: 40×)

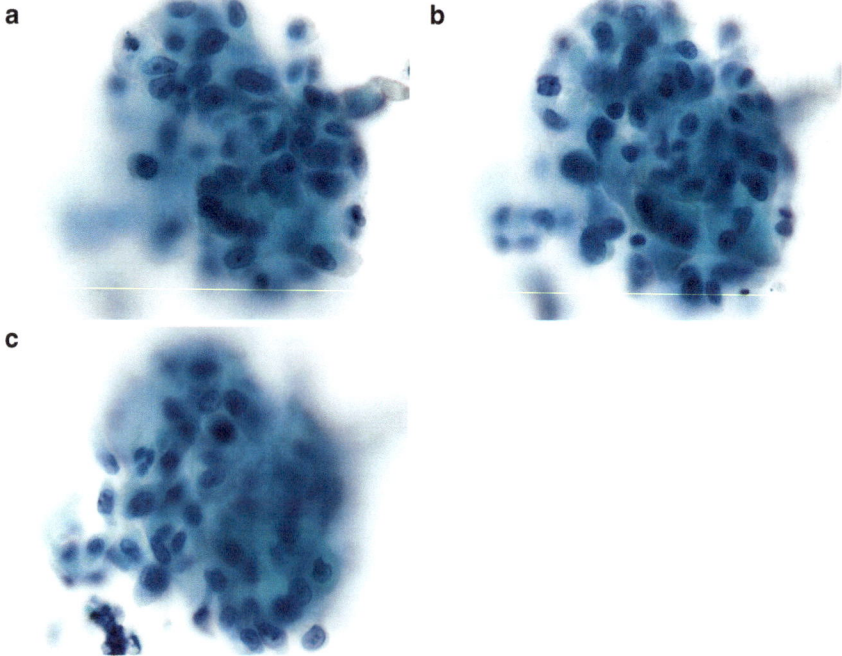

Fig. 13.12 ATEC-AE (71.4%). Late 50s. More than three layers could be identified by focusing up and down (**a**, **b** and **c**). Nuclear enlargement, irregular array, prominent nucleoli were observed in each cell layer. Metaplastic change is identified in the cytoplasm. In this case, because no typical EGBD findings were not observed, many OSG members choose ATEC-AE. Following endometrial biopsy could not be available because collected specimen was not enough. Repeat cytological test was "Negative for malignancy". (Papanicolaou stain, original magnification **a**, **b**, **c**: 80×)

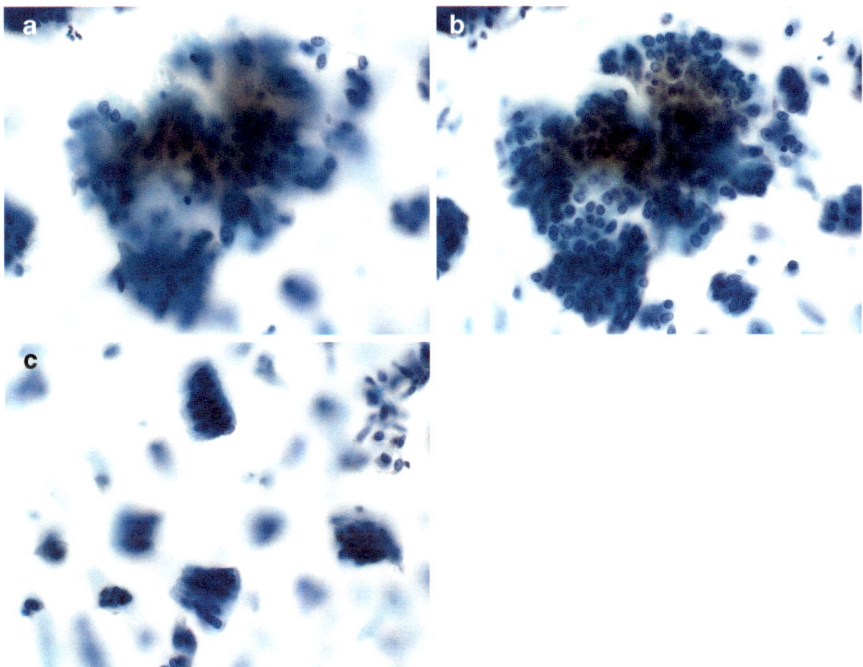

Fig. 13.13 ATEC-AE (57.1%) vs Malignant tumor (35.7%). Late 50s. Many abnormal cell clusters consisted of atypical cells and more than three layers were seen in the specimen. As no necrotic background was identified, possible interpretations include atypical endometrial hyperplasia and early, high-grade endometrioid carcinoma. In this case, however, columnar cells showing palisade arrangement were also observed. Because this finding is not typical in endometrioid carcinoma, many OSG members could not decide cytological result as "Malignant tumor." Final histological diagnosis obtained from operative material was endometrioid carcinoma grade 2, which shows protruded lesion (8 × 6 × 3 mm in size) accompanied with pseudomyxoma peritonei (**a**, **b** and **c** are the same field). (Papanicolaou stain, original magnification **a**, **b**, **c**: 40×)

References

1. The National Cancer Institute Workshop. The 1988 Bethesda system for reporting cervical/vaginal cytologic diagnoses. Acta Cytol. 1989;33:567–74.
2. Solomon D, Davey D, Kurman R, et al. The 2001 Bethesda system. Terminology for reporting results of cervical cytology. JAMA. 2002;287:2114–9.
3. Yanoh K, Hirai Y, Sakamoto A, et al. New terminology for intrauterine endometrial samples: a group study by the Japanese society of clinical cytology. Acta Cytol. 2012;56:233–41.
4. Margari N, Pouliakis A, Anoinos D. Et al. a reporting system for endometrial cytology.: cytomorphologic criteria – implied risk of malignancy. Diag Cytopathol. 2016;44:888–901.
5. Fulciniti F, Yanoh K, Karakitsos P, et al. The Yokohama system for reporting directly sampled endometrial cytology: The quest to develop a standardized terminology. Diagn Cytopathol. 2018;46:400–12.

Cell Block Techniques for Endometrial Cytology Technical Procedures, Role of Immunocytochemistry, Advantages, Applications

14

Niki Margari, Alessia Di Lorito, and Ioannis G. Panayiotides

«Ἡ χαρίεν ἔστ· ἄνθρωπος, ἂν ἄνθρωπος ᾖ».
"How delightful is a human being, if he really is human".
(Menandros [Greek writer, 342–292 B.C.])
This chapter is dedicated to the loving memory of Professor Petros Karakitsos, MD, PhD, former Head of the Department of Diagnostic Cytopathology, National and Kapodistrian University of Athens Medical School ("Attikon" University Hospital), who passed away suddenly on June 26, 2017: a dedicated doctor, a Pythagorean teacher, an outstanding scientist, and a cherished friend and collaborator.

14.1 Cell Block Preparation and Techniques

14.1.1 Cell Block Preparation

Cell block (CB) preparation is a well-established technique for cytological material preparation [1, 2]. Residual material can be used to obtain cell blocks as a complementary technique to cytological slides (Figs. 14.1 and 14.2). Preparation of CB·s in addition to liquid based cytology increased the ability to diagnose malignancy by 67% when compared to liquid based cytology alone [3].

N. Margari (✉)
Private Cytopathology Laboratory, Ex Scientific Collaborator of Department of Cytopathology at National and Kapodistrian University of Athens, Athens, Greece
e-mail: nikimarg@gmail.com

A. Di Lorito
Pathology Department, SS Annunziata Hospital, Chieti, Italy

I. G. Panayiotides
Professor of Pathology at University of Athens, Chair, 2nd Department of Pathology UoA, "Attikon" University Hospital, Athens, Greece

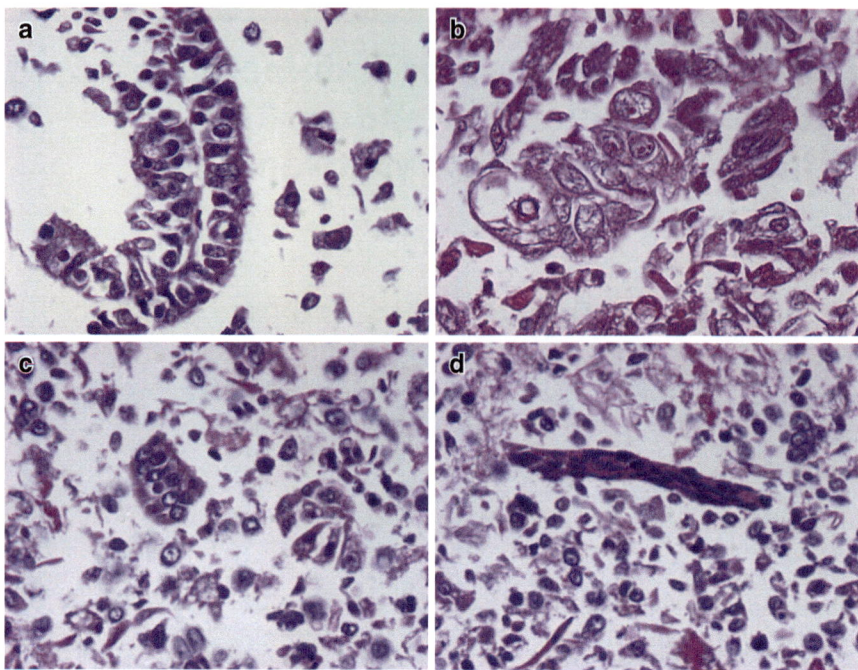

Fig. 14.1 Case examples of endometrial lesions prepared with the cell block technique (particularly the Cellient technique) and their histologic diagnoses. H&E stain, ×40. (**a**) a case of atypical hyperplasia, (**b**) a case of serous carcinoma showing also characteristics of clear cell carcinoma, (**c** and **d**) a case of proliferative endometrium and polyp on histology

The overall diagnostic accuracy of endometrial cytology is 96% and 100% for benign lesions and adenocarcinoma, respectively, when cell blocks are prepared from the residual material, whereas the diagnostic accuracy exceeds 90% in cases of hyperplasia [4]. Zhang et al. examined CB preparations on residual cytological material; in their study the diagnostic accuracy was 95.1%, with a sensitivity of 82.8% and specificity of 98.3%. In the same study combination of liquid based cytology and CB presented a diagnostic accuracy of 95.8%, a sensitivity of 89.7%, and specificity of 97.4% [5].

CB preparation on residual cytological material offers the advantages of valuable diagnostic evidence, maintenance of architecture, preservation of nuclear and cytoplasmic characteristics, and the presence of tissue fragments that cannot be obtained by cytology alone. The increasing demand of the clinicians for the application of ancillary tests in order to determine the profile of the lesions and their optimal treatment calls for the development of accurate techniques on cytological material. In addition to their significant role in diagnosis, many ancillary techniques may be performed out of CB preparations, such as immunocytochemistry and molecular techniques, including polymerase chain reaction (PCR), fluorescent in situ hybridization (FISH) determining the biological behavior of the lesion [6, 7] and, recently, artificial intelligence [8].

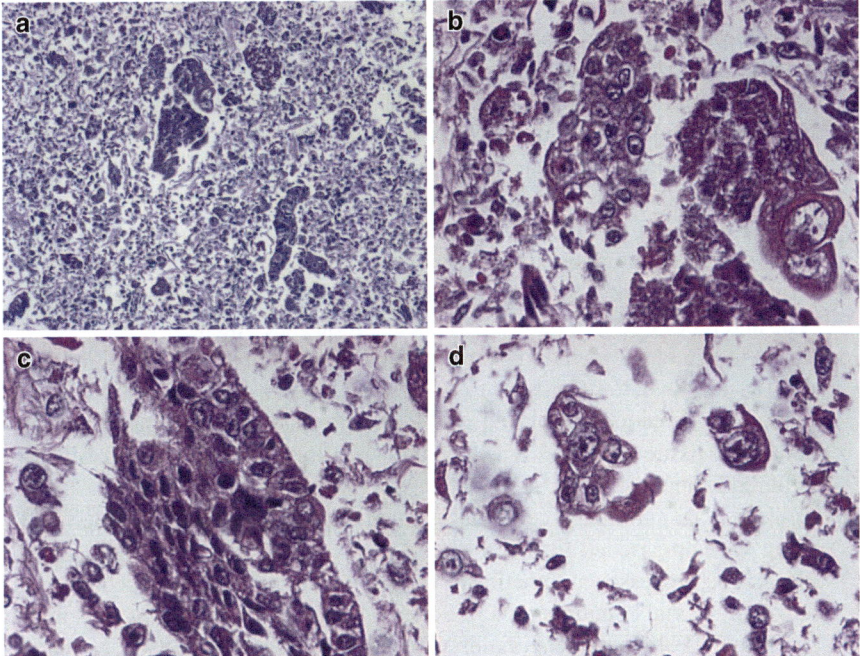

Fig. 14.2 A case prepared with the cell block technique diagnosed as endometrioid carcinoma grade 3 on histology (×40, H&E staining)

The first step for the preparation of a CB is the collection of cells by centrifugation to form a cell pellet, which is subsequently immersed into a fixative solution. Following material processing, according to the technique used, the pellet is transferred to a cassette for paraffin embedding and sectioning.

Different fixatives may be used for material collection prior to CB preparation, such as buffered formalin, ethanol, methanol, picric acid, or Bouin's solution. It is important to validate the fixative prior to use, since both the type of fixative and time of fixation may affect immunocytochemistry and molecular studies.

14.1.2 Cell Block Techniques

Several techniques have so far been described:

(a) Fixed sediment method: the material is mixed with 96% ethyl alcohol and 4% buffered formalin and centrifuged for 10 min. The supernatant is then discarded and the sediment is placed into a filter paper. The wrapped filter paper is subsequently placed into a cassette and processed as a paraffin-embedded tissue specimen.

(b) Bacterial Agar method: the material is centrifuged and the supernatant discarded. Solid 3% agar medium within a tube is melted in boiling water, according to manufacturer's instructions, and 4 ml of molten agar medium is added to a sterile tube, from which it is poured within another tube containing the sediment; stirring and cooling of the mixture follows. Excess of agar should be gently removed. The cell button is subsequently placed into a cassette and processed as a paraffin-embedded tissue specimen. Lege artis solidification of the agar medium prevents heating related artifacts, such as bubbles or shrunken cells.

(c) Plasma-thrombin clot method: the material is centrifuged and the supernatant discarded. The sediment is mixed with the same amount of blood plasma and thrombin, whereupon a clot is formed. This is subsequently placed into a cassette and processed as a paraffin-embedded tissue specimen.

(d) Shandon cytoblock method is a manual cell block preparation technique using the Thermo Shandon Cytospin to concentrate cells within the cell block. The mixture is subsequently placed into a cassette and processed as a paraffin-embedded tissue specimen. The technique requires phosphate-free fixatives or buffered formalin.

(e) Cellient is an automated cell block system applied on ThinPrep material [1], using vacuum-assisted filtration to concentrate available cells within the cell block. The processor automatically aspirates material from the specimen collected in the vial containing the PreservCyt solution, an alcohol-based solution (methanol). Following dehydration with isopropyl alcohol (99.8%) and dispensation of xylene (98.5% minimum), the sample is heated and paraffin is added. Serial sectioning of the cell block provides several slides: one is stained with Hematoxylin and Eosin (H&E), whereas the remaining may be used for immunocytochemistry and other molecular techniques.

(f) Other cell block techniques include Simplified cell block technique, compact cell block technique, microwave technique for rapid processing of cell blocks, cell block from Millipore [1].

14.2 Immunocytochemistry

14.2.1 Application of Immunocytochemistry in Endometrial Cancer

Immunocytochemistry may be performed in cell block preparations to increase diagnostic accuracy and facilitate differential diagnosis, as well as to assess the status of markers related to prognosis and/or response to various treatment modalities.

Endometrioid adenocarcinomas typically co-express cytokeratins 8/18 and vimentin. Vimentin is positive in more than 90% of endometrial carcinomas, with a wide range of expression from focal to diffuse [9]. CEA immunostaining is usually negative or may show a focally positive cytoplasmic staining in endometrioid carcinomas. Mucinous variants, as well as areas with squamous differentiation may be positive. Endometrial carcinomas show a positive cytoplasmic immunostaining for CA-125. The combination of low CEA and high CA-125 is more compatible with an endometrial or ovarian origin [1].

PTEN is a tumor suppressor gene that antagonizes the PI3K-AKT (phosphatidylinositol3-kinase/protein kinase B) signaling pathway. It is frequently altered in many cancers such as in endometrial cancer. For the evaluation of PTEN immunoreactivity, a staining reaction of brown granules in the cytoplasm and nucleus can be considered positive, and the loss of staining is considered negative. In addition, PTEN signals in stromal cells or in vascular endothelial cells can be used as internal positive controls. In fact, normal endometrial epithelium and stroma present strong immunoreactivity for PTEN protein during the proliferative phase, whereas it is decreased during the secretory phase. PTEN mutations have been described by Mutter et al. in 83% of endometrioid adenocarcinomas and in 55% of precancerous lesions [10]. In the same study, cases of proliferative endometrium and hyperplasia without atypia were positive for PTEN immunostaining; hyperplasia without atypia showed a less homogenous staining pattern, with a few negative hyperplastic glands among positive ones. Maxwell et al. found PTEN mutations in 20% of endometrial hyperplasias, suggesting that its inactivation occurs in early stages of carcinogenesis [11]. Norimatsu et al. examined immunocytochemical expression of PTEN in liquid based preparations [12]; they demonstrated that a cutoff value of 50% was necessary for accurate diagnosis of endometrial carcinoma. Djordjevic et al. examined immunohistochemical expression of PTEN in endometrial carcinomas, using stromal cells as internal positive controls; cases with diffuse positive cytoplasmic and nuclear staining in more than 90% of malignant cells were considered positive. They also confirmed that PTEN mutations occur more frequently in endometrioid carcinomas than in other types, with 51% and 28%, respectively [13].

p53 gene is activated by cell stress and its protein accumulates in the nucleus, resulting in nuclear positive stain, although cytoplasmic staining with concomitant variable nuclear staining has been described in some cases. Complete loss of staining ("null pattern") may also be associated with p53 mutations. p53 positive immunostain is typically found in high-grade endometrioid carcinomas and serous carcinomas [14]. Only a small number of low-grade endometrioid carcinomas present p53 mutation-type immunoreactivity. It is an important marker, because mutation-type p53 null staining is associated with an unfavorable survival outcome. Garg et al. examined p53 overexpression using immunohistochemistry. They considered the marker to be overexpressed only in cases with strong and diffuse staining in at least 75% of tumor cells, although other studies define a cutoff value of 50% of cells [15]. In a recent work, Norimatsu et al. reported the evaluation of intensity of nuclear staining and the nuclear labeling index (N-LI). The intensity of nuclear staining was scored as negative (0), weak (1), moderate (2), or strong (3). The N-LI was scored as <10% (0), from 10% to 25% (1), from 26% to 50% (2), or > 50% (3). The final p53 score was calculated by the summation of both scores. If total scores were > 1, then the sample was considered positive [12].

14.2.1.1 Beta-Catenin

Catenin beta-1, also known as ß-catenin, is a protein encoded by the CTNNB1 gene. It is a dual function protein, involved in regulation and coordination of cell adhesion and gene transcription. ß-catenin is a subunit of the cadherin protein complex and acts as an intracellular signal transducer in the Wnt signaling pathway.

Beta-catenin is widely expressed in many tissues. Mutations and overexpression of **ß**-catenin are associated with many cancers. Presence or absence of nuclear staining is evaluated and percentage of tumor demonstrating nuclear staining should be recorded. For the evaluation of ß-catenin immunoreactivity, a staining reaction of brown granules in cells should be considered positive, whereas the loss of staining is considered negative. Positive immunoreactivity can usually be membranous, cytoplasmic, and nuclear. If both the cytoplasm and the nucleus are stained, cell is classified as positive. Membranous staining for vascular endothelial cells and epithelial cells can be utilized as an internal positive control [16].

In endometrial cytology, a panel made up of these biomarkers can be used for EIN diagnosis. Norimatsu et al. reported loss of PTEN and positive nuclear staining of ß-catenin frequently seen in EIN but not in normal proliferative cases. The combination of PTEN negative/ß-catenin-positive results can be applied for detecting EIN. Because the overexpression of gene products of types I and II ECs correlates with clinico-pathological factors and prognosis, it is also important to evaluate the immunocytochemical expression of endometrial malignancy markers including precursor lesions on LBC samples. In literature the usefulness of PTEN staining was also confirmed by different studies. Both nuclear and cytoplasmic ß-catenin should be underexpressed in type II carcinomas. In a study, Kosmas et al. reported that the ICC positive expression of p53 on imprint cytologic smears was correlated with surgical-pathological stage, histological grade, and lymph node metastases [17].

Endometrioid tumors stage 1 and 2 according to FIGO show strong and diffuse positivity for estrogen receptors (ER) and progesterone receptors (PR) in 80% of cases, whereas in FIGO 3 stage tumors positivity is found to be 15% to 50% [18]. Kuramoto et al. examined the immunohistochemical expression of ER and PR in endometrial carcinomas, showing this to decrease with increase of tumor grade and/ or depth of myometrial invasion [19]. Hormone receptor (HR) expression is also required by clinicians for treatment in order to predict response to endocrine therapy and it is the most validated prognostic biomarker for endometrial cancer, being associated with favorable survival outcome [18]. According to the ASCO CAP guidelines, as for breast cancer diagnosis, HR immunoreactivity is evaluated by counting the number of positively staining neoplastic nuclei (expressed as a percentage) and by scoring the degree of positive immunoreactivity, respectively, as weak or strong). The staining intensity can be affected by the amount of protein present, by the clone of the antibody used, and the antigen retrieval system used. In most cancers, there is heterogeneous immunoreactivity with weakly to darkly staining nuclei. Guan et al. proposed a cutoff value of 1% for the interpretation of ER and PR staining, although a cutoff point of 10% was set by other authors [19]. They suggested that the values between 1% and 9% for ER positive cancer cells might have been underestimated. Moreover, patients whose carcinomas had a percentage of 1% or more positive nuclei showed a better survival related to patients with negative values [19]. ER/PR expression in serous carcinoma is similar to FIGO grade 3 endometrioid carcinomas and the staining rate and intensity of positive cases are lower than in endometrioid well differentiated carcinomas. Indeed, in the carcinosarcoma cases, it should be remembered that the epithelial component rarely expresses ER/PR as in clear cell carcinomas.

Endometrioid carcinomas typically present either a patchy p16 staining or no staining at all. However, weak focal positive staining has been described in a few cases of FIGO 1 and 2 endometrioid carcinomas and stronger staining has been found in 25% of grade 3 carcinomas [9]. Carcinomas with squamous differentiation may have a more diffuse positive staining (Fig. 14.3).

Serous carcinomas most commonly present a diffuse and strong p53 expression and a high percentage of Ki-67 labeling, whereas hormone receptor immunostaining is usually negative, seldom weak or positive. However, some cases may present a phenotype with negative p53 expression and positive hormone receptors. Yemelyanova et al. examined the value of p16 in differential diagnosis of serous carcinomas; serous carcinomas showed a diffuse, either moderate or strong p16

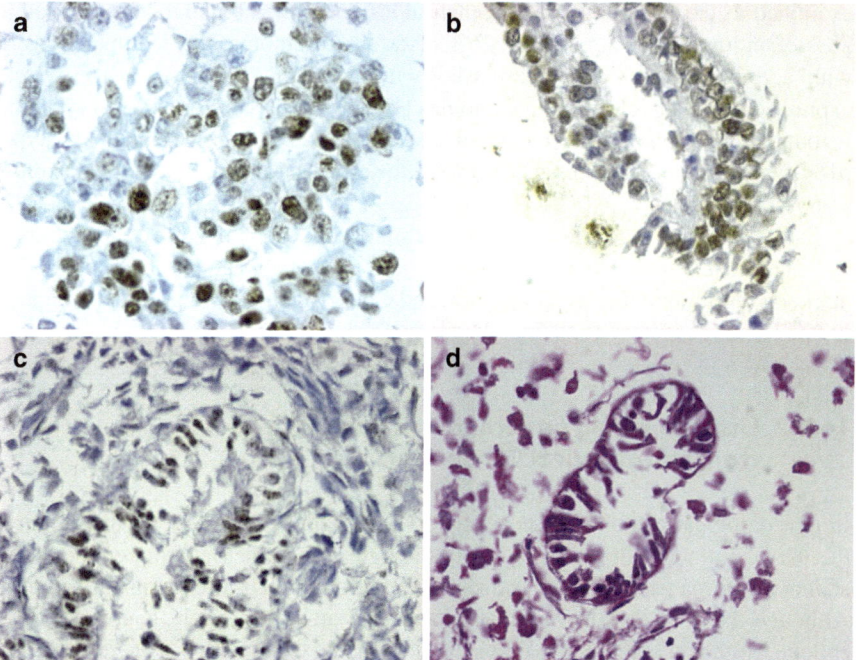

Fig. 14.3 Case examples of immunocytochemical staining in endometrial lesions prepared with the cell block technique (particularly the Cellient technique) and their histologic diagnoses. H&E stain, ×40. (**a**) Ki-67, a case of endometrioid carcinoma grade 3, (**b**) ER, a case of endometrioid carcinoma grade 1, (**c** and **d**) p53 and H&E, a case of endometrioid carcinoma grade 1

expression, whereas in cases of endometrioid carcinomas expression was less diffuse and less intense. In 92% of serous carcinomas there is a diffuse immunoreactivity for p16, in contrast to low- and high-grade endometrioid carcinomas, where p16 immunostaining is found in 7% and 25%, respectively [20, 21]. PTEN expression is also retained in serous carcinomas; H-J An et al. found most serous carcinomas to have an intense PTEN expression, in contrast to endometrioid carcinomas, where PTEN loss or decreased expression is seen in 66% of cases [22]. Serous carcinomas more commonly present a different immunohistochemical profile compared to grade 3 endometrioid carcinomas: the former are positive for PTEN, negative for ER/PR, diffusely positive for p16, and have an aberrant mutation-type expression for p53, whereas the latter are negative for PTEN, positive for ER/PR, focally positive for p16, and present a wild type absence of staining for p53 [23].

As endometrioid and serous subtypes, clear cell carcinoma expresses pan-cytokeratins, CK7, CA125, and vimentin. Clear cell carcinoma is usually negative for ER/PR, though positive staining may be found in a small percentage of cases with a weak and focal staining. They are also typically negative for CK20 and WT1. Napsin A is a cytoplasmic marker found to be positive in 67%–88% of clear cell carcinomas, in contrast to 0–0.5% of endometrioid carcinomas [21]. HNF1B is a useful nuclear marker typically presenting a strongly positive staining in clear cell carcinomas; however, it has a high sensitivity but a low specificity. Němejcová et al. examined its expression in 320 endometrial lesions; it was found to be expressed in 28% of endometrioid carcinomas, 26% of serous carcinomas, 88% of hyperplasias with atypia, and 91% of hyperplasias without atypia, though it was strongly expressed mostly in clear cell carcinomas [24]. Expression of mutational type p53 is found in up to one third of clear cell carcinomas [25]. In this carcinoma subtype, p16 is less commonly expressed compared to serous carcinomas and PTEN expression is also retained [22].

Indeed, neuroendocrine carcinomas and small cell carcinomas have been reported and classified separately. In addition to morphological features, neuroendocrine markers (chromogranin, synaptophysin, and CD56 together with proliferative index as Ki67) are mandatory for a final diagnosis [26].

14.2.2 Distinguishing Endometrioid Carcinoma from Endocervical Carcinoma

In gynecological pathology, the most common challenging diagnosis that requires the use of immunocytochemistry is to distinguish endometrial from endocervical adenocarcinoma especially in endocervical and endometrial cytological samples. Although the most common carcinoma subtypes are, respectively, endometrioid in endometrium and mucinous in cervix, endometrioid carcinoma can arise in the cervical tract and mucinous adenocarcinomas can be found in endometrium. This differentiation is clinically important, because it indicates to the gynecological surgeons the surgical planning and subsequent treatment approach. In this situation, it is really important to consider the patient·s age and the clinical features [27].

In fact, in general, samples from post-menopausal patients with endometrial hyperplasia and/or squamous morules indicate an endometrial origin, while samples from premenopausal patients with cervical squamous dysplasia favor an endocervical lesion. However, in the cases in which morphology alone is not conclusive, the use of ancillary techniques is mandatory. The use of a panel of markers, instead of a single biomarker, is encouraged because any analysis can produce unexpected positive or negative (aberrant) staining reactions using a single biomarker but generally this is not possible with a panel of markers. In routine practice, a few selected biomarkers can help to distinguish most cases of low-grade endometrial endometrioid adenocarcinoma and high-risk HPV-related endocervical adenocarcinoma. Carcinoembryonic antigen (CEA) and vimentin have traditionally been used for this differential diagnosis, while the most useful markers, currently recommended, are p16 and hormone receptors (estrogen and progesterone) [27].

Endometrioid carcinomas typically express vimentin and estrogen receptors (ER) with a diffuse and strong staining, in contrast to endocervical adenocarcinomas that are negative for vimentin and ER although some cases may be focally weakly positive for the latter.

CEA is diffusely positive in most cases of endocervical adenocarcinomas with a cytoplasmic or membranous staining. In contrast, it is negative or focally positive in endometrioid carcinomas. In literature, there are discordant studies showing that both endometrial and endocervical mucinous carcinomas can express CEA. However, the type of antibody used (polyclonal vs monoclonal) has some significance. According to previous data, although endometrial adenocarcinomas were largely negative for CEA, the rate of CEA positivity varied in endocervical adenocarcinomas, with significantly higher rates observed with monoclonal CEA as compared to polyclonal CEA [28].

Vimentin is a type III intermediate filament protein that is expressed in mesenchymal cells. Vimentin characteristically shows cytoplasmic expression in endometrial cancer. However it has a heterogeneous distribution and staining patterns, also seen in many carcinomas. Indeed, a number of endometrial cancer cases showed only weak and focal vimentin staining. In small biopsy or in cell blocks, its value as a discriminatory marker for gynecologic carcinomas could represent a problem, so that it is recommended to include this biomarker in a panel of different antibodies [28].

p16 is positive in high-risk HPV-related endocervical adenocarcinomas. It is expressed in a diffuse pattern, either moderately or strongly; its expression is related to Rb tumor suppressor pathway and triggered by the E7 HPV protein. On the other hand, endometrioid carcinomas express p16 in a weak and patchy pattern. The pattern of staining is critical for interpretation. High-risk HPV-related endocervical adenocarcinomas have diffuse and moderate to strong p16 expression (essentially all tumor cells are positive, "block positivity") because it is correlated with HPV-mediated molecular alterations, resulting in p16 overexpression. So, diffuse moderate to strong p16 staining can be used as a surrogate marker for high-risk HPV in this setting. However, it should be also noted that diffuse and strong p16 expression can also be found in serous and undifferentiated carcinomas, as well as in rare cases of endometrioid carcinomas [29].

Table 14.1 Distinguishing endometrial carcinoma from endocervical carcinoma

Markers	HPV-related endocervical adenocarcinoma	Non-HPV-related endocervical adenocarcinoma	Low-grade endometrial adenocarcinoma	High-grade endometrial adenocarcinoma	Endometrial serous adenocarcinoma
ER/PR	Negative (rarely weakly positive only focally)	Negative	Positive	Usually positive (decreases as the grade increases)	Negative (rarely positive)
p53	Wild type	Wild type or mutation type	Wild type	Usually wild type	Mutation type
p16	Positive (diffuse)	Negative or focally positive	Usually negative or patchy staining	Variable positive	Positive (diffuse)
High-risk HPV testing	Positive	Negative	Negative	Negative	Negative

HPV testing has a critical role in the diagnosis of cervical adenocarcinoma, its 16, 18, and 45 genotypes being the most frequent ones. Nevertheless, one should remember that not all cervical adenocarcinomas are HPV related. Also in these cases, most of the times, p16 immunocytochemistry is more readily available and constitutes a reasonable surrogate to HPV testing [30].

p53 expression in endocervical carcinomas is usually patchy and weak, in contrast to strong expression in serous and high-grade endometrioid carcinomas (Table 14.1).

14.2.2.1   NON-HPV-Related Carcinomas.

Some cervical adenocarcinomas are unrelated to high-risk HPV. These include gastric-type adenocarcinomas, mesonephric adenocarcinomas, most clear cell carcinomas. In distinguishing these tumor types from a primary endometrial adenocarcinoma, different panels of markers need to be evaluated which depend on the subtype of the cervical and endometrial adenocarcinoma considered. These non-HPV-related cervical adenocarcinomas are almost always p16-negative or focally positive. There are also rare cases with diffuse positivity, related to non-HPV causative mechanisms. Gastric-type cervical adenocarcinomas are mucinous adenocarcinomas. Gastric-type adenocarcinomas exhibit negative hormone receptor expression as high-risk HPV-related endocervical adenocarcinomas. Rare subtypes display aberrant (mutation-type) p53 expression, while usual-type HPV-related adenocarcinomas have a heterogeneous p53 expression. A challenging differential diagnosis is between endometrial mucinous adenocarcinoma and gastric-type cervical adenocarcinoma. In this situation, hormone receptor assessment can be of use because endometrial mucinous adenocarcinomas or adenocarcinomas with mucinous differentiation are ER/PR positive whereas gastric-type cervical adenocarcinomas are negative.

The rare form of mesonephric adenocarcinoma can be identified by morphology, negativity for ER and PR, and positivity for other markers as GATA3. Cervical adenocarcinomas (HPV-related and HPV-non-related cancers) such as endometrioid adenocarcinoma are usually GATA3 negative. Exceptions have been described in high-grade tumors, including grade 3 endometrioid, serous and clear cell carcinomas, that can also express this marker [31].

14.2.3 Distinguishing Endometrial Carcinoma from Colonic Adenocarcinoma

Differentiation of colorectal adenocarcinoma from adenocarcinoma arising at other sites can sometimes be challenging. Carcinoembryonic antigen (CEA) is diffusely expressed, depending on the type of antibody used, from 83.8% (CEA D14) to 90.3% (mCEA) of colonic carcinomas, but is negative or focally positive in endometrial carcinoma [31].

Vimentin is typically positive in endometrioid carcinomas and negative or rarely positive in colonic adenocarcinomas [31].

Endometrial carcinomas are most commonly CK 7 positive and CK 20 negative, whereas the reverse is true for 75% of colorectal carcinomas. However, CK 7 (−)/CK 20 (−) or CK 7 (+)/CK 20 (+) colorectal carcinomas occur in 15% and 10%, respectively (Koss, p1663). The stain is cytoplasmic and/or membranous.

Colonic adenocarcinomas are also positive for CA 19–9 (83%), CA 15–3 (85%), EMA (88%) [1].

Pax8 is a nuclear marker typically expressed in carcinomas of Müllerian origin [32]. It is positive in endometrial carcinoma and commonly negative in colorectal carcinomas. CDX2 is a useful marker for colorectal carcinomas, with nuclear immunoreactivity; however, it may be positive in endometrial carcinomas, especially in cases with mucinous metaplasia. Finally, ER/PR are usually found in endometrial carcinomas.

14.2.4 Distinguishing Endometrial Carcinoma from Ovarian Carcinoma

Most ovarian carcinomas present a diffuse and strong positive reaction for WT-1 [33]. WT-1 is usually negative or only focally positive in serous endometrial carcinoma and negative in endometrioid carcinoma. However, some studies have found a significant percentage of endometrial serous carcinomas (up to 44%) to be positive for WT-1. In the same study they also found WT-1 to be a marker with a prognostic value, its expression being associated with a shorter disease-free survival [34]. Fadare et al. examined the combination of ER, PR, and WT-1 immunostaining for the differential diagnosis of ovarian serous from endometrial serous carcinomas; they concluded that a ER(−), PR(−), WT1(−) profile favors an endometrial origin, while a ER(+), PR(+), WT1(+) profile favors an ovarian origin [35]. However, a simultaneous involvement of the uterine corpus and one or both ovaries by an otherwise not otherwise specified adenocarcinoma may occur. Most commonly, it is an endometrioid subtype but sometimes it can be serous. With endometrioid adenocarcinomas involving the uterus and one or both ovaries, immunohistochemistry is of little or no value to establish the relationship between carcinomas also because the immunophenotype of a primary ovarian and uterine endometrioid adenocarcinoma is essentially identical and comprehensive molecular studies showed that these tumors are clonally related in almost all cases irrespectively of their clinico-pathological features [36].

References

1. Koss LG, Melamed MR, Koss LG. Koss' diagnostic cytology and its histopathologic bases. 5th ed. Philadelphia: Lippincott Williams & Wilkins; 2006.
2. Rabban J, Soslow R, Zaloudek C. Immunohistology of the female genital tract. In: Dabbs D, editor. Diagnostic immunohistochemistry: theranostic and genomic applications. 3rd ed. Philadelphia, PA: Saunders/Elsevier; 2010. p. 690–720.
3. Jain D, Mathur SR, Iyer VK. Cell blocks in cytopathology: a review of preparative methods, utility in diagnosis and role in ancillary studies. Cytopathology. 2014;25:356–71. https://doi.org/10.1111/cyt.12174.
4. Maksem J, Robboy S, Bishop J, et al. Endometrial cytology with tissue correlations. New York: Springer US; 2009.
5. Zhang H, Wen J, Xu PL, et al. Role of liquid-based cytology and cell block in the diagnosis of endometrial lesions. Chin Med J (Engl). 2016;129:1459–63. https://doi.org/10.4103/0366-6999.183431.
6. Satturwar S, Malenie R, Sutton A, et al. Validation of immunohistochemical tests performed on cytology cell block material: practical application of the College of American pathologists' guidelines. Cytojournal. 2019;16:6. https://doi.org/10.4103/cytojournal.cytojournal_29_18.
7. Briffod M, Hacene K, Le Doussal V. Immunohistochemistry on cell blocks from fine-needle cytopunctures of primary breast carcinomas and lymph node metastases. Modern Pathol. 2000;13:841–50. https://doi.org/10.1038/modpathol.3880149. An official journal of the United States and Canadian Academy of Pathology, Inc.
8. Pouliakis A, Margari N, Karakitsou E, et al. Artificial intelligence via competitive learning and image analysis for endometrial malignancies: discriminating endometrial cells and lesions. Int J Reliab Qual E-Healthc (IJRQEH). 2019;8:38–54. https://doi.org/10.4018/IJRQEH.2019100102.
9. Reid-Nicholson M, Iyengar P, Hummer AJ, et al. Immunophenotypic diversity of endometrial adenocarcinomas: implications for differential diagnosis. Mod Pathol. 2006;19:1091–100. https://doi.org/10.1038/modpathol.3800620.
10. Mutter GL, Lin MC, Fitzgerald JT, et al. Altered PTEN expression as a diagnostic marker for the earliest endometrial precancers. J Natl Cancer Inst. 2000;92:924–30. https://doi.org/10.1093/jnci/92.11.924.
11. Maxwell GL, Risinger JI, Gumbs C, et al. Mutation of the PTEN tumor suppressor gene in endometrial hyperplasias. Cancer Res. 1998;58:2500–3.
12. Norimatsu Y, Miyamoto M, Kobayashi TK, et al. Diagnostic utility of phosphatase and tensin homolog, beta-catenin, and p53 for endometrial carcinoma by thin-layer endometrial preparations. Cancer. 2008;114:155–64. https://doi.org/10.1002/cncr.23495.
13. Djordjevic B, Hennessy BT, Li J, et al. Clinical assessment of PTEN loss in endometrial carcinoma: immunohistochemistry outperforms gene sequencing. Mod Pathol. 2012;25:699–708. https://doi.org/10.1038/modpathol.2011.208.
14. Kobel M, Ronnett BM, Singh N, et al. Interpretation of P53 Immunohistochemistry in Endometrial Carcinomas: Toward Increased Reproducibility. Int J Gynecol Pathol. 2019;38:S123–S31. https://doi.org/10.1097/PGP.0000000000000488.
15. Garg K, Leitao MM Jr, Wynveen CA, et al. p53 overexpression in morphologically ambiguous endometrial carcinomas correlates with adverse clinical outcomes. Mod Pathol. 2010;23:80–92. https://doi.org/10.1038/modpathol.2009.153.
16. Norimatsu Y, Moriya T, Kobayashi TK, et al. Immunohistochemical expression of PTEN and beta-catenin for endometrial intraepithelial neoplasia in Japanese women. Ann Diagn Pathol. 2007;11:103–8.
17. Kosmas K, Stamoulas M, Marouga A, et al. Expres-sion of p53 in imprint smears of endometrial carcinoma. Diagn Cytopathol. 2014;42:416–22.

18. Wei JJ, Paintal A, Keh P. Histologic and immunohistochemical analyses of endometrial carcinomas: experiences from endometrial biopsies in 358 consultation cases. Arch Pathol Lab Med. 2013;137:1574–83. https://doi.org/10.5858/arpa.2012-0445-OA.
19. Kuramoto H, Hata H, Kato Y, et al. Mechanism of action of endocrine therapy of endometrial carcinoma. Gan To Kagaku Ryoho. 1989;16:1851–7.
20. Zhang Y, Zhao D, Gong C, et al. Prognostic role of hormone receptors in endometrial cancer: a systematic review and meta-analysis. World J Surg Oncol. 2015;13:208. https://doi.org/10.1186/s12957-015-0619-1.
21. Guan J, Xie L, Luo X, et al. The prognostic significance of estrogen and progesterone receptors in grade I and II endometrioid endometrial adenocarcinoma: hormone receptors in risk stratification. J Gynecol Oncol. 2019;30:e13. https://doi.org/10.3802/jgo.2019.30.e13.
22. Yemelyanova A, Ji H, Shih Ie M, et al. Utility of p16 expression for distinction of uterine serous carcinomas from endometrial endometrioid and endocervical adenocarcinomas: immunohistochemical analysis of 201 cases. Am J Surg Pathol. 2009;33:1504–14. https://doi.org/10.1097/PAS.0b013e3181ac35f5.
23. Deavers M, Coffey D. Precision molecular pathology of uterine cancer. New York: Springer; 2017.
24. An HJ, Lee YH, Cho NH, et al. Alteration of PTEN expression in endometrial carcinoma is associated with down-regulation of cyclin-dependent kinase inhibitor, p27. Histopathology. 2002;41:437–45. https://doi.org/10.1046/j.1365-2559.2002.01455.x.
25. Gilks CB, Oliva E, Soslow RA. Poor interobserver reproducibility in the diagnosis of high-grade endometrial carcinoma. Am J Surg Pathol. 2013;37:874–81. https://doi.org/10.1097/PAS.0b013e31827f576a.
26. Nemejcova K, Ticha I, Kleiblova P, et al. Expression, epigenetic and genetic changes of HNF1B in endometrial lesions. Pathol Oncol Res. 2016;22:523–30. https://doi.org/10.1007/s12253-015-0037-2.
27. Murali R, Davidson B, Fadare O, et al. High-grade endometrial carcinomas: morphologic and immunohistochemical features, diagnostic challenges and recommendations. Int J Gynecol Pathol. 2019;38:S40–63. https://doi.org/10.1097/PGP.0000000000000491.
28. Hu R, Jiang J, Song G, et al. Mixed large and small cell neuroendocrine carcinoma of the endometrium with serous carcinoma: A case report and literature review. Medicine (Baltimore). 2019;98:e16433.
29. Lee S, Rose MS, Sahasrabuddhe VV, et al. Tissue-based immunohistochemical biomarker accuracy in the diagnosis of malignant glandular lesions of the uterine cervix: a systematic review of the literature and meta-analysis. Int J Gynecol Pathol. 2017;36:310–22.
30. Malpica A. How to approach the many faces of endometrioid carcinoma. Mod Pathol. 2016;29:S29–44. https://doi.org/10.1038/modpathol.2015.142.
31. Stewart CJR, Crum CP, McCluggage WG, et al. Guidelines to aid in the distinction of endometrial and endocervical carcinomas, and the distinction of independent primary carcinomas of the endometrium and adnexa from metastatic spread between these and other sites. Int J Gynecol Pathol. 2019;38:S75–92. https://doi.org/10.1097/PGP.0000000000000553.
32. Mahajan A. Practical issues in the application of p16 immunohistochemistry in diagnostic pathology. Hum Pathol. 2016;51:64–74.
33. Rabban JT, Longacre TA. Chapter 18: Immunohistology of the female genital tract. In: Dabbs DJ, editor. Diagnostic immunohistochemistry theranostic and genomic applications. 4th ed. Philadelphia, PA: Saunders Elsevier; 2014. p. 653–709.
34. Laury AR, Perets R, Piao H, et al. A comprehensive analysis of PAX8 expression in human epithelial tumors. Am J Surg Pathol. 2011;35:816–26. https://doi.org/10.1097/PAS.0b013e318216c112.
35. Al-Hussaini M, Stockman A, Foster H, et al. WT-1 assists in distinguishing ovarian from uterine serous carcinoma and in distinguishing between serous and endometrioid ovarian carcinoma. Histopathology. 2004;44:109–15. https://doi.org/10.1111/j.1365-2559.2004.01787.x.

36. Hedley C, Sriraksa R, Showeil R, et al. The frequency and significance of WT-1 expression in serous endometrial carcinoma. Hum Pathol. 2014;45:1879–84. https://doi.org/10.1016/j.humpath.2014.05.009.
37. Fadare O, James S, Desouki MM, et al. Coordinate patterns of estrogen receptor, progesterone receptor, and Wilms tumor 1 expression in the histopathologic distinction of ovarian from endometrial serous adenocarcinomas. Ann Diagn Pathol. 2013;17:430–3. https://doi.org/10.1016/j.anndiagpath.2013.04.011.
38. Hájková N, Tichá I, Hojný J, et al. Synchronous endometrioid endometrial and ovarian carcinomas are biologically related: a clinico-pathological and molecular (next generation sequencing) study of 22 cases. Oncology Letters. 2019;17:2207–14.

Molecular Pathology of Endometrial Carcinoma: A General Appraisal

15

Alessia Di Lorito, Fernando Schmitt, Milo Frattini,
Luca Mazzucchelli, and Franco Fulciniti

15.1 Introduction

Traditionally, endometrial carcinomas (EC) have long been classified into two groups (types I and II) as Bockman reported in 1983 [1]. This first classification was based on metabolic, clinic and endocrine features of epithelial tumors. Type I carcinomas (70%) have been correlated with hyperlipidemia, diabetes, caused by hyperestrogenism due to anovulatory uterine bleeding, infertility or late onset of menopause and have a good prognosis. These latter usually of low grade, moderately/highly differentiated, arise from hyperplasia and show immunoreactivity for estrogen receptors.

In contrast, type II ECs arise from atrophic endometrium, are independent of metabolic or endocrine pathologies and are usually found in non-obese women. This type of EC consists of poorly differentiated tumors that are characterised by a

A. Di Lorito
Pathology Department, SS Annunziata Hospital, Chieti, Italy

F. Schmitt
Department of Pathology, Medical Faculty of Porto University, Porto, Portugal

Unit of Molecular Pathology, Institute of Molecular Pathology and Immunology, Porto University, Porto, Portugal

RISE, Health Research Network, Porto, Portugal

M. Frattini (✉)
Laboratory of Molecular Pathology, Institute of Pathology, Locarno, Switzerland

Istituto Cantonale di Patologia, Service of Molecular Pathology, Locarno, Switzerland
e-mail: milo.frattini@eoc.ch

L. Mazzucchelli
Istituto Cantonale di Patologia, Locarno, Switzerland

F. Fulciniti
Clinical Cytology Service, Istituto Cantonale dì Patologia, Ente Ospedaliero Cantonale, Locarno, Switzerland

non-favourable prognosis. Type II tumors are diagnosed in the advanced stages of the disease, are clinically aggressive, with a tendency to give metastases. On the contrary, type I tumors are more likely to be diagnosed in the early stage. Type I carcinomas generally arise from a sequence of premalignant conditions ranging from complex endometrial hyperplasia with atypia to EIN (endometrial intraepithelial neoplasia) while type II are preceded by EIC (endometrial serous and clear cell intraepithelial carcinoma) [2].

During the years, the first subtype was mainly represented by the endometrioid subtype, while the second group consists of serous and clear cells carcinomas [2].

In the past, therapeutic options were based only on histology, grade and stage based on surgical procedures, and women were stratified in risk group staging, these latter indicating which women should receive adjuvant therapy [radiation (vaginal or external beam), chemotherapy or both]. Additional features, such as patient age, depth of myometrial invasion, endocervical stromal involvement, lymph node status, and LVSI (lymphovascular space invasion), are increasingly being incorporated in pathological reports [3].

However, in the last years, the morphological and clinical differences between the two types are also well mirrored by different molecular changes, so data from literature established that type I and II cancers carry mutations in independent sets of genes.

15.1.1 Molecular Alterations in Endometrioid Carcinoma

A series of studies have reported that endometrioid carcinomas are highly mutated tumors, especially in *PTEN, PIK3CA, CTNNB1, ARID1A and KRAS genes. Furthermore, a non-negligible number of cases is characterised by* microsatellite instability (MSI) [4, 5].

The downstream receptor tyrosine kinases pathways, such as the MAP kinase pathway and the PI3K/mTOR pathway, are often dysregulated in endometrioid carcinoma, with the second one playing the most relevant role in endometrioid carcinogenesis. In the PI3K pathway, in fact, two actors are frequently altered: PTEN and PIK3CA, the last one encoding for the p110alpha subunit of PI3K protein. The expression of PTEN, which is a negative regulator of the PI3K/AKT/mTOR pathway leading to the hyperstimulation of the PI3K/mTOR axis, is frequently lost in endometrioid carcinomas following mutations inactivating the protein, and it is frequently altered in endometrial hyperplasia as well. As a consequence, this event is widely considered an initial and early step in the pathogenesis of endometrioid carcinoma [5].

PIK3CA mutations, causing the hyperactivation of PI3K and, then, Akt and finally mTOR, leading to deregulation of cell proliferation and apoptosis, are observed in 20% of endometrioid carcinomas.

The other pathway, the MAP kinase axis, is constitutively activated at the RAS level, typically after the occurrence of KRAS mutations. KRAS encodes for the small GTPase K-Ras protein, which is an oncoprotein involved in the activation of several signalling pathways, including the PI3K pathway, and in the activation of other small GTPases, such as RalA. KRAS mutations, resulting in Ras proteins with constitutively bound GTP, consequently activating these downstream oncogenic pathways [6]; they have been described, although at low frequencies, in both EC types.

Another alteration detected in endometrial carcinogenesis is represented by mutations in the CTNNB1 gene, which causes the Wnt/β-catenin pathway dysregulation. *CTNNB1* gene encodes for β-catenin, the key regulator enzyme of the whole pathway. Activation of the pathway leads to the accumulation of β-catenin and its migration into the nucleus, which leads to upregulation of many target genes such as cyclin D, *VEGF, Myc, and* E-*cadherin [7]*. CTNNB1 mutations have been observed almost exclusively in endometrioid carcinomas.

Finally, ARID1A encodes for the BAF250A tumor suppressor gene and is functionally involved in the SWI/SNF chromatin-remodelling complex. ARID1A mutations are commoner in type I than in type II ECs. Interestingly, ARID1A mutations can result in PI3K pathway activation via downregulation of PI3K interacting protein 1 (PIK3IP1). It is noteworthy that, at the preclinical level, it has been demonstrated that inhibition of the EZ2H methylesterase in ARID1A mutated ovarian cancer cells results in synthetic lethality, suggesting EZ2H as a potential new therapeutic target for ARID1A mutated tumors [8].

The last alteration frequently associated with endometrial carcinogenesis (detected in approximately 20–30% of cases) is microsatellite instability (MSI), which occurs secondary to sporadic mutations or epigenetic silencing in genes encoding for the DNA mismatch repair (MMR) enzymes. Microsatellites are short segments of repetitive DNA sequences found predominantly in non-coding DNA of the genome. MSI causes an increased possibility to develop changes in the number of repeat elements as compared with normal tissue. This is caused by DNA repair errors occurring during replication. These events have two possible causes. The first one is sporadic or germline mutations in at least one of the DNA mismatch repair enzymes (MMR), the second one is an epigenetic silencing due to *MLH1* gene promoter hypermethylation [9, 10]. The system of MMR proteins plays a key role in maintaining genomic stability, recognizing and repairing base–base mismatches and deletion/insertion of DNA generated during replication and recombination. Defects in MMR genes are associated with genome-wide instability and the progressive accumulation of mutations, especially in regions of microsatellites, leading to MSI. When the MMR system is altered, tumor develops through the selection of cancer-promoting mutations in pathways involved in apoptosis and cell growth. In addition to sporadic alterations, MMR genes can be altered through germline mutations, thus conferring a genetic predisposition to cancers referred to as Hereditary Non-Polyposis Colorectal Cancer (HNPCC) Syndrome or Lynch Syndrome. Women with this such an inherited disorder have an increased risk to develop ECs as well as colorectal carcinomas and other cancer types as stomach, small bowel, lower urinary tract and ovary.

15.2   Molecular Alterations in Non-endometrioid Carcinomas

Non-endometrioid carcinomas are characterised by different molecular alterations, such as p53 mutations (90%), inactivation of p16 (40%) and E-cadherin
(80–90%), c-erbB2 gene amplification (30%), alterations in genes involved in
the regulation of the mitotic spindle checkpoint (STK-15) and loss of heterozy-
gosity at multiple loci, reflecting the presence of chromosomal instability. *TP53*
gene encodes for the tumor suppressor protein p53, which plays a fundamental
role in conserving stability in the human genome. TP53 mutations decrease cell
ability to repair damage to DNA before entry to S-phase, leading to a higher
possibility that mutations will be present in the genome and passed to succes-
sive generations of cells. TP53 mutations have been detected in up to 90% of
serous carcinomas and in 10–20% of endometrioid carcinomas, which are
mostly grade 3 tumors.

Inactivation of the cell cycle regulator p16 is another frequent alteration, reported
in 40% of non-endometrioid carcinomas. Reduced expression of E-cadherin is frequent in EC and may be caused by LOH (loss-of-heterozygosity) or promoter
hypermethylation. In fact, LOH at16q22.1 is observed in almost 60% of non-
endometrioid carcinomas. Moreover, c-erbB2 protein overexpression (HER2) and/
or gene amplification are also associated more frequently with non-endometrioid
carcinomas (43% and 29%) than endometrioid ones. HER2 is encoded by the
c-erbB2 gene and is a member of the human epidermal growth factor receptor
(HER/EGFR/ERBB) family [2, 11] for which targeted therapies are available in
other tumor types, such as breast and gastric cancers.

The most typical molecular feature found in non-endometrioid tumor samples is
chromosomal instability. This phenomenon is due to the presence of widespread
chromosomal gains and losses, which reflect the presence of aneuploidy. Moreover,
numerous up-regulated genes (STK-15, BUB1, CCNB2) are also reported in this
group that are involved in the functioning of the mitotic spindle checkpoint. One of
them, STK-15 is essential for chromosome segregation and centrosome functions
and it is frequently amplified [2, 11].

15.3   New Perspectives after TCGA Classification

At the present time, the improvement of polymerase chain reaction-based clonal
assays and new molecular high-sensitivity technologies as Next Generation
Sequencing platforms (NGS) have led to the discovery of relevant biomarkers, facil-
itating a molecular approach to cancer diagnosis and treatment. In fact, by using
integrated genomics and epigenomic, transcriptomic, and proteomic techniques,
during the last years, the Cancer Genome Atlas (TCGA) has published the results
from an integrated endometrial cancer classification. The TCGA project demonstrated that EC can be classified into four categories based on distinct molecular
subgroups recognized by the number of gene alterations and tumor mutational bur-
den. The four groups are Polymerase Epsilon (POLE) ultramutated, Microsatellite

Instability hypermutated, copy-number low, and copy-number high serous-like carcinomas [12].

The first molecular subgroup showed copy number-stable and ultra-mutated carcinomas with recurrent mutations in the exonuclease domain of DNA polymerase epsilon (*POLE*). This gene is involved in nuclear DNA replication and repair. The tumors belonging to the POLE group are characterized by very high somatic mutation frequencies, exceeding 100 mutations per megabase (Mb) in the majority of cases. Subsequent studies showed that, at the histological point of view, *POLE*-mutant ECs are typically high-grade endometrioid carcinomas with a superficial broad front of invasion pattern and the presence of prominent tumor-infiltrating lymphocyes (TILs) with tumor giant cells. However, it has also been noted that *POLE* mutant carcinomas are morphologically difficult to be classified.

The second molecular EC subgroup reported by the TCGA are hypermutated, MSI-high carcinomas caused by alterations of mismatch repair proteins. The predominant underlying mechanism is epigenetic silencing of *MLH1* by promoter hypermethylation, but tumors associated with Lynch syndrome also fall into this group.

MMR system deficiency leads to high mutation frequencies, usually exceeding 10 mutations per Mb. Carcinomas with MSI or with an MMR deficient system, in general, are mostly endometrioid; however, non-endometrioid subtypes have been also reported. In this group of tumors, the presence of TILs and a microcystic elongated and fragmented (MELF) pattern of invasion and LVSI have been described.

The third molecular subgroup is genomically relatively stable, MMR-proficient carcinomas. They are characterized by having a moderate number of mutations, mostly within the PI3K/Akt and Wnt signalling pathways. This subgroup is almost exclusively composed of neoplasms expressing both estrogen- (ER) and progesterone receptor (PR).

The fourth subgroup has high somatic copy number alterations, very similar to high-grade serous ovarian carcinomas, and shows frequent *TP53* mutations (92%). Morphologically, these carcinomas are of high-grade (grade 3), including most serous endometrial carcinomas, and they are commonly reported as 'serous-like cancers' [12, 13].

15.4 Technical Analyses

The TCGA-based classification can be easily performed using the NGS approach. Indeed, by using large panels (e.g.: Foundation One from Roche, Oncomine Comprehensive assay version 4 Plus from Thermo Fisher, TSO500 from Illumina), a combined analysis of gene mutations, copy number variations and microsatellite instability evaluation can be done in a single experiment, with progressively decreasing costs with the passage of time. However, these large panels may show failure in small biopsies for two main reasons: an insufficient number of cancer cells or an insufficient percentage of tumor cells in the tissue available for molecular characterization. In these cases, smaller, customized panels are very useful to perform molecular characterization even in these difficult cases.

On the other hand, standard and "old" techniques such as Sanger sequencing and Fragment analysis/immunohistochemistry, for assessing gene mutations and the status of the MMR repair system may be used as well, with good results. However, if Sanger sequencing is time consuming even if the analysis is limited to POLE gene mutations (due to the large spectrum of POLE mutations that must be analysed), the combination of fragment analysis for assessing MSI and immunohistochemistry for the evaluation of expression of the proteins of the MMR system still represents the gold standard method: indeed, the international guidelines recommend to use these two approaches, by applying recognized panels or molecular antibodies. As it concerns protein expression, normal expression indicates that a protein is functional and, if all the four proteins are expressed as in the internal controls, the overall MMR system is considered intact. When at least one protein is absent or heavily underexpressed, the MMR shows a deficiency in a member and is overall inactive, with the following evaluation criteria: when the downregulation is limited to PMS2 or MSH6, it means that the corresponding gene is altered; on the contrary, since MLH1 dominates over PMS2 and MSH2 over MSH6, when the genetic inactivation is at MLH1 or MSH2 level, pathologists observe the absence of a couple of proteins, MLH1/PMS2 and MSH2/MSH6, respectively. The precise identification of the altered proteins is of absolute relevance for the identification of the majority of Lynch-syndrome-associated cases: in fact, PMS2, MLH2 and MSH6 are almost exclusively altered as a result of inherited genetic mutations, and therefore only in Lynch syndrome patients. On the contrary, MLH1 can be deregulated in both inherited and sporadic cases, always showing protein downregulation. However, molecular analyses can provide a substantial help: promoter hypermethylation has been observed in a substantially exclusive association with sporadic cases.

In general, immunohistochemistry may, therefore, provide the precise identification of the member of the MMR system that is altered in a given case and, in the majority of patients, may also subdivide inherited from sporadic cases. However, this technique is often challenging, because of some technical problems due to the intrinsic features of the proteins belonging to the MMR system: these proteins are located in the nucleus and antigen retrieval or accessibility to the reagents cannot be efficient always. The molecular analysis of MSI may bypass this problem, because MSI evaluation represents a more objective technique, although it is not able to identify which protein is precisely altered. A tumor is typically classified as having MSI if changes are detected in at least 2 of 5 loci using the PCR method. The use of NCI (BAT-25, BAT-26, D2S123, D5S346 and D17S250) and Promega (BAT-25, BAT-26, NR-21, NR-24 and MONO-27) panels in PCR and the use of MLH1, MSH2, MSH6 and PMS2 proteins in IHC are currently the gold standard approaches [10].

15.5   Clinical Relevance of Molecular Characterization

The molecular characterization of EC is clinically relevant at the prognostic level: the four groups are associated with a different course of the disease. POLE mutant patients, irrespectively on MSI status (in rare cases both POLE mutations and MSI have been observed), display the best follow-up period with nearly 100% of patients alive 5 years after the first diagnosis. In POLE wild-type, MSI patients, the course of the disease is still very good, with more than 80% of the 5-year survival rate. Starting from the third group, the possibility to be alive 5 years after disease diagnosis decreases significantly, with the worse situation observed in the fourth group.

In parallel, the molecular classification has opened new findings of the possibility to use new therapies in advanced/metastatic and high aggressive cancers [12]. Indeed, among patients with metastatic cancer, cancers with MSI may be efficiently addressed to the administration of immune checkpoint inhibitors, including anti-PD-1 antibody and anti-PD-L1 antibody immunotherapies, especially with pembrolizumab. In fact, the approval of immune checkpoint inhibitor pembrolizumab (anti-PD-1), for all solid tumors with defective DNA MMR system, could lead to a benefit for 20–30% of patients with advanced EC [14].

In addition, due to the role of the PI3K/mTOR pathway in endometrial carcinogenesis, it has been proposed, and now it is under evaluation in several clinical trials, the use of mTOR inhibitors [13].

Finally, the application of an NGS approach may lead to the discovery of new potential molecular targets, in a subgroup of EC patients: in this way, it has been proven that patients with relevant genomic alterations may respond to other targeted therapies such as BGJ-98 (in FGFR2 altered cases) or pazopanib (in POLE mutant cases) [15]. Molecular analyses may be also successfully performed on Liquid Based Cytology (LBC) samples [16] (See also Chap. 16 of this book for further information).

References

1. Bokhman JV. Two pathogenetic types of endometrial carcinoma. Gynecol Oncol. 1983;15:10–7.
2. Llobet D, Pallares J, Yeramian A, et al. Molecular pathology of endometrial carcinoma: practical aspects from the diagnostic and therapeutic viewpoints. J Clin Pathol. 2009;62:777–85.
3. Colombo N, Creutzberg C, Amant F, et al. Sessa C; ESMO-ESGO-ESTRO endometrial consensus conference working group. ESMO-ESGO-ESTRO consensus conference on endometrial Cancer: diagnosis, treatment and follow-up. Ann Oncol. 2016;27:16–41.
4. Prendergast EN, Liu AJ, Fahey JN, et al. Genomic alteration patterns and clinical actionability of comprehensive genomic profiling for the management of endometrial carcinoma. Gynecol Oncol. 2017;145:2–220.
5. Di Lorito A, Rosini S, Falò E, et al. Molecular alterations in endometrial archived liquid-based cytology. Diagn Cytopathol. 2013;41:492–6.
6. Sideris M, Emin EI, Abdullah Z, et al. The role of KRAS in endometrial Cancer: a Mini-review. Anticancer Res. 2019;39:533–9.
7. Kurnit KC, Kim GN, Fellman BM, et al. CTNNB1 (beta-catenin) mutation identifies low grade, early stage endometrial cancer patients at increased risk of recurrence. Mod Pathol. 2017;30:1032–41.

8. Takeda T, Banno K, Okawa R, et al. ARID1A gene mutation in ovarian and endometrial cancers (review). Oncol Rep. 2016;35:607–13.
9. Yamashita H, Nakayama K, Ishikawa M, et al. Microsatellite instability is a biomarker for immune checkpoint inhibitors in endometrial cancer. Oncotarget. 2017;9:5652–64.
10. Kunitomi H, Banno K, Yanokura M, et al. New use of microsatellite instability analysis in endometrial cancer. Oncol Lett. 2017;14:3297–301. https://doi.org/10.3892/ol.2017.6640.
11. Remmerie M, Janssens V. Targeted therapies in type II endometrial cancers: too little, but not too late. Int J Mol Sci. 2018;19:2380.
12. Cancer Genome Atlas Research Network, Kandoth C, Schultz N, et al. Integrated genomic characterization of endometrial carcinoma. Nature. 2013;497:67–73.
13. McAlpine J, Leon-Castillo A, Bosse T. The rise of a novel classification system for endometrial carcinoma; integration of molecular subclasses. J Pathol. 2018;244:538–49.
14. Lorenzi M, Amonkar M, Zhang J, et al. Epidemiology of Microsatellite Instability High (MSI-H) and Deficient Mismatch Repair (dMMR) in Solid Tumors: a structured literature review. J Oncol. 2020;2020., Article ID 1807929:17.
15. Vermij L, Smit V, Nout R, et al. Incorporation of molecular characteristics into endometrial cancer management. Histopathology. 2020;76:52–63.
16. Fulciniti F, Yanoh K, Karakitsos P, et al. The Yokohama system for reporting directly sampled endometrial cytology: the quest to develop a standardized terminology. Diagn Cytopathol. 2018;46:400–12.

Molecular Pathology of Endometrial Carcinoma on LBC Samples and Cell Blocks

16

Diana Martins, Fernando Schmitt, Milo Frattini, and Franco Fulciniti

16.1 Introduction

Stratification of Endometrial Carcinoma (EC) is an essential criterion for the accurate evaluation of prognosis, and its ultimate goal is to improve the outcome of patients through the optimization of treatment guidelines. There are presently two kinds of stratification systems, the conventional pathology assignment and the emerging molecular classification proposed by The Cancer Genome Atlas (TCGA) [1, 2]. In the past, endometrial cancer diagnosis and treatment has been driven by histopathological

D. Martins
I3S, Instituto de Investigação e Inovação em Saúde, University of Porto, Porto, Portugal

Polytechnic Institute of Coimbra, ESTESC-Coimbra Health School, Department of Biomedical Laboratory Sciences, Coimbra, Portugal

University of Coimbra, Coimbra Institute for Clinical and Biomedical Research (iCBR) Area of Environment Genetics and Oncobiology (CIMAGO), Biophysics Institute of Faculty of Medicine, Coimbra, Portugal

F. Schmitt (✉)
Department of Pathology, Medical Faculty of Porto University, Porto, Portugal

Unit of Molecular Pathology, Institute of Molecular Pathology and Immunology, Porto University, Porto, Portugal

RISE, Health Research Network, Porto, Portugal
e-mail: fschmitt@ipatimup.pt

M. Frattini
Laboratory of Molecular Pathology, Institute of Pathology, Locarno, Switzerland

Istituto Cantonale di Patologia, Service of Molecular Pathology, Locarno, Switzerland

F. Fulciniti
Clinical Cytology Service, Istituto Cantonale dì Patologia, Ente Ospedaliero Cantonale, Locarno, Switzerland

© The Author(s), under exclusive license to Springer Nature Singapore Pte Ltd. 2022
Y. Hirai, F. Fulciniti (eds.), *The Yokohama System for Reporting Endometrial Cytology*, https://doi.org/10.1007/978-981-16-5011-6_16

subtype, grade, stage, myometrial invasion, and lymphovascular invasion [1]. Although histopathologic prognostic factors are associated with the risk of recurrence and the response to adjuvant therapy in EC, none of the current risk stratification systems has a satisfactory predictive power for the identification of patients at risk for recurrence or metastatic disease [3]. Inaccurate EC histopathologic classification, imprecise risk stratification, and diverse treatment strategies [4] are the main causes of the poor predictive power of this prognostic system [5]. Furthermore, heterogeneity of EC is ignored in this traditional system [6], resulting in cancers of the same stage and histology having very distinct molecular and genomic profiles. Translational research is progressing rapidly and EC specific precision medicine is evolving. Molecular characterization of EC is advancing rapidly, and understanding how a specific subset of mutations or combinations of molecular characteristics can lead to targeted therapeutics may have an utmost importance in the treatment of advanced or recurrent EC [7]. Molecular prognostic factors, such as POLE mutation, copy number variation (CNV), and genetic instability related to an abnormal expression of mismatch repair proteins, were able to classify EC into four molecular subtypes: POLE-mutant, microsatellite instability (MSI), low copy number variation (CNV-L), and high copy number variation (CNV-H) [2]. Molecular characterization of EC into subgroups has prognostic and therapeutic implications. Further development of an integrated molecular risk profile may identify patients who could benefit from additional treatment because of a higher risk of recurrence [3].

16.2   Liquid-Based Cytology and Cell Blocks in Endometrial Cancer Diagnosis

To diagnose EC, transvaginal ultrasound, hysteroscopy, endometrial biopsy, dilatation, and curettage are the most commonly used strategies. Although endometrial hysteroscopy and/or curettage remains the most common diagnostic modality, it is anyhow an invasive procedure with the risk of infection and perforation [8]. The cytological diagnosis of endometrial lesions offers a valuable alternative to biopsy and it may be useful in obtaining material to study prognostic and predictive markers [9, 10]. Endometrial cytology allows cytomorphologic interpretation of a wide range of lesions, using a lesser amount of cells [11], showing a sensitivity of 79% and a specificity of 99% [12, 13]. However, its main limitations are represented by artefacts associated with conventional smear quality, the presence of excess blood, and presence of 3D dense overlapping cell clusters [14]. The introduction of liquid-based cytology (LBC) and the experience acquired in specimen collection, processing, and interpretation represents an opportunity to reconsider endometrial cytology as a means of obtaining morphological diagnosis [14]. LBC represents a great advantage in cell preservation and clearance of background, critical for an accurate diagnosis [15]. Since satisfactory sampling is one of the major factors affecting the quality of the cellular sample in endometrial cytology, several endometrial samplers have been developed, such as Endoflower, the Tao brush, and Endocyte [16, 17]. A recent study compared the effectiveness between Li brushes and Tao Brushes for the diagnosis of endometrial lesions. The results suggest that cytologic samples obtained by both Li brushes and Tao brushes have a high accuracy with optimal cyto-histologic correlation in detecting EC and the Li brush may be a better endometrial cell collector [18]. Endometrial cytology seems to be a reliable approach for evaluating endometrium with a lower insufficient sample rate; anyhow the sensitivity of LBC for the detection of endometrial cancer is between 70% and 96% and, hence, still unsatisfactory [13]. Moreover, in the endometrial cytologic samples, the evaluation of glandular architecture relies indirectly on the morphology of cell clusters and LBC alone may not be sufficient for the detection [13] of early EC, hence the development of new diagnostic tools is urgently needed.

Cell blocks or Cytoblocks (CBs) are an important diagnostic tool in cytopathology and are already a well-established technique for cytological material preparation [11]. They are a type of preparation by which cytological material is compacted into a pellet, fixed in buffered formalin, and embedded in paraffin blocks for further processing as a histological specimen [19]. Using residual cytological material offers the advantage of resolving the complexity of dense cellular clusters on thin-layer preparations and adds valuable diagnostic evidence by producing tissue fragments that cannot be obtained by cytology alone and has the further diagnostic advantage of permitting architectural study [19]. Some optimization is required due to the diversity of cell block techniques: fixed sediment method, Agar embedding method, plasma-thrombin clot method, or even automated method. However, it is self-evident that they offer many advantages over other cytological preparations and represent their invaluable complement [20]. Multiple serial sections may be obtained by cell blocks for performing special stains, immunocytochemistry (ICC), and molecular tests (MT), utilizing formalin-fixed paraffin-embedded (FFPE) tissue sections, allowing microscopic examination of the tumor and also quantification of tumor fraction. They can also be used for microarray construction in the study of immunomarkers [21]. In fact, an immunohistochemistry antibody panel has been exploited to perform differential diagnosis of endometrial lesions using: ER, PR, Phosphate and Tensin homologue (PTEN), p53, Ki-67, p16 between endometrial and endocervical adenocarcinoma and of endometrioid carcinoma from serous and clear cell carcinoma [19].

16.3 Molecular Tests in Liquid-Based Cytology and Cell Blocks: A New ERA

Molecular cytopathology is a rapidly evolving field of modern cytopathology, which underlines the effective interplay between genomics and cytology [22]. The development of personalized medicine, including the cancer gene testing on cytological samples from patients with surgically unresectable, high-stage, locally advanced, and metastatic malignancies is crucial [23].

DNA extracted from cytology preparations is, in general terms, better in quality than histology due to lack of stroma with resultant increase in the cancer cells fraction and, consequently, lower amounts of DNA may be sufficient for routine molecular tests [24]. CBs are most commonly used for molecular tests [19, 25], although smears and LBC allow comparable amounts of DNA [26, 27].

However, CBs present nucleic acid preservation issues similar to those of histological FFPE material. Neutral buffered formalin, the most commonly used fixative for tissue preservation, induces methylene bridging of bases and the formation of cross-links between nucleic acids and available proteins [27]. With regard to RNA extraction, the addition of proteinase K digestion followed by heating steps is employed in an effort to break the methylene bridges and optimize the quality. Even with optimized digestion and heating steps, however, it is not always possible to remove all modifications such as residual methyl groups from FFPE-extracted RNA [28].

Cellularity is also an important issue in CBs; it is evaluated through the examination of an H&E-stained section prepared from the CB. The percentage of tumor cells for molecular testing is extrapolated based on the H&E section, which constitutes a disadvantage, especially when the cellularity is low [29]. In a CB with low tumor content, some practices including additional dedicated steps for cellular enrichment are needed [30, 31], due to the fact that the analytical sensitivity of molecular diagnostic assays depends on the percentage of tumor cells vs. the total number of cells in the preparation [26].

Several studies have described using LBC specimens for molecular analysis [26, 32]. The issues regarding LBC mainly concern the DNA preservation/RNA preservation and the amount of tumor cells [26]. The alcohol-based fixative has a direct effect on DNA preservation. For example, LBC samples fixed with CytoRich Red have shown worse DNA preservation due to the presence of a small amount ($<1\%$) of formaldehyde that may cause DNA degradation and modification [29]. Also, material from CytoLyt samples has demonstrated optimal RNA integrity being suitable for nucleic acid isolation and subsequent analysis by reverse transcription-PCR (RT-PCR) [33]. In a study, the authors assessed LBC samples to verify the concordance with the FFPE DNA results. The results obtained were concordant; the wild type cases in FFPE were also wild type in LBC and vice versa for the mutated case [26].

When a low-sensitive direct sequencing method is used, neoplastic cell enrichment is mandatory [34]. However, manual or laser microdissection on LCB slides remains difficult and time consuming [26]. Alternatively, molecular techniques, such as real-time PCR methods, can be used directly on the DNA extracted from the preserving solution, without slide preparation [35], suggesting LBC as a valuable method for molecular testing.

16.4   NGS (Next-Generation Sequencing)

More recently, cytological specimens have also been validated for next-generation sequencing (NGS) assays to simultaneously screen different types of mutations in multiple genes and in multiple patient samples using small amounts of material [36]. NGS has also enabled the detection of somatic mutations at low frequencies in cancer tissues, and the identification of tumor DNA derived from cancer tissues (ctDNA) in circulating peripheral blood. Genomic profiles of ctDNA have been shown to closely associate with those of the malignant tumor, suggesting that ctDNA could contribute to the diagnosis and personalized treatment of cancer [37]. NGS and fully automated platforms may need specific sample requirements. In fact, establishing the number of cells needed to perform NGS from a cytology sample is crucial. The studies that applied NGS to cytological material had usually needed/exploited at least 20% neoplastic cellularity [38]. Moreover, sample requirements depend on target capture, gene panel, and platform types. For example, NGS assays from illumin a usually require more cells and/or higher DNA input than Ion Torrent NGS, due to the different chemistry that is the base of the two methodologies (Hybrid capture vs amplicon-based) [38]. The success of molecular testing on cytology is strongly dependent on standardized pre-analytical protocols. Besides specific analytical issues, dependent on the given molecular technique, appropriate test request, specimen collection, fixation, processing, staining, tumor fraction enrichment, DNA quality/quantity assessment, and storage conditions are also crucial [26]. In fact, a recent study compared the efficacy of LBC with liquid-based genetic diagnosis (LBGDx) by amplicon sequencing of five genes including PTEN, PIK3CA, CTNNB1, KRAS, and TP53 in LBC subjects who underwent endometrial screening [26]. The results show that LBGDx could diagnose EC at a sensitivity of 85%. Among the EC cases diagnosed by LBGDx, a few were cytologically negative, corroborating the fact that it is difficult to diagnose EC by cytology alone [39]. The study suggests that, if the cytological examination was combined with LBGDx, all ECs could be correctly diagnosed and additional histological examination is not necessary for five of the six suspicious cases on cytological examination alone [13]. Moreover it was prospected that, by increasing the number of genes studied, the sensitivity of the genetic diagnosis could, in turn, further be increased, suggesting that the combination of LBC and LBGDx was able to diagnose all EC cases, supporting the assertion that LBGDx is a useful strategy to improve the sensitivity of screening of EC by LBC [13]. A recent study proposed chromosomal instability (CIN),(i.e.: an increase in rate of addition or loss of entire chromosomes or sections thereof) as a biological mechanism for the evolution of different histopathologic and molecular pathologic prognostic factors and may be applicable to the risk stratification system of EC [40]. The results suggest that CIN was associated with unfavourable prognosis in EC. Moreover, the combination of pathology, CIN signatures and molecular classification proposed by TCGA, leads to a further refinement of prognostic scores, with potential clinical utility in EC. This integrated risk model holds promise to reduce both overtreatment and undertreatment [40].

Based on the above, and although further studies are needed to support this important studies, it is expected that the implementation of genetic analysis in LCB and CB for screening of endometrial cancer will contribute to reduce the mortality in patients with EC by early detection of neoplastic lesions.

References

1. Colombo N, Creutzberg C, Amant F, et al. ESMO-ESGO-ESTRO consensus conference on endometrial Cancer: diagnosis, treatment and follow-up. Ann Oncol [Internet]. 2016;27:16–41. Available from: https://linkinghub.elsevier.com/retrieve/pii/S0923753419353372.
2. Levine DA. Integrated genomic characterization of endometrial carcinoma. Nature [Internet]. 2013;497:67–73. Available from: http://www.nature.com/articles/nature12113.
3. Chang Z, Talukdar S, Mullany SA, et al. Molecular characterization of endometrial cancer and therapeutic implications. Curr Opin Obstet Gynecol [Internet]. 2019;31:24–30. Available from: http://journals.lww.com/00001703-201902000-00006.
4. Bendifallah S, Canlorbe G, Collinet P, et al. Just how accurate are the major risk stratification systems for early-stage endometrial cancer? Br J Cancer [Internet]. 2015;112:793–801. Available from: http://www.nature.com/articles/bjc201535.
5. Clarke BA, Gilks CB. Endometrial carcinoma: controversies in histopathological assessment of grade and tumour cell type. J Clin Pathol [Internet]. 2010;63:410–5. Available from: http://jcp.bmj.com/lookup/doi/10.1136/jcp.2009.071225.
6. Stelloo E, Nout RA, Osse EM, et al. Improved risk assessment by integrating molecular and Clinicopathological factors in early-stage endometrial Cancer—combined analysis of the PORTEC cohorts. Clin Cancer Res [Internet]. 2016;22:4215–24. Available from: http://clincancerres.aacrjournals.org/lookup/doi/10.1158/1078-0432.CCR-15-2878.
7. Arend RC, Jones BA, Martinez A, et al. Endometrial cancer: molecular markers and management of advanced stage disease. Gynecol Oncol [Internet]. 2018;150:569–80. Available from: https://linkinghub.elsevier.com/retrieve/pii/S0090825818309028.
8. Tokuda H, Nakago S, Kato H, et al. Bleeding in the retroperitoneal space under the broad ligament as a result of uterine perforation after dilatation and curettage: report of a case. J Obstet Gynaecol Res. 2017;43:779–82.
9. Buccoliero AM, Castiglione F, Gheri CF, et al. Liquid-based endometrial cytology: its possible value in postmenopausal asymptomatic women. Int J Gynecol Cancer [Internet]. 2007;17:182–7. Available from: https://clsjournal.ascls.org/lookup/doi/10.1111/j.1525-1438.2006.00757.x.
10. Kondo E, Tabata T, Koduka Y, et al. What is the best method of detecting endometrial cancer in outpatients?-endometrial sampling, suction curettage, endometrial cytology. Cytopathology [Internet]. 2007;19:28–33. Available from: http://doi.wiley.com/10.1111/j.1365-2303.2007.00509.x.
11. Fulciniti F, Yanoh K, Karakitsos P, et al. The Yokohama system for reporting directly sampled endometrial cytology: the quest to develop a standardized terminology. Diagn Cytopathol [Internet]. 2018;46:400–12. Available from: http://doi.wiley.com/10.1002/dc.23916.
12. Yanoh K, Hirai Y, Sakamoto A, et al. New terminology for intrauterine endometrial samples: a group study by the Japanese Society of Clinical Cytology. Acta Cytol [Internet]. 2012;56:233–41. Available from: https://www.karger.com/Article/FullText/336258.
13. Matsuura M, Yamaguchi K, Tamate M, et al. Efficacy of liquid-based genetic diagnosis of endometrial cancer. Cancer Sci [Internet]. 2018;109:4025–32. Available from: https://onlinelibrary.wiley.com/doi/abs/10.1111/cas.13819.
14. Di Lorito A, Rosini S, Falò E, et al. Molecular alterations in endometrial archived liquid-based cytology. Diagn Cytopathol [Internet]. 2013;41:492–6. Available from: http://doi.wiley.com/10.1002/dc.22869.
15. Lv S, Wang R, Wang Q, et al. A novel solution configuration on liquid-based endometrial cytology. PLoS One [Internet]. 2018;13:1–13. Available from: https://doi.org/10.1371/journal.pone.0190851.
16. Iavazzo C, Vorgias G, Mastorakos G, et al. Uterobrush method in the detection of endometrial pathology. Anticancer Res [Internet]. 2011;31:3469–74. Available from: http://www.ncbi.nlm.nih.gov/pubmed/21965763.
17. Yang X, Ma K, Chen R, et al. Liquid-based endometrial cytology associated with curettage in the investigation of endometrial carcinoma in a population of 1987 women. Arch Gynecol

Obstet [Internet]. 2017;296:99–105. Available from: http://link.springer.com/10.1007/s00404-017-4400-2.

18. Lv S, Wang Q, Li Y, et al. A clinical comparative study of two different endometrial cell samplers for evaluation of endometrial lesions by Cytopathological diagnosis. Cancer Manag Res [Internet]. 2020;12:10551–7. Available from: https://www.dovepress.com/a-clinical-comparative-study-of-two-different-endometrial-cell-sampler-peer-reviewed-article-CMAR.

19. Nambirajan A, Jain D. Cell blocks in cytopathology: an update. Cytopathology [Internet]. 2018;29:505–24. Available from: http://doi.wiley.com/10.1111/cyt.12627.

20. Saqi A. The state of cell blocks and ancillary testing: past, present, and future. Arch Pathol Lab Med [Internet]. 2016;140:1318–22. Available from: http://meridian.allenpress.com/aplm/article/140/12/1318/194165/The-State-of-Cell-Blocks-and-Ancillary-Testing.

21. El Hag MI, Ha J, Farag R, et al. Utility of GATA-3 in the work-Up of breast adenocarcinoma and its differential diagnosis in serous effusions. Diagn Cytopathol [Internet]. 2016;44:731–6. Available from: http://doi.wiley.com/10.1002/dc.23521.

22. Idowu MO. Epidermal growth factor receptor in lung cancer: the amazing interplay of molecular testing and cytopathology. Cancer Cytopathol [Internet]. 2013;121:540–3. Available from: http://doi.wiley.com/10.1002/cncy.21321.

23. Clark DP. Seize the opportunity. Cancer Cytopathol [Internet]. 2009;117:289–97. Available from: http://doi.wiley.com/10.1002/cncy.20045.

24. Gailey MP, Stence AA, Jensen CS, et al. Multiplatform comparison of molecular oncology tests performed on cytology specimens and formalin fixed, paraffin-embedded tissue. Cancer Cytopathol [Internet]. 2015;123:30–9. Available from: http://doi.wiley.com/10.1002/cncy.21476.

25. da Cunha Santos G, Saieg MA. Preanalytic specimen triage: smears, cell blocks, cytospin preparations, transport media, and cytobanking. Cancer Cytopathol [Internet]. 2017;125:455–64. Available from: http://doi.wiley.com/10.1002/cncy.21850.

26. Bellevicine C, Malapelle U, Vigliar E, et al, Troncone G. How to prepare cytological samples for molecular testing. J Clin Pathol [Internet]. 2017;70:819–26. Available from: http://jcp.bmj.com/lookup/doi/10.1136/jclinpath-2017-204561.

27. Doxtader EE, Cheng Y-W, Zhang Y. Molecular testing of non–small cell lung carcinoma diagnosed by endobronchial ultrasound–guided Transbronchial fine-needle aspiration: the Cleveland Clinic experience. Arch Pathol Lab Med [Internet]. 2019;143:670–6. Available from: http://meridian.allenpress.com/aplm/article/143/6/670/10030/Molecular-Testing-of-NonSmall-Cell-Lung-Carcinoma.

28. Bridge JA. Reverse transcription-polymerase chain reaction molecular testing of cytology specimens: pre-analytic and analytic factors. Cancer Cytopathol [Internet]. 2017;125:11–9. Available from: http://doi.wiley.com/10.1002/cncy.21762.

29. Roy-Chowdhuri S, Aisner DL, Allen TC, et al. Biomarker testing in lung carcinoma cytology specimens: a perspective from members of the pulmonary pathology society. Arch Pathol Lab Med [Internet]. 2016;140:1267–72. Available from: http://meridian.allenpress.com/aplm/article/140/11/1267/65644/Biomarker-Testing-in-Lung-Carcinoma-Cytology.

30. Crapanzano JP, Heymann JJ, Monaco S, et al. The state of cell block variation and satisfaction in the era of molecular diagnostics and personalized medicine. Cytojournal [Internet]. 2014;11:7. Available from: https://cytojournal.com/the-state-of-cell-block-variation-and-satisfaction-in-the-era-of-molecular-diagnostics-and-personalized-medicine/.

31. Bellevicine C, Malapelle U, de Luca C, et al. EGFR analysis: current evidence and future directions. Diagn Cytopathol [Internet]. 2014;42:984–92. Available from: http://doi.wiley.com/10.1002/dc.23142.

32. Malapelle U, de Rosa N, Rocco D, et al. EGFR and KRAS mutations detection on lung cancer liquid-based cytology: a pilot study. J Clin Pathol [Internet]. 2012;65:87–91. Available from: http://jcp.bmj.com/lookup/doi/10.1136/jclinpath-2011-200296.

33. Malapelle U, de Rosa N, Bellevicine C, et al. EGFR mutations detection on liquid-based cytology: is microscopy still necessary? J Clin Pathol [Internet]. 2012;65:561–4. Available from: http://jcp.bmj.com/lookup/doi/10.1136/jclinpath-2011-200659.

34. Malapelle U, Bellevicine C, Zeppa P, et al. Cytology-based gene mutation tests to pre-dict response to anti-epidermal growth factor receptor therapy: a review. Baloch Z, editor. Diagn Cytopathol [Internet]. 2011;39:703–10. Available from: http://doi.wiley.com/10.1002/dc.21512.
35. Bellevicine C, Malapelle U, Vigliar E, et al. Epidermal growth factor receptor test performed on liquid-based cytology lung samples: experience of an academic referral center. Acta Cytol [Internet]. 2014;58:589–94. Available from: https://www.karger.com/Article/FullText/369756.
36. Vigliar E, Malapelle U, de Luca C, et al. Challenges and opportunities of next-generation sequencing: a cytopathologist's perspective. Cytopathology [Internet]. 2015;26:271–83. Available from: http://doi.wiley.com/10.1111/cyt.12265.
37. Siravegna G, Marsoni S, Siena S, et al. Integrating liquid biopsies into the management of can-cer. Nat Rev Clin Oncol [Internet]. 2017;14:531–48. Available from: https://doi.org/10.1038/nrclinonc.2017.14.
38. Roy-Chowdhuri S, Stewart J. Preanalytic variables in cytology: lessons learned from next-generation sequencing—the MD Anderson experience. Arch Pathol Lab Med [Internet]. 2016;140:1191–9. Available from: http://meridian.allenpress.com/aplm/article/140/11/1191/65706/Preanalytic-Variables-in-Cytology-Lessons-Learned.
39. Fujiwara H, Takahashi Y, Takano M, et al. Evaluation of endometrial cytology: Cytohistological correlations in 1,441 Cancer patients. Oncology [Internet]. 2015;88:86–94. Available from: https://www.karger.com/Article/FullText/368162.
40. Li Y, Li J, Guo E, et al. Integrating pathology, chromosomal instability and mutations for risk stratification in early-stage endometrioid endometrial carcinoma. Cell Biosci [Internet]. 2020;10:122. Available from: https://cellandbioscience.biomedcentral.com/articles/10.1186/s13578-020-00486-0.

Future Challenges and Perspectives of Endometrial Cytology

17

Yasuo Hirai, Tadao K. Kobayashi, Yoshiaki Norimatsu, Jun Watanabe, Tetsuji Kurokawa, Akiko Shinagawa, Akira Mitsuhashi, and Akihiko Kawahara

17.1 Background

In recent years, liquid-based cytology (LBC) has replaced conventional smears as the method of choice for clinical cytological evaluation of various organs. This approach has much improved the cytomorphological evaluation of endometrial cells in health and disease and has permitted to use residual cellular material to prepare

Y. Hirai (✉)
Department of Obstetrics and Gynecology, Faculty of Medicine, Dokkyo Medical University, Tochigi, Japan

PCL Japan Pathology and Cytology Center, PCL Inc., Saitama, Japan
e-mail: yhirai-ind@umin.ac.jp

T. K. Kobayashi
Division of Health Sciences, Cancer Education and Research Center, Osaka University Graduate School of Medicine, Osaka, Japan

Y. Norimatsu
Department of Medical Technology, Faculty of Health Sciences, Ehime Prefectural University of Health Sciences, Tobe-cho, Iyo-gun, Ehime, Japan

J. Watanabe
Department of Bioscience and Laboratory Medicine, Hirosaki University Graduate School of Health Science, Hirosaki, Japan

T. Kurokawa · A. Shinagawa
Department of Gynecology and Obstetrics, Faculty of Medical Sciences, University of Fukui, Fukui, Japan

A. Mitsuhashi
Department of Reproductive Medicine, Graduate School of Medicine, Chiba University, Chiba, Japan

A. Kawahara
Department of Diagnostic Cytopathology, Kurume University Hospital, Fukuoka, Japan

© The Author(s), under exclusive license to Springer Nature Singapore Pte Ltd. 2022
Y. Hirai, F. Fulciniti (eds.), *The Yokohama System for Reporting Endometrial Cytology*, https://doi.org/10.1007/978-981-16-5011-6_17

cell blocks for subsequent morphological and immunocytochemical studies as well as to examine the molecular pathological changes associated to endometrial malignancies.

However, an additional possibility with LBC is that to exploit the advantages of computer-based image analysis technology for assistance in the interpretation of various types of cytologic samples. To this concern, the experience with cervical cytology has matured sufficiently to enter routine clinical laboratory use. Such interpretation support devices are being manufactured by several companies and are already in use in large cytology laboratories around the world. They are indispensable for maintaining the accuracy of daily cytological interpretation. Up to now, the commonest application of image analysis has been cervical cytology, due to the large number of specimens available.

The present chapter deals with the current states and future perspectives of digital cytology and the morphometric approach to endometrial cytology interpretation. Appropriate technological progress in these areas will significantly improve the future utility of endometrial cytology. Subsequently, we will discuss the current state of flow cytometry (LC-1000)-assisted endometrial cytology interpretation. This topic is rarely discussed elsewhere in this book. Finally, we will discuss the latest advances in the value of cell-free DNA samples in the follow-up of endometrial malignancies.

17.2   Current Status and Future Perspectives of Digital Cytology

Digital cytology is a technique for converting microscopic images of cytological specimens into digital data and observing them using a computer. The breakdown of the technology includes still images, moving images, and whole slide imaging (WSI), which can virtually freely observe fields in the magnification and Z-axis directions by converting the entire field of view into digital data. These technologies are developing rapidly and various apparatuses are being developed.

Applications of digital cytology include cytodiagnosis, remote cytodiagnostic, conferences, control surveys, education, remotely administered e-learning and quantification of morphological images, objective differential diagnosis [1, 2] by digital image analysis (DIA).

Among future perspectives in this field, automatic DIA by artificial intelligence (AI) based on the image analysis developed with the spread of liquid-based cytology (LBC), automated interpretation and reporting of immunocytochemical staining genetic diagnosis by in situ hybridization (FISH, CISH ISH, etc.), network construction of remote cytodiagnosis and international standardization of cytodiagnosis, etc., are expected.

17.3   The Morphometric Approach to Endometrial Cytology Interpretation

According to the International Federation of Gynecology and Obstetrics (FIGO) histologic grading system [3], nuclear atypia should be also evaluated in endometrial cytology samples, as its recognition and grading are critical for a diagnosis. However, the assessment of nuclear atypia is highly subjective in the FIGO histologic grading [4]. Lax et al. [5] reported an interobserver agreement of 35% (0.22), while Sagae et al. [6] showed a 49% agreement (0.23).

In several studies evaluating endometrial cytology, evaluation of nuclear atypia includes several objective variables as: evaluation of the nuclear area and nuclear roundness, nuclear pleomorphism (anisonucleosis), nuclear size, chromatin structure and distribution, size of nucleoli, and mitoses [7–9].

However, to avoid the negative impact of subjectivity in endometrial cytological samples, it would be useful, at least for some of these latter, like nuclear area and nuclear roundness, to use objective data obtained from morphometric evaluation when applying the FIGO grading system.

Since the liquid-based cytology (LBC) technique—originally developed for gynecologic cervical smears—permits a standardized specimen preparation, the application of morphometric analysis on these samples becomes a real possibility.

Recently, Norimatsu et al. [10] investigated whether an objective evaluation of nuclear findings in endometrial cytology could be more accurate than the simple cytomorphological evaluation. Nuclear image morphometry for geometric and texture features (area, grey value, aspect ratio, internuclear distance, nucleolar diameter) was performed using the ImageJ software (version 1.51) by the U.S. National Institutes of Health (http://rsbweb.nih.gov/ij/).

Consequently, it became obvious that image morphometry is more useful than simple microscopic evaluation to score nuclear atypia in LBC of directly sampled endometrial mucosa.

In this regard, Pouliakis et al. [11] have attempted to reduce the human subjectivity, by exploiting a more objective methodology using a classification and regression trees (CARTs) system. The proposed methodology gave encouraging results in the discrimination between the nuclei from cases of hyperplasia with cytological atypia and carcinoma, from those of benign endometrial cases and cases without cytological atypia.

Pouliakis et al. also underlined that LBC facilitates the application of objective criteria since it permits a standardized procedure of fixation, transfer, and cytological preparation of the material which can be easily deployed in any cytopathology laboratory, and, moreover, enables the application of ancillary techniques.

In retrospect, according to Baak's definition [12], morphometry is a quantitative description of the form, and unlike other morphological methods, it enables the findings to be numerically expressed. The method is inexpensive and technically simple, thus being applicable on the material processed by standard procedure while morphometric parameters are objective and reproducible.

Meanwhile, AI, currently a hot and controversial topic, has been introduced into many aspects of our everyday life, including medicine. Compared with other applications in the treatment of diseases, AI is more likely to enter the diagnostic disciplines based on image analysis such as pathology, included cellular pathology, ultrasound, radiology, and skin disease diagnosis. Among these applications, the implementation of AI in cellular pathology represents a special challenge due to the complexity and great responsibility of the cytopathologic diagnosis of endometrial precancerous and cancerous lesions.

17.4   Current Status and Future Perspectives of Flow Cytometry (LC-1000)-Assisted Endometrial Cytology Interpretation

The LC-1000 (Japanese notification number: 28B1X10014000038) exfoliated cell analyzer developed by the Sysmex Corporation is a medical device that calculates the cell proliferation index (CPIx). This is a unique value reflecting proliferative capacity as a proxy for the neoplastic tendency of exfoliated (e.g., endometrial) cells suspended in an alcohol-based preservative solution. Efficacy of the LC-1000 for cervical cancer screening has recently been reported by a Japanese Society of Clinical Cytology study [13] and also from Korea [14].

Currently, endometrial cancer screening by means of endometrial cytology is rarely performed in Western countries other than Japan; such countries favor instead endometrial aspiration biopsy only in high-risk groups. Recently a pilot study on efficiency of the LC-1000 in the detection of endometrial cancer in directly sampled endometrial cells has been conducted by ourselves with the cooperation of Sysmex corporation. The result data are currently being prepared for publication. The preliminary results were as follows:

I. The manufacturer's recommended ultrasonic dispersion setting was effective for measuring LC-1000 endometrial cell samples.

II. According to the results of the measurement using LC-1000 of endometrial cell samples taken from clinical cases, the detection sensitivity of cancer cases was 100%, specificity was 92%, positive predictive value was 73%, and negative predictive value was 100%.

III. Although the number of samples measured was limited, measuring endometrial cell samples using LC-1000 was suggested to be useful in screening cancer groups from non-cancer groups.

As mentioned above, LC-1000 is suggested to provide supplemental information for detecting endometrial cancer based on a principle different from that of endometrial cytology. Because only a small volume of endometrial cell suspension is required for analysis by the LC-1000, a single endometrial cell collection is sufficient for both LBC and LC-1000 analysis.

A large-scale study is currently underway in Japan to evaluate the utility of LC-1000 by comparing its performance in the detection of both endometrial precursor and cancerous lesion with that of endometrial aspiration biopsy or cytology. If the large-scale study supports above mentioned our results of LC-1000 study, LC-1000 analysis in screening for endometrial cancer in a high-risk group combined with endometrial cytology is likely to significantly improve the detection efficiency of endometrial cancer.

17.5 Value of Cell-Free DNA Samples in the Follow-Up of Endometrial Malignancies

One of the most important developments in therapeutic medicine in the last decade has been the routine introduction of genomic detection and analysis technologies, which have further widened the already known complexities of molecular changes in human neoplasms. Cell-free DNA (cf-DNA) is released into the peripheral blood by tumor cells and cells in the tumor microenvironment that undergo apoptosis or necrosis. A number of recent publications have documented the utility and feasibility of mutational analyses in cf-DNA in lung cancer [15], pancreatic cancer [16], and colorectal cancer [17]. Recently, the FDA (Food and Drug Administration) has also approved the use of cell-free extracts for the study of EGFR mutations as a guide to the treatment of Non-small cell lung cancer (NSCLC) patients with EGFR inhibitors [18]. Nonetheless, while the detection of known mutations or of resistance mutations is feasible by using cell-free extracts from plasma and other biological fluids, the study of cf-DNA extracts cannot yet be applied to the primary molecular diagnosis of tumors, due to the very low amount of cf-DNA in biological fluids, requiring targeted enrichment for accurate detection. Several previous reports have used malignant pleural supernatant for EGFR mutation analysis, and Liu et al. report that malignant pleural supernatants may be a potential substitute for metastatic pleural tumor tissue study in EGFR mutational testing [19]. In 2018, Kawahara et al. [20] have had the opportunity to assess the EGFR mutational status using pleural effusion cytology samples. In his study he concluded that the combined test using both cell sediment DNA and supernatant cf-DNA of the same samples increases the concordance rate of EGFR mutation between primary tumor as determined in tissue and its corresponding pleural localizations. Supernatant cf-DNA may help in this way to compensate for the lack of sufficient DNA in EGFR mutational analysis, allowing to detect EGFR mutations in a similar way to cancer cell sediments.

In recent years, the concept of "liquid biopsy" has been rapidly applied also to the field of gynecologic oncology [21–23]. In 2017, Cicchillitti et al. [24] demonstrated that significantly elevated cf-DNA levels were detected in an endometrial cancer group compared with controls, with higher levels in high-grade groups (G2 and G3). An earlier investigation by Dobyrzycka et al. [25] has shown that a considerable difference exists between the mean level of cf-DNA in patients with type 1 and type 2 endometrial cancer. Recently, Vizza et al. [26]

adopted a special calculation method which is called Alu-quantitative real-time PCR (qPCR) analysis to investigate the role of cf-DNA in patients with endometrial carcinoma. This method compares two variables, of which one is a quantitative parameter (Alu 115) (total cf-DNA) and another is a quantitative one (Alu 247), which represents the degree of integrity of cf-DNA. They found a significantly increased ratio of total cf-DNA content in high-grade endometrial cancer patients, with a parallel decrease of the DNA integrity index in patients with lympho-vascular invasion or unfavorable prognostic features (obesity or hypertension).

The detection of cf-DNA has also been carried out in non-blood-derived samples such as pleural fluid, urine, and uterine lavage [22, 23, 27]. This latter samples were used not only in the management of endometrial cancer but also in ovarian cancer patients [22, 23, 26]. In 2019, Ponti et al. noted that compared to bloodstream, the levels of cf-DNA detected in non-blood-derived fluids can be higher in certain cancer types. However, more experience and research is needed to improve the reliability of these data for cancer molecular profiling in clinical pathology and oncology [28].

17.5.1 Molecular Study of Liquid-Based Pap Samples and of cf-DNA in Uterine Lavage Fluid

Recently, genetic analysis on Papanicolaou test (Pap smear) samples using liquid-based cytology rather than traditional Pap smears showed potential advantages in early detection of both ovarian carcinoma and endometrial cancer patients. In 1943 Papanicolaou and Traut's published their landmark monograph entitled "Diagnosis of Uterine Cancer by the Vaginal Smear" [29]. This was undoubtedly a foundational work in gynecologic cytopathology. It is noteworthy that already at that time they suggested that endocervical sampling could in theory be used to detect not only cervical cancers but also other cancers arising in the female reproductive tract, including endometrial and ovarian carcinomas [30]. Accumulated recent research moves us much closer to that goal. In honor of Papanicolaou's pioneering contribution to the field of early cancer detection, Kinde et al. also have named the approach described as the "PapGene" test in his recent report [31]. In their study, these authors used liquid-based cytologic samples (LBC) Pap (test) samples, (which are routinely used for human papillomavirus detection) and showed that cancerous cells shedding from the upper gynecological tract can be present in the uterine cervix [31]. Massive parallel sequencing for tumor-specific mutations was performed on DNA from LBC Pap smear specimens. This technique was successfully applied to 24 (100%) of 24 patients with endometrial cancer, in which cancer-associated mutations were identified in all cases. However, the sensitivity was lower in ovarian cancer patients, with mutations identified in 9/22 patients (41%). It is now clear enough that, in the future, molecular genetic analysis of the Pap test using LBC samples, may help in the early detection of endometrial cancer and ovarian cancer patients [31–33].

Uterine fluid is the term used to describe the fluid found in the uterine cavity; its cellular composition has been extensively investigated by Casslen et al. [27]. A similar diagnostic approach to LBC samples of Pap smears has already been applied for the demonstration of genomic mutations using uterine lavage fluid or uterine aspirates in order to detect early cancers, in addition to cyto/histopathological examination [34–36]. Tumor cells which shed from both ovarian and endometrial cancer can be in fact collected through lavage of uterine cavity. Maritschnegg et al. [35] performed lavage of the uterine cavity to obtain samples from 65 patients with ovarian cancer ($n = 30$), endometrial cancer ($n = 5$), other malignancies ($n = 3$), or benign lesions of the gynecologic organs ($n = 27$). They used next-generation sequencing to examine these samples along with the corresponding tumor tissue for the presence of somatic mutations. Molecular pathologic examination of uterine lavage samples led to the detection of ovarian cancer, endometrial cancer, and clinically occult ovarian cancer, suggesting its potential usage as an additional diagnostic tool for an early diagnosis. They also emphasized that although they analyzed uterine lavage samples of only five patients with endometrial cancer, the diagnostic potential of the uterine lavage technique is obvious.

Currently, the above described research field is extraordinarily active and it is warmly expected that genetic profiling techniques will significantly increase the diagnostic specificity and sensitivity of precision medicine in gynecologic oncology.

References

1. Washiya K, Nakamura M, Mizuki Y, et al. Discriminating analysis of atypical squamous cells of undetermined significance of the uterine cervix using nuclear three-dimensional analysis. Acta Cytol. 2014;58:96–102.
2. Yoshioka H, Herai A, Oikawa S, et al. Fractal analysis method for the complexity of cell cluster staining on breast FNAB. Acta Cytol. 2021;65:1–12.
3. FIGO committee on gynecologic oncology. Revised FIGO staging for carcinoma of the vulva, cervix, and endometrium. Int J Gynecol Obstet. 2009;105:103–4.
4. Scholten AN, Smit VT, Beerman H, et al. Prognostic significance and interobserver variability of histologic grading systems for endometrial carcinoma. Cancer. 2004;100:764–72.
5. Lax SF, Kurman RJ, Pizer ES, et al. A binary architectural grading system for uterine endometrial endometrioid carcinoma has superior reproducibility compared with FIGO grading and identifies subsets of advance-stage tumors with favorable and unfavorable prognosis. Am J Surg Pathol. 2000;24:1201–8.
6. Sagae S, Saito T, Satoh M, et al. The reproducibility of a binary tumor grading system for uterine endometrial endometrioid carcinoma, compared with FIGO system and nuclear grading. Oncology. 2004;67:344–50.
7. Nishimura Y, Watanabe J, Jobo T, et al. Cytologic scoring of endometrioid adenocarcinoma of the endometrium. Cancer. 2005;105:8–12.
8. Yamaguchi T, Kawahara A, Hattori S, et al. Cytological nuclear atypia classification can predict prognosis in patients with endometrial cancer. Cytopathology. 2015;26:157–66.
9. Kato R, Hasegawa K, Torii Y, et al. Cytological scoring and prognosis of poorly differentiated endometrioid adenocarcinoma. Acta Cytol. 2015;59:83–90.
10. Norimatsu Y, Irino S, Maeda Y, et al. Nuclear morphometry as an adjunct to cytopathologic examination of endometrial brushings on LBC samples: a prospective approach to combined evaluation in endometrial neoplasms and look alikes. Cytopathology. 2021;32:65–74.

11. Pouliakis A, Margari C, Margari N, et al. Using classification and regression trees, liquid-based cytology and nuclear morphometry for the discrimination of endometrial lesions. Diagn Cytopathol. 2014;42:582–91.
12. Baak JP. Further evaluation of the practical applicability of nuclear morphometry for the prediction of the outcome of atypical endometrial hyperplasia. Anal Quant Cytol Histol. 1986;8:46–8.
13. Nakamura M, Ueda M, Iwata T, et al. A clinical trial to Verify the efficiency of the LC-1000 exfoliative cell analyzer as a new method of cervical cancer screening. Acta Cytol. 2019;63:391–400.
14. Heo I, Kwak HJ, Nah EH, et al. Evaluation of the LC-1000 flow cytometry screening system for cervical Cancer screening in routine health checkups. Acta Cytol. 2018;62:279–87.
15. Rolfo C, Mack PC, Scagliotti GV, et al. Liquid biopsy for advanced non-small cell lung cancer (NSCLC): a statement paper from the IASLC. J Thrac Oncol. 2018;13:1248–68.
16. Piietrasz D, Pecuchet N, Garlan F, et al. Plasma circulating tumor DNA in pancreatic cancer is a prognostic marker. Clin Cancer Res. 2017;23:116–23.
17. Strickler JH, Loree JM, Ahronian LG, et al. Genomic landscape of cell-free DNA in patients with colorectal cancer. Cancer Discov. 2018;8:164–73.
18. Kwapisz D. The first liquid biopsy test approved. Is it a new era of mutation testing for non-small cell lung cancer? Ann Transl Med. 2017;5:46. https://doi.org/10.21037/atm.01.32.
19. Liu X, Lu Y, Zhu G, et al. The diagnostic accuracy of pleural effusion and plasma samples versus tumour tissue for detection of EGFR mutation in patient with advanced non-small cell lung cancer: comparison of methodologies. J Clin Pathol. 2013;66:1065–9.
20. Kawahara A, Fukumitsu C, Azuma K, et al. A combined test using both cell sediments and supernatant cell-free DNA in pleural effusion shows increased sensitivity in detecting activating EGFR mutation in lung cancer patients. Cytopathology. 2018;29:150–5. https://doi.org/10.1111/cyt.12517.
21. Cheng X, Zhang L, Chen Y, et al. Circulating cell-free DNA and circulating tumor cells, the "liquid biopsies" in ovarian cancer. J Ovarian Res. 2017;10:75. https://doi.org/10.1186/s13048-017-0369-5.
22. Muinelo-Romay L, Casas-Arozamena AM. Liquid biopsy in endometrial cancer: new opportunities for personalized oncology. Int J Mol Sci. 2018;19:2311. https://doi.org/10.3390/ijms19082311.
23. Chen Q, Zhhang Z-H, Wang S, et al. Circulating cell-free DNA or circulating tumor DNA in the management of ovarian and endometrial cancer. Onco Targets Therapy. 2019;12:11517–30.
24. Cicchillitti L, Corrado G, De Angeli M, et al. Circulating cell-free DNA content as blood based biomarker in endometrial cancer. Oncotarget. 2017;8:115230–43.
25. Dobrzycka B, Terlikowski SJ, Mazurek A. Et a. circulating free DNA, p53 antibody and mutations of KRAS gene in endometrial cancer. Int J Cancer. 2010;127:612–21.
26. Vizza E, Corrado G, De Angeli M, et al. Serum DNA integrity index as a potential molecular biomarker in endometrial cancer. J Exp Clin Res. 2018;37:16. https://doi.org/10.1186/s13046-018-0688-4.
27. Casslen B, Kobayashi TK, Stormby N. Cyclic variation of the cellular components in human uterine fluid. J Reprod Fertil. 1982;66:213–8.
28. Ponti G, Manfredini M, Tomasi A. Non-blood source of cell-free DNA for cancer molecular profiling in clinical pathology and oncology. Crit Rev Oncol Hematol. 2019;141:36–42. https://doi.org/10.1016/jcritrevone.2019.06.005.
29. Papanicolaou GN, Traut HF. Diagnosis of uterine cancer by the vaginal smear. New York: The Commonwealth Fund; 1943.
30. Carmichael DE. Diagnosis by vaginal smear. In: The Pap Smear: Life of George N. Papanicolaou. Illinois: Charles C Thomas Publisher; 1973. p. 54–61.
31. Kinde I, Bettegowda C, Wang Y, et al. Evaluation of DNA from the Papanicolaou Test to detect ovarian and endometrial cancers. Sci Transl Med. 2013;5:167ra4. https://doi.org/10.1126/scitranslmed.3004952.

32. Wang Y, Li L, Douville D, et al. Evaluation of liquid from the Papanicolaou test and other liquid biopsies for the detection of endometrial and ovarian cancers. Sci Transl Med. 2018;10:eaap8793. https://doi.org/10.1126/scitranslmed.aap8793.
33. Matsuura M, Yamaguchi K, Tamate M, et al. Efficacy of liquid-based genetic diagnosis of endometrial cancer. Cancer Sci. 2018;109:4025–32.
34. Nair N, Camacho-Vanegas RD, et al. Genomic analysis of uterine lavage fluid detects early endometrial cancers and reveals a prevalent landscape of driver mutations in women without histopathologic evidence of cancer: a prospective cross-sectional study. PLoS Med. 2016;13:e1002206. https://doi.org/10.1371/journal.pmed.1002206.
35. Maritschnegg E, Wang Y, Pecha N, et al. Lavage of the uterine cavity for molecular detection of Mullerian duct carcinomas: a proof-of-concept study. J Clin Oncol. 2015;33:4293–300.
36. Casas-Arozamena DE, Moiola CP, et al. Genomic profiling of uterine aspirates and cfDNA as an integrative liquid biopsy strategy in endometrial cancer. J Clin Med. 2020;9:585. https://doi.org/10.3390/jcm9020585.

The Future Direction in Endometrial Oncology through the Liquid Biopsy

18

Natalia Malara, Tadao K. Kobayashi, Akihiko Kawahara, Alarice C. Lowe, and Arrigo Capitanio

18.1   Introduction

N. Malara

Liquid Biopsy is defined as a non-invasive and repeatable test that evaluates biomarkers released by diseased tissue and found in peripheral blood samples or in other liquid matrices which may be cerebrospinal fluid, urine, secretions, effusions, etc. In clinical practice, the currently validated liquid biopsy is based on the analysis of circulating molecular biomarkers as analysis of circulating tumor DNA (ct-DNA) and of cellular biomarkers as circulating tumor cells (CTCs) [1, 2]. Oncological research is focused on identifying and validating further circulating biomarkers, alone or in combination, including molecular ones such as transcriptomics (i.e., specific miRNA, etc.) [3], proteomics (i.e., CEA, etc.) [4], metabolomics (i.e., methylglyoxal, etc.) [5, 6] and cellular ones as circulating endothelial cells [7, 8], in order to improve early diagnosis and personalized monitoring of the cancer patient through liquid biopsy.

N. Malara (✉)
Department of Experimental and Clinical Medicine, University Magna Graecia, Bionem Lab, Catanzaro, Italy

T. K. Kobayashi
Division of Health Sciences, Cancer Education and Research Center, Osaka University Graduate School of Medicine, Osaka, Japan

A. Kawahara
Department of Diagnostic Cytopathology, Kurume University Hospital, Fukuoka, Japan

A. C. Lowe
Department of Pathology, Stanford University Hospital, Stanford, CA, USA

A. Capitanio
Department of Clinical Pathology, Linköping University Hospital, Linköping, Sweden

235

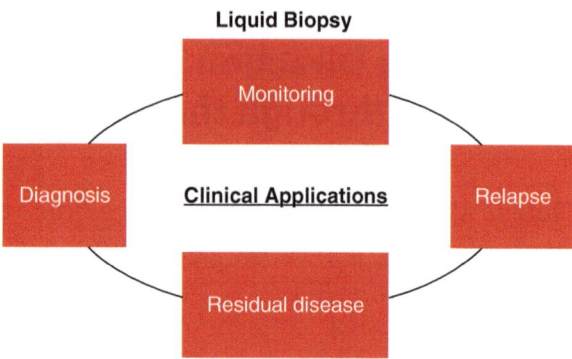

The circulating biomarkers pertaining to endometrial cancer (EC) can be found mainly in two types of liquid matrices, i.e. in samples of peripheral blood and uterine fluid, respectively. In both matrices, it is possible to collect molecules and cells released by the endometrial tumor tissue. In particular, in the peripheral blood samples the circulating biomarkers collected are representative of primary and secondary EC while those collected from uterine fluid are representative only of primary EC disease. The main clinical applications of liquid biopsy are diagnosis, monitoring of minimal residual disease, risk assessment, and unmasking patients at high-risk of recurrence [9, 10] (Fig. 18.1).

18.2 Circulating Tumor DNA (ct-DNA) Evaluation for Monitoring Endometrial Cancer

T. K. Kobayashi, A. Kawahara, and N. Malara

One of the most important developments in therapeutic medicine in the last decade has been the routine introduction of molecular genetic detection and analysis technologies, which have further widened the already known complexities of molecular changes in human neoplasms.

Cell-free DNA (cf-DNA) originates from any type of cell found in body fluids and refers to extracellular DNA molecules (double-stranded DNA and mitochondrial DNA). It was detected in the blood of patient and healthy individuals by Mandel and Metais in 1948 [11]. In prenatal diagnostics, cf-DNA analysis represents now a routine clinical practice [12]. Cf-DNA is prevalently released by the cells of the hematopoietic system and it still remains a challenge to determine the relative contribution to the overall amount of cf-DNA of different tissues in health and pathological conditions [13]. Cells release cf-DNA into the fluids of the proximal microenvironment in free form or as inclusions in extracellular vesicles, by apoptosis or necrosis, senescence, ferroptosis, neutrophil extracellular trap (NETosis), phagocytosis, active secretion, expulsion of mature nuclei by erythroblasts, management of mitochondrial DNA or Vital NETosis [14].

Cf-DNA is passively transported by local fluids to reach the blood circulation, where it is subjected to extra- and intracellular mechanisms regulating its catabolism. Fragmented cf-DNA can be eliminated directly from the renal emunctorium, by becoming a substrate of enzymatic excision mechanisms or by degradation by the hepatocytic filter, especially when it is organized into nucleosomal complexes [15]. A number of recent publications have documented the utility and feasibility of mutational analyses in cf-DNA in lung cancer [16], pancreatic cancer [17], and colorectal cancer [18]. Recently, the FDA has also approved the use of cell free extracts for the study of EGFR mutations as a guide to the treatment of Non-Small Cell Lung Cancer (NSCLC) patients with EGFR inhibitors [19]. Nonetheless, while the detection of known mutations or of resistance mutations is feasible by using cell free extracts from plasma and other biological fluids, the study of cf-DNA extracts cannot yet be applied to the primary molecular diagnosis of tumors. In particular, the quantitative interpretation in the peripheral blood of the cf-DNA of tumor origin is not yet clear. In fact, the kinetic of cf-DNA does not depend exclusively from dead cancer cells, being also strongly influenced by the tumor volume, metabolism, and proliferation rate [20]. Furthermore, it is not yet entirely clear how cancer cells surviving chemotherapy can modify their active secretion of extracellular vesicles incorporating circulating tumor-DNA (ct-DNA) [21].

Further studies need to define the clinical potential and diagnostic sensitivity of cf-DNA, also in function of the anatomic tumor site. This variable can influence the concordance between plasma and tumor data by causing lower amounts of ct-DNA into the bloodstream. In these cases the sources of ct-DNA, such as effusions, secretions, can represent precious sources to obtain a molecular profile [22]. Several previous reports have used supernatant from malignant pleural effusions for EGFR mutation analysis, and Liu et al. report that malignant pleural effusion supernatants may be a potential substitute for metastatic pleural tumor tissue study in EGFR mutational testing [23]. In 2018, Kawahara et al. [24] have had the opportunity to assess the EGFR mutational status using pleural effusion cytology samples. In his study they concluded that the combined test using both cell sediment DNA and supernatant cf-DNA of the same samples increases the concordance rate of EGFR mutation between primary tumor as determined in tissue and its corresponding pleural localizations. Supernatant cf-DNA may help in this way to compensate for the lack of sufficient DNA in EGFR mutational analysis, allowing to detect EGFR mutations in a similar way to cancer cell sediments.

In recent years, the concept of "liquid biopsy" has been rapidly applied also to the field of gynecologic oncology [25–27]. In 2017, Cicchillitti et al. [28] demonstrated that significantly elevated cf-DNA levels were detected in an endometrial cancer group compared with controls, with higher levels in high-grade groups (G2 and G3) (Fig 18.2). An earlier investigation by Dobyrzycka et al. [29] has shown that a considerable difference exists between the mean level of cf-DNA in patients with type 1 and type 2 endometrial cancer. Recently, Vizza et al. [30] adopted a special calculation method which is called Alu-quantitative real-time PCR (qPCR) analysis to investigate the role of cf-DNA in patients with endometrial carcinoma. This method compares two variables, of which one is a quantitative parameter (Alu

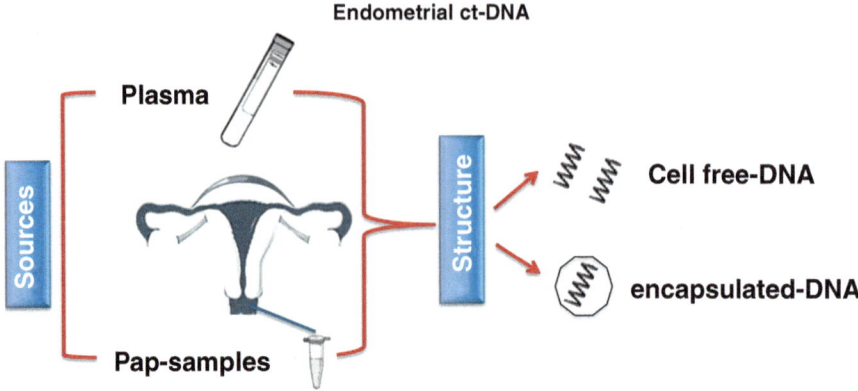

Fig. 18.2 Schematic overview on circulating tumor-DNA biomarkers in endometrial cancer. Prevalent matrices used are plasma and uterine fluid (Pap-sample). It can be in two alternative formats, free (non-encapsulated) or encapsulated (within extracellular membrane)

115) (total cf-DNA) and another is a qualitative one (Alu 247), which represents the degree of integrity of cf-DNA. They found a significantly increased ratio of total cf-DNA content in high-grade endometrial cancer patients, with a parallel decrease of the DNA integrity index in patients with lympho-vascular invasion or unfavorable prognostic features (obesity or hypertension). The detection of cf-DNA has also been carried out in non-blood-derived samples such as pleural fluid, urine, and uterine lavage. This latter samples were used not only in the management of endometrial cancer but also in ovarian cancer patients. In 2019, Ponti et al. noted that compared to bloodstream, the levels of cf-DNA detected in non-blood-derived fluids can be higher in certain cancer types. However, more experience and research is needed to improve the reliability of these data for cancer molecular profiling in clinical pathology and oncology [31].

18.3 Liquid-Based Pap Samples and ct-DNA in Uterine Fluid

N. Malara and T. K. Kobayashi

Recently, genetic analysis on Papanicolaou test (Pap smear) samples using liquid-based cytology rather than traditional Pap smears showed potential advantages in early detection of both ovarian carcinoma and endometrial cancer patients. In 1943 Papanicolaou and Traut published their landmark monograph entitled "Diagnosis of Uterine Cancer by the Vaginal Smear" [32]. This was undoubtedly a foundational work in gynecologic cytopathology. It is noteworthy that already at that time they suggested that endocervical sampling could in theory be used to detect not only cervical cancers but also other cancers arising in the female reproductive tract, including endometrial and ovarian carcinomas [33]. Accumulated recent research moves us much closer to that goal. In honor of Papanicolaou's pioneering contribution to the field of early cancer detection, Kinde et al. also have named the approach described as the "PapGene" test in his recent report [34]. In their study, these authors [34] used liquid-based cytologic samples (LBC) Pap (test) samples, (which are routinely used for human papillomavirus detection) and showed that cancerous cells shedding from the upper gynecological tract can be found in cervical smears.

Massive parallel sequencing for tumor-specific mutations was performed on cf-DNA from LBC Pap smear specimens. This technique was successfully applied to 24 (100%) of 24 patients with endometrial cancer, in which cancer-associated mutations were identified in all cases. However, the sensitivity was lower in ovarian cancer patients, with mutations identified in 9/22 patients (41%). It is now clear enough that, in the future, molecular genetic analysis of the Pap test using LBC samples may help in the early detection of endometrial cancer and ovarian cancer patients [34, 35].

Uterine fluid is the term used to describe the fluid found in the uterine cavity; its cellular composition has been extensively investigated by Casslen et al. [36]. A similar diagnostic approach to LBC samples of Pap smears has already been applied for the demonstration of mutations using uterine lavage fluid or uterine aspirates in order to detect early cancers, in addition to cyto/histopathological examination [37–39]. Tumor cells which shed from both ovarian and endometrial cancer can be in fact collected through lavage of uterine cavity. Maritschnegg et al. [38] performed lavage of the uterine cavity to obtain samples from 65 patients with ovarian cancer ($n = 30$), endometrial cancer ($n = 5$), other malignancies ($n = 3$), or benign lesions of the gynecologic organs ($n = 27$). They used next-generation sequencing to examine these samples along with the corresponding tumor tissue for the presence of somatic mutations. Molecular pathologic examination of uterine lavage samples led to the detection of ovarian cancer, endometrial cancer, and clinically occult ovarian cancer, suggesting its potential usage as an additional diagnostic tool for an early diagnosis. They also emphasized that, although they analyzed uterine lavage samples of only five patients with endometrial cancer, the diagnostic potential of the uterine lavage technique is obvious. Ct-DNA can be present in the uterine fluid also in encapsulated form within extracellular vesicles. Extracellular vesicles are small cell-derived vesicles containing a range of molecular information comprised in DNA, RNA, and proteins. Depending on their size and biological role generally, the extracellular vesicles are referred to as microvesicles, ectosomes, exosomes, apoptotic bodies, shedding vesicles, or microparticles among others [39].

Recently, the isolation and analysis of exosomes from uterine fluid in pregnant women has opened new possibilities for the non-invasive gynecological cancer diagnostics. The analysis of the content of the isolated exosomes allowed access to molecular information of the host organism. The molecules of DNA, RNA, and proteins encapsulated within the exosomal lipid bilayer were protected from degradation [40].

Similarly in the oncology field, this procedure has opened up new possibilities to access directly to molecular and biological profile of gynecological diseases, and in particular endometrial carcinomas, for screening and cancer risk monitoring [41].

Currently, the above described research field is extraordinarily active and it is warmly expected that genetic profiling techniques will significantly increase the diagnostic specificity and sensitivity of precision medicine in gynecologic oncology [42].

18.4 Circulating Tumor Cells (CTCs) Evaluation for Monitoring Endometrial Cancer Patients

N. Malara and A. C. Lowe

The circulating tumor cells, (CTCs) are released from the tumor tissue into the proximal microenvironment through passive and active mechanisms. The passive mechanism is closely linked to the high proliferative rate, hallmark of cancer [43]. The high proliferation favors the detachment of the endometrial cancer cells and their release into the uterine cavity or into the bloodstream. The passive release in the systemic circulation is permitted by the fenestrated endothelial wall of the intra-lesional vessels [44]. The active mechanism presupposes the activation of signaling pathways that make the cancer cell capable of intra and extravasation and invasion [45].

The CTCs are cells structurally similar to the resident tumor cells in the neoplastic tissue. When in the blood, CTCs are detectable as single elements with a globular shape and/or as multi-cellular groups or clusters characterized by a 3D spherical structure. The destiny of single CTC is to undergo a form of programmed cell death occurring in anchorage-dependent cells after loss of contact with extracellular matrix, defined anoikisis [45]. Moreover, CTC can be destroyed by the lymph nodes or tissue immune system, return to the primary lesion or display bone marrow homing [46]. The endometrial CTCs display bone marrow homing remaining as dormant Disseminated Tumor Cells (DTCs). The DTCs exhibit a quiescent state for months or years and are not detectable. The metastatic CTCs have active signaling pathways to sustain tumorigenic abilities to re-form a tumor similar to the primary one [47].

Tumorigenic CTCs are also defined as cancer stem cells because they have molecular characteristics similar to normal stem cells, are prevalently characterized by a hybrid phenotype, and induce epithelial–mesenchymal transition [48]. Based on these concepts, it is easy to understand why, in order to monitor and to prognostically define the patient with EC, the isolation and analysis of CTCs directly from peripheral blood is needed, rather than that of CTCs floating in the uterine fluid. The detection in the peripheral blood of CTCs in patients with EC is an indication of myometrial and/or lymph node invasion up to the presence of distant metastases [49]. The detection of CTCs in uterine fluids is suggestive of the presence of endometrial cancer disease and therefore has a greater diagnostic than prognostic value.

The problem of isolating CTCs from peripheral blood samples of patients with EC is similar for all types of solid tumor, and is due to the rarity of this population in the blood with respect to the hematological cell population [50].

The US Food and Drug Administration (FDA) has approved the CellSearch® system (Janssen Diagnostics), a validated biomarker assay as an analytical test for clinical use to detect CTCs in peripheral blood. The CTCs are recognized and captured according to the expression of antigens expressed on the surface that identify their epithelial nature. In particular, for patients with EC the CTCs are identified by the CellSearch® using Epcam expression [51]. The limit of this methodological approach is related to the immuno-selection, which guarantees a specific but not very sensitive approach. The circulating tumor population is a non-homogenous population, which is, nonetheless, representative of the systemic dimension of the tumor, being inclusive of information derived by all cancer lesions present in the patient's body at the time of the liquid biopsy collection. In this context, selecting the CTCs by using an epithelial specific antibody restricts the view on the heterogeneous circulating cell population, only partially represented by the epithelial phenotype. To this concern, the choice of the right methodology to capture CTCs by preserving their natural heterogeneity permits an optimal exploitation of liquid biopsy in order to minimize the two main causes of therapeutic cancer failure: i.e.: the intra-patient heterogeneity (due to the heterogeneous cellular composition of primary and metastatic lesions) and by the inter-patient heterogeneity (due to the different clinical outcome in patients with same type and stage of the disease) [52].

The study by Dockery and co-workers collects CTCs from blood with the ISET (isolation by size of epithelial tumor cells) device for early identification of the 20–25% of stage I EC patients who will develop extrauterine pathology. The study demonstrates that CTCs are useful indicators of poor post-surgical prognosis [53]. Another study uses the polymerase chain reaction directly on the blood to identify the presence of CTCs in patients with EC and, subject with the results, gives an indication to candidate these patients for an adjuvant therapy [54].

Studies specifically conducted on EC DTCs in bone marrow have confirmed the close correlation between DTC with stage and recurrence rate [55]. Finally, in advanced EC cases, monitoring with CTCs, and the use of Veridex has given satisfactory results [56]. In this direction, methodologies able to isolate CTCs in respect to their heterogeneity, as the Charactex protocol [49], represent promising solutions to enhance the quality for monitoring advanced EC through CTCs evaluation.

18.5   Circulating Tumor Cells (CTCs) Evaluation for Early Cancer Detection of Endometrial Cancer

N. Malara and A. C. Lowe

Endometrial cancer releases cancer cells through the cervical canal into the genital tract. In fact, in 45% of women with endometrial cancer, cancer cells were diagnosed on routine cervical cytology. Endometrial cytology offers many advantages including low cost, quick turn-around times with personalization of tumor surveillance and prevention times, but is currently based at least on an endometrial mucosa direct brushing. It has recently been shown that it is possible to find cancer cells in the lower urogenital tract. In particular, an early diagnosis of endometrial cancer could even be

Fig. 18.3 Short-time cultured CTCs, by applying the Charactex protocol, were identified through immunocytochemistry with a monoclonal antibody against pan-CK (magnification 20×)

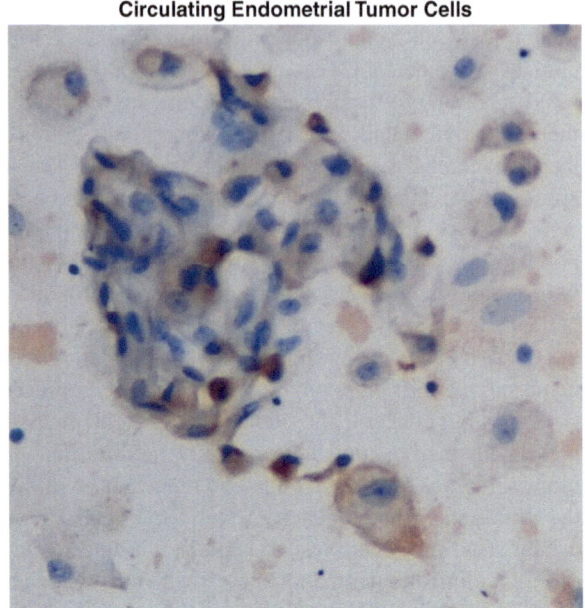

Circulating Endometrial Tumor Cells

possible by the cytological examination of avoided urine and non-invasive vaginal samples (i.e., post-menopausal bleeding), offering a potential liquid biopsy with high diagnostic accuracy [57]. Furthermore, subclinical ECs can release tumor cells directly into the peripheral blood in the absence of clinically documented lesions, as also reported in other types of solid tumors [49, 58]. Figure 18.3 reports a case of subclinical EC in which isolated and short-time cultured CTCs were identified by applying the Charactex protocol and surface immunocytochemical expression for pan-cytokeratin (pan) CK, one year before the clinical manifestation of the disease. Further studies are needed to verify whether non-invasive endometrial cytology has the same diagnostic sensitivity of traditional invasive procedures and if it can be considered as a screening examination able to identify high-risk EC women and to candidate them for a subsequent invasive endometrial diagnostic procedure.

18.6   Digital Imaging and Deep-learning Model and Liquid Biopsy

A. Capitanio and N. Malara

Studies on circulating tumor cells (CTCs) morphology and on their molecular landscape variations are paramount to access real-time information on this heterogeneous population of cells. This forms the basis for early diagnosis of new or relapsing lesions [49, 59]. The ability to define cellular characteristics of CTCs, in order to eliminate the need for costly and time-consuming labeling protocols, requires sufficient statistical power and sensitivity, nowadays, achievable and depending only on current sophisticated technologies. Combination between deep neural network and transfer learning demonstrates the power of delineating the physical and functional signatures to detect rare cell populations in heterogeneous cytological samples. This technique could guide in the next future, the development of cellular imaging assays that stratify cytological endometrial derived-preparation, preferring those cytology samples obtainable by non-invasive procedures such as from peripheral blood or uterine fluid/brushing samples. The goal is to identify selected cytological phenotypes linked to a high probability of primary-recurrent cancer diagnosis; on this basis, the patient selection process would be more specifically oriented and directed towards patients necessitating further instrumental and even invasive procedures to confirm and establish the stage and grade of endometrial disease.

The operational tools to achieve the results mentioned above are based on the application of digital imaging technologies. Nowadays, the so-called digital pathology is based on digital Whole Slide Imaging (WSI). Digital scanning of histological preparations in everyday use in many centers for teaching, research, documentation and, increasingly, primary diagnosis [60–62]. The same cannot be said for cytopathology. Indeed, many teaching and research programs make extensive use of digital cytological preparations [63], but diagnostic applications are still limited [64–66]. The reason for this is to be found in the very nature of the cytological preparation. When observed under the microscope, a histological section does not require extensive micro-focusing as the structural, morphological elements are more or less all on the same focal plane.

On the other hand, the cytological preparation consists of cells dispersed in the mounting medium and arranged on several focal planes. This three-dimensional structure, to be reproduced digitally, requires scans on multiple z-stack focal planes. The digital files resulting from this scan mode are much larger than those obtained by scanning a single focal plane. Furthermore, the time required for multiple-level screening becomes incompatible with the need for rapid diagnostic.

The final solution to this problem will be achieved by developing hardware capable of performing simultaneous scans of different z-stack levels. There are already examples of digital scanners for microscopy that allow this result [67, 68], but their use and diffusion are minimal.

An acceptable non-hardware related solution is the use of software that analyzes the different z-stack levels, choosing the perfectly focused portions of the image in each level and reconstructing with them a new image where all the cells appear in focus [65]. In practice, the software splits the digital image of each z-stack layer into small tiles. Then, it examines the corresponding tiles in each level by calculating the parameter—usually, the Laplacian Variance Analysis or the Gray Level Co-Occurrence Matrix (GLCM contrast analysis) [69–73]—which allows deciding which tile expresses the best focus. At the end of the process, these tiles build up the new digital image where all the cytological details are sharp.

This brief excursus on digital cytology techniques, of which the liquid biopsy is an expression, is essential to frame the perspectives introduced by Artificial Intelligence (AI) in cytology and, therefore, in liquid biopsy.

According to Britannica Encyclopedia, AI is defined as "...the ability of a digital computer or computer-controlled robot to perform tasks commonly associated with intelligent beings." In our specific field, we ask AI to recognize and classify cancer cells in digital images of liquid biopsies by practically replicating the mental process we apply to our observations under the microscope. Therefore, it is intuitive that the previously exposed methods aimed at producing excellent digital images form the basis of every AI procedure. It is equally intuitive that if the purpose of AI is to replicate human capabilities, possibly improving them, the steps to be taken are in order:

1. Learn from reality to observe the objects' characteristics to be recognized (in our case, the cancer cells).
2. Memorize and process knowledge.
3. Apply the models and categories of knowledge resulting from the previous step to recognize the diagnostic elements autonomously.

Machine Learning (ML) is an AI method founded on the analysis of sets of known data (known characteristics and features of tumor cells) and using the extracted pieces of information to make decisions at the analysis of new unknown data [74, 75].

Deep Learning (DL) is a kind of ML based on neural networks and, mainly, on the so-called Convolutional Neural Network (CNN). In the last twenty years, DL became a tool of paramount importance for image classification [76].

The first step of the DL-CNN process is the so-called training dataset. It consists of sets of digital images with known input (they are cells) and known output (they are tumor cells). Filters and convolutional algorithms are applied to each image to allow the network to learn which cell features are associated with the wanted output.

An intermediate step between the training and the final use of the network for diagnostic decision is the use of validation datasets. It means that the network is tested with datasets with unknown output during the training to verify and calibrate its performances. This calibration is necessary because in some circumstances (for example, in small training sets) the network over fits the training data but often fails the final decision test.

In liquid biopsies and, more generally, in cytology, the amount of data available for training sets can be small. In some cases, such as liquid biopsies, this is because there are few diagnostic cells. In other cases, it may be rare diseases or diseases rarely investigated with cytology. In all these cases, the DL can be performed using "data augmentation"methods. The data augmentation technique involves increasing the data present in training set by slightly modifying the existing data. For example, in digital images, it is possible to change the resolution, size, and orientation of the images themselves [77]. In doing so, it may appear that falsehoods and bias are being introduced into the data under consideration. In reality, if the changes are not to distort the meaning of the data, new variables are simply introduced, which increase the robustness of the dataset.

In conclusion, the methods briefly mentioned above can potentially constitute a powerful tool in cytological diagnostics, especially in diagnostic difficulty areas such as liquid biopsy. At the moment, however, the experience is still minimal. With the sole exception of studies published in the field of cervicovaginal cytology [78–84], the studies published in recent years illustrating AI's diagnostic applications with DL and CNN are few [85–91]. But they are increasing, as is the awareness among pathologists that such methods can achieve excellent reliability levels.

References

1. Ignatiadis M, Sledge GW, Jeffrey SS. Liquid biopsy enters the clinic – implementation issues and future challenges. Nat Rev Clin Oncol. 2021;18:297–312. https://doi.org/10.1038/s41571-020-00457-x.
2. Chen D, Xu T, Wang S, et al. Liquid biopsy applications in the clinic. Mol Diagn Ther. 2020;24:125–32. https://doi.org/10.1007/s40291-019-00444-8.
3. Drula R, Ott LF, Berindan-Neagoe I, et al. MicroRNAs from liquid ebiopsy derived extracellular vesicles: recent advances in detection and characterization methods. Cancers (Basel). 2020;12:2009. https://doi.org/10.3390/cancers12082009.
4. Martinez-Garcia E, Lesur A, Devis L, et al. Targeted proteomics identifies proteomic signatures in liquid biopsies of the endometrium to diagnose endometrial cancer and assist in the prediction of the optimal surgical treatment. Clin Cancer Res. 2017;23:6458–67. https://doi.org/10.1158/1078-0432.CCR-17-0474.
5. Coluccio ML, Gentile F, Presta I, et al. Tailoring chemometric models on blood-derived cultures secretome to assess personalized cancer risk score. Cancers (Basel). 2020;12:1362. https://doi.org/10.3390/cancers12061362.
6. Singhal S, Rolfo C, Maksymiuk AW, et al. Liquid biopsy in lung cancer screening: the contribution of metabolomics. Results of a pilot study. Cancers (Basel). 2019;11:1069. https://doi.org/10.3390/cancers11081069.
7. Lanuti P, Rotta G, Almici C, et al. Endothelial progenitor cells, defined by the simultaneous surface expression of VEGFR2 and CD133, are not detectable in healthy peripheral and cord blood. Cytometry A. 2016;89:259–70. https://doi.org/10.1002/cyto.a.22730.
8. Lanuti P, Simeone P, Rotta G, et al. A standardized flow cytometry network study for the assessment of circulating endothelial cell physiological ranges. Sci Rep. 2018;8:5823. https://doi.org/10.1038/s41598-018-24234-0.
9. Pantel K, Alix-Panabières C. Liquid biopsy and minimal residual disease - latest advances and implications for cure. Nat Rev Clin Oncol. 2019;16:409–24. https://doi.org/10.1038/s41571-019-0187-3.
10. Babayan A, Pantel K. Advances in liquid biopsy approaches for early detection and monitoring of cancer. Genome Med. 2018;10:21. https://doi.org/10.1186/s13073-018-0533-6.
11. Mandel P, Metais P. Les acides nucléiques du plasma sanguin chez l'homme. C R Acad Sci Paris. 1948;142:241–3.
12. Lo YMD, Chan KCA, Sun H, et al. Maternal plasma DNA sequencing reveals the genome-wide genetic and mutational profile of the fetus. Sci Transl Med. 2010;2:61ra91. https://doi.org/10.1126/scitranslmed.3001720.
13. Abbou SD, Shulman DS, SG DB, Crompton BD. Assessment of circulating tumor DNA in pediatric solid tumors: the promise of liquid biopsies. Pediatr Blood Cancer. 2019;66:e27595. https://doi.org/10.1002/pbc.27595.

14. Keller L, Belloum Y, Wikman H, et al. Clinical relevance of blood-based ctDNA analysis: mutation detection and beyond. Br J Cancer. 2021;124:345–58. https://doi.org/10.1038/s41416-020-01047-5.
15. Lehmann-Werman R, Magenheim J, Moss J, et al. Monitoring liver damage using hepatocyte-specific methylation markers in cell-free circulating DNA. JCI Insight. 2018;3:e120687. https://doi.org/10.1172/jci.insight.120687.
16. Rolfo C, Mack PC, Scagliotti GV, et al. Liquid biopsy for advanced non-small cell lung cancer (NSCLC): a statement paper from the IASLC. J Thrac Oncol. 2018;13:1248–68.
17. Piietrasz D, Pecuchet N, Garlan F, et al. Plasma circulating tumor DNA in pancreatic cancer is a prognostic marker. Clin Cancer Res. 2017;23:116–23.
18. Strickler JH, Loree JM, Ahronian LG, et al. Genomic landscape of cell-free DNA in patients with colorectal cancer. Cancer Discov. 2018;8:164–73.
19. Kwapisz D. The first liquid biopsy test approved. Is it a new era of mutation testing for non-small cell lung cancer? Ann Transl Med. 2017;5:46. https://doi.org/10.21037/atm.01.32.
20. Mair R, Mouliere F, Smith CG, et al. Measurement of plasma cell-free mitochondrial tumor DNA improves detection of glioblastoma in patient-derived orthotopic xenograft models. Cancer Res. 2019;79:220–30.
21. Lazaro-Ibanez E, Lasser C, Shelke GV, et al. DNA analysis of low- and high-density fractions defines heterogeneous subpopulations of small extracellular vesicles based on their DNA cargo and topology. J Extracell Vesicles. 2019;8:1656993. https://doi.org/10.1080/2001307 8.2019.1656993.
22. Ribeiro IP, de Melo JB, Carreira IM. Head and neck cancer: searching for genomic and epigenetic biomarkers in body fluids—the state of art. Mol Cytogenet. 2019;12:33. https://doi.org/10.1186/s13039-019-0447-z.
23. Liu X, Lu Y, Zhu G, et al. The diagnostic accuracy of pleural effusion and plasma samples versus tumor tissue for detection of EGFR mutation in patient with advanced non-small cell lung cancer: comparison of methodologies. J Clin Pathol. 2013;66:1065–9.
24. Kawahara A, Fukumitsu C, Azuma K, et al. A combined test using both cell sediments and supernatant cell-free DNA in pleural effusion shows increased sensitivity in detecting activating EGFR mutation in lung cancer patients. Cytopathology. 2018;29:150–5. https://doi.org/10.1111/cyt.12517.
25. Cheng X, Zhang L, Chen Y, et al. Circulating cell-free DNA and circulating tumor cells, the "liquid biopsies" in ovarian cancer. J Ovarian Res. 2017;10:75. https://doi.org/10.1186/s13048-017-0369-5.
26. Muinelo-Romay L, Casas-Arozamena AM. Liquid biopsy in endometrial cancer: new opportunities for personalized oncology. Int J Mol Sci. 2018;19:2311. https://doi.org/10.3390/ijms19082311.
27. Chen Q, Zhhang Z-H, Wang S, et al. Circulating cell-free DNA or circulating tumor DNA in the management of ovarian and endometrial cancer. Onco Targets Therapy. 2019;12:11517–30. https://doi.org/10.2147/OTT.S227156.
28. Cicchillitti L, Corrado G, De Angeli M, et al. Circulating cell-free DNA content as blood based biomarker in endometrial cancer. Oncotarget. 2017;8:115230–43.
29. Dobrzycka B, Terlikowski SJ, Mazurek A, et al. Circulating free DNA, p53 antibody and mutations of KRAS gene in endometrial cancer. Int J Cancer. 2010;127:612–21.
30. Vizza E, Corrado G, De Angeli M, et al. Serum DNA integrity index as a potential molecular biomarker in endometrial cancer. J Exp Clin Res. 2018;37:16. https://doi.org/10.1186/s13046-018-0688-4.
31. Ponti G, Manfredini M, Tomasi A. Non-blood source of cell-free DNA for cancer molecular profiling in clinical pathology and oncology. Crit Rev Oncol Hematol. 2019;141:36–42. https://doi.org/10.1016/jcritrevone.2019.06.005.
32. Papanicolaou GM, Traut HF. Diagnosis of uterine cancer by the vaginal smear. New York: The Commonwealth Fund; 1943.
33. Carmichael DE. The Pap Smear: life of George N. Papanicolaou. Illinois: Charles C Thomas Publisher; 1973.

34. Kinde I, Bettegowda C, Wang Y, et al. Evaluation of DNA from the Papanicolaou Test to detect ovarian and endometrial cancers. Sci Transl Med. 2013;5:167ra4. https://doi.org/10.1126/scitranslmed.3004952.

35. Wang Y, Li L, Douville D, et al. Evaluation of liquid from the Papanicolaou test and other liquid biopsies for the detection of endometrial and ovarian cancers. Sci Transl Med. 2018;10:eaap8793. https://doi.org/10.1126/scitranslmed.aap8793.

36. Casslen B, Kobayashi TK, Stormby N. Cyclic variation of the cellular components in human uterine fluid. J Reprod Fertil. 1982;66:213–8.

37. Nair N, Camacho-Vanegas RD, et al. Genomic analysis of uterine lavage fluid detects early endometrial cancers and reveals a prevalent landscape of driver mutations in women without histopathologic evidence of cancer: a prospective cross-sectional study. PLoS Med. 2016;13:e1002206. https://doi.org/10.1371/journal.pmed.1002206.

38. Maritschnegg E, Wang Y, Pecha N, et al. Lavage of the uterine cavity for molecular detection of Mullerian duct carcinomas: a proof-of-concept study. J Clin Oncol. 2015;33: 4293–300.

39. György B, Szabó TG, Pásztói M, et al. Membrane vesicles, current state-of-the-art: emerging role of extracellular vesicles. Cell Mol Life Sci. 2011;68:2667–88.

40. Keller S, Ridinger J, Rupp AK, et al. Body fluid derived exosomes as a novel template for clinical diagnostics. J Transl Med. 2011;9:86. https://doi.org/10.1186/1479-5876-9-86.

41. Malentacchi F, Sgromo C, Antonuzzo L, Pillozzi S. Liquid biopsy in endometrial cancer. J Cancer Metastasis Treat. 2020;6:34. https://doi.org/10.20517/2394-4722.2020.34.

42. Casas-Arozamena DE, Moiola CP, et al. Genomic profiling of uterine aspirates and cfDNA as an integrative liquid biopsy strategy in endometrial cancer. J Clin Med. 2020;9:585. https://doi.org/10.3390/jcm9020585.

43. Fares J, Fares MY, Khachfe HH, et al. Molecular principles of metastasis: a hallmark of cancer revisited. Sig Transduct Target Ther. 2020;5:28. https://doi.org/10.1038/s41392-020-0134-x.

44. Hashizume H, Baluk P, Morikawa S, et al. Openings between defective endothelial cells explain tumor vessel leakiness. Am J Pathol. 2000;156:1363–80. https://doi.org/10.1016/S0002-9440(10)65006-7.

45. Aguirre-Ghiso JA, Sosa MS. Emerging topics on disseminated cancer cell dormancy and the paradigm of metastasis. Annu Rev Cancer Biol. 2018;2:377–93.

46. Leone K, Poggiana C, Zamarchi R. The Interplay between circulating tumor cells and the immune system: From immune escape to cancer immunotherapy. Diagnostics (Basel). 2018;30:8. https://doi.org/10.3390/diagnostics8030059.

47. Ayob AZ, Ramasamy TS. Cancer stem cells as key drivers of tumour progression. J Biomed Sci. 2018;25:20. https://doi.org/10.1186/s12929-018-0426-4.

48. Luo YT, Cheng J, Feng X, et al. The viable circulating tumor cells with cancer stem cells feature, where is the way out? J Exp Clin Cancer Res. 2018;26:37. https://doi.org/10.1186/s13046-018-0685-7.

49. Malara N, Trunzo V, Foresta U, et al. Ex-vivo characterization of circulating colon cancer cells distinguished in stem and differentiated subset provides useful biomarker for personalized metastatic risk assessment. J Transl Med. 2016;14:133. https://doi.org/10.1186/s12967-016-0876-y.

50. Bogani G, Liu MC, Dowdy SC, et al. Detection of circulating tumor cells in high-risk endometrial cancer. Anticancer Res. 2015;35:683–7.

51. Ni T, Sun X, Shan B, et al. Detection of circulating tumour cells may add value in endometrial cancer management. Eur J Obstet Gynecol Reprod Biol. 2016;207:1–4. https://doi.org/10.1016/j.ejogrb.2016.09.031.

52. Miyauchi T, Yaguchi T, Kawakami Y. Inter-patient and intra-tumor heterogeneity in the sensitivity to tumor-targeted immunity in colorectal cancer. Nihon Rinsho Meneki Gakkai Kaishi. 2017;40:54–9. https://doi.org/10.2177/jsci.40.54.

53. Dockery LE, Tanaka TT, Zhang R, et al. Incidence and implications of circulating tumor cells in endometrial cancer. Gynecol Oncol. 2019;154:1–252. https://doi.org/10.1016/j.ygyno.2019.03.217.

54. Kölbl AC, Victor LM, Birk AE, et al. Quantitative PCR marker genes for endometrial adeno-carcinoma. Mol Med Rep. 2016;14:2199–205. https://doi.org/10.3892/mmr.2016.5483.
55. Fehm T, Becker S, Bachmann C, et al. Detection of disseminated tumor cells in patients with gynecological cancers. Gynecol Oncol. 2006;103:942–7. https://doi.org/10.1016/j.ygyno.2006.05.049.
56. Malentacchi F, Sgromo C, Antonuzzo L, et al. Liquid biopsy in endometrial cancer. J Cancer Metastasis Treat. 2020;6:34. https://doi.org/10.20517/2394-4722.2020.34.
57. O'Flynn H, Ryan NAJ, Narine N, et al. Diagnostic accuracy of cytology for the detection of endometrial cancer in urine and vaginal samples. Nat Commun. 2021;12:1–8. https://doi.org/10.1038/s41467-021-21257-6.
58. Malara N, Gentile F, Coppedè N, et al. Superhydrophobic lab-on-chip measures secretome protonation state and provides a personalized risk assessment of sporadic tumour. NPJ Precis Oncol. 2018;2:26. https://doi.org/10.1038/s41698-018-0069-7.
59. Coluccio ML, Presta I, Greco M, et al. Microenvironment molecular profile combining glyca-tion adducts and cytokines patterns on Secretome of short-term blood-derived cultures during tumour progression. Int J Mol Sci. 2020;21:4711. https://doi.org/10.3390/ijms2113471.
60. Araújo ALD, Arboleda LPA, Palmier NR, et al. The performance of digital microscopy for pri-mary diagnosis in human pathology: a systematic review. Virchows Arch 2019;474:269–287. doi: https://doi.org/10.1007/s00428-018-02519-z. Epub 2019 Jan 26.
61. Williams BJ, Treanor D. Practical guide to training and validation for primary diagno-sis with digital pathology. J Clin Pathol 2020;73:418–422. doi: https://doi.org/10.1136/jclinpath-2019-206319. Epub 2019 Nov 29.
62. Borowsky AD, Glassy EF, Wallace WD, et al. Digital whole slide imaging compared with light microscopy for primary diagnosis in surgical pathology. Arch Pathol Lab Med. 2020;144:1245–53. https://doi.org/10.5858/arpa.2019-0569-OA.
63. Wilbur DC. Digital pathology and its role in cytology education. Cytopathology. 2016;27:325–30. https://doi.org/10.1111/cyt.12377.
64. Capitanio A, Dina RE, Treanor D. Digital cytology: a short review of technical and meth-odological approaches and applications. Cytopathology 2018;29:317–325. doi: https://doi.org/10.1111/cyt.12554. Epub 2018 May 28.
65. Hanna MG, Pantanowitz L. Feasibility of using the Omnyx digital pathology system for cytology practice. J Am Soc Cytopathol 2019;8:182–189. doi: https://doi.org/10.1016/j.jasc.2019.01.003. Epub 2019 Jan 18.
66. Wilbur DC. Digital cytology: current state of the art and prospects for the future. Acta Cytol 2011;55:227–238. doi: https://doi.org/10.1159/000324734. Epub 2011 Apr 27.
67. Li N, Lv T, Sun Y, et al. High throughput slanted scanning whole slide imaging system for digi-tal pathology. J Biophotonics 2021:e202000499. doi: https://doi.org/10.1002/jbio.202000499. Epub ahead of print.
68. Liao J, Wang Z, Zhang Z, et al. Dual light-emitting diode-based multichannel microscopy for whole-slide multiplane, multispectral and phase imaging. J Biophotonics. 2018;11. https://doi.org/10.1002/jbio.201700075. https://doi.org/10.1002/jbio.201700075. Epub 2017 Aug 7. PMID: 28700137; PMCID: PMC5766431.
69. Paunovic J, Vucevic D, Radosavljevic T, et al. Gray-level co-occurrence matrix analysis of chromatin architecture in periportal and perivenous hepatocytes. Histochem Cell Biol 2019;151:75–83. doi: https://doi.org/10.1007/s00418-018-1714-5. Epub 2018 Aug 24.
70. Utino FL, Garcia M, Velho PENF, et al. Second-harmonic generation imaging analysis can help distinguish sarcoidosis from tuberculoid leprosy. J Biomed Opt. 2018;23:1–7. https://doi.org/10.1117/1.JBO.23.12.126001.
71. Kono K, Hayata R, Murakami S, et al. Quantitative distinction of the morphological char-acteristic of erythrocyte precursor cells with texture analysis using gray level co-occurrence matrix. J Clin Lab Anal. 2018;32:e22175. https://doi.org/10.1002/jcla.22175. Epub 2017 Feb 21. PMID: 28220972; PMCID: PMC6816968.

72. Yang J, Zhu G, Shi YQ. Analyzing the effect of JPEG compression on local variance of image intensity. IEEE Trans Image Process. 2016;25:2647–56. https://doi.org/10.1109/TIP.2016.2553521.

73. Liao B, Jiang Y, Liang W, et al. Gene selection using locality sensitive Laplacian score. IEEE/ACM Trans Comput Biol Bioinform. 2014;11:1146–56. https://doi.org/10.1109/TCBB.2014.2328334.

74. LeCun Y, Bengio Y, Hinton G. Deep learning. Nature. 2015;521:436–44. https://doi.org/10.1038/nature14539.

75. Chartrand G, Cheng PM, Vorontsov E, et al. Deep learning: a primer for radiologists. Radiographics. 2017;37:2113–31. https://doi.org/10.1148/rg.2017170077.

76. Krizhevsky BA, Sutskever I, Hinton GE. Imagenet classification with deep convolutional neural networks. Adv Neural Inf Process Syst. 2012;25:1097–105. https://doi.org/10.1145/3065386.

77. Gonzalez D, Dietz RL, Pantanowitz L. Feasibility of a deep learning algorithm to distinguish large cell neuroendocrine from small cell lung carcinoma in cytology specimens. Cytopathology 2020;31:426–431. doi: https://doi.org/10.1111/cyt.12829. Epub 2020 May 20.

78. Gonzalez D, Dietz RL, Pantanowitz L. Feasibility of a deep learning algorithm to distinguish large cell neuroendocrine from small cell lung carcinoma in cytology specimens. Cytopathology 2020;31:426 431. doi: https://doi.org/10.1111/cyt.12829. Epub 2020 May 20.

79. Bao H, Bi H, Zhang X, et al. Artificial intelligence-assisted cytology for detection of cervical intraepithelial neoplasia or invasive cancer: a multicenter, clinical-based, observational study. Gynecol Oncol 2020;159:171–178. doi: https://doi.org/10.1016/j.ygyno.2020.07.099. Epub 2020 Aug 16.

80. Zhu XH, Li XM, Zhang WL, et al. [Application of artificial intelligence-assisted diagnosis for cervical liquid-based thin-layer cytology]. Zhonghua Bing Li Xue Za Zhi. 2021;50:333–338. Chinese. https://doi.org/10.3760/cma.j.cn112151-20201013-00780.

81. Wentzensen N, Lahrmann B, Clarke MA, et al. Accuracy and efficiency of deep-learning-based automation of dual stain cytology in cervical cancer screening. J Natl Cancer Inst. 2021;113:72–79. https://doi.org/10.1093/jnci/djaa066. PMID: 32584382; PMCID: PMC7781458.

82. Tang HP, Cai D, Kong YQ, et al. Cervical cytology screening facilitated by an artificial intelligence microscope: a preliminary study. Cancer Cytopathol 2021. https://doi.org/10.1002/cncy.22425. Epub ahead of print.

83. Holmström O, Linder N, Kaingu H, et al. Point-of-care digital cytology with artificial intelligence for cervical cancer screening in a resource-limited setting. JAMA Netw Open. 2021;4:e211740. https://doi.org/10.1001/jamanetworkopen.2021.1740. PMID: 33729503; PMCID: PMC7970338.

84. Sanyal P, Barui S, Deb P, Sharma HC. Performance of a convolutional neural network in screening liquid based cervical cytology smears. J Cytol. 2019;36:146–151. https://doi.org/10.4103/JOC.JOC_201_18. PMID: 31359913; PMCID: PMC6592125.

85. Sanyal P, Mukherjee T, Barui S, et al. Artificial intelligence in cytopathology: a neural network to identify papillary carcinoma on thyroid fine-needle aspiration cytology smears. J Pathol Inform. 2018;9:43.

86. Guan Q, Wang Y, Ping B, et al. Deep convolutional neural network VGG-16 model for differential diagnosing of papillary thyroid carcinomas in cytological images: a pilot study. J Cancer. 2019;10:4876–82.

87. Zhang L, Le L, Nogues I, et al. DeepPap: deep convolutional networks for cervical cell classification. IEEE J Biomed Health Inform. 2017;21:1633–43.

88. Sornapudi S, Brown GT, Xue Z, et al. Comparing deep learning models for multi-cell classification in liquid- based cervical cytology image. AMIA Annu Symp Proc. 2019;2020:820–7.

89. Teramoto A, Tsukamoto T, Kiriyama Y, et al. Automated classification of lung Cancer types from cytological images using deep convolutional neural networks. Biomed Res Int. 2017;2017:4067832.
90. Wu M, Yan C, Liu H, et al. Automatic classification of cervical cancer from cytological images by using convolutional neural network. Biosci Rep. 2018;38:BSR20181769.
91. Wu M, Yan C, Liu H, Liu Q. Automatic classification of ovarian cancer types from cytological images using deep convolutional neural networks. Biosci Rep. 2018;38:BSR20180289.

Correction to: Endometrial Cell Sampling Procedure to Obtain Cytologic Specimens

Tetsuji Kurokawa, Toshimichi Onuma, Akiko Shinagawa, and Yoshio Yoshida

---

Correction to:
Chapter 4 in: Y. Hirai, F. Fulciniti (eds.), *The Yokohama System for Reporting Endometrial Cytology*,
https://doi.org/10.1007/978-981-16-5011-6_4

Reference citation in line 7 of page 25 was published with incorrect reference. This has now been corrected to [12] from [13]. The caption of figures 4.1, 4.2, and 4.3 were initially published with errors as it lacked the reference citation from which they were used. Figure 4.1 was reused from reference 10 and figures 4.2 and 4.3 were reused from reference 12. Page numbers were missing in reference 12. This was corrected and page numbers 13–20 are added to reference 12 in this version.

The updated version of this chapter can be found at
https://doi.org/10.1007/978-981-16-5011-6_4

Index